Advanced Courses in Mathematics
CRM Barcelona

Centre de Recerca Matemàtica

Managing Editor:
David Romero i Sànchez

More information about this series at http://www.springer.com/series/5038

Karen R. Strung

An Introduction to C*-Algebras and the Classification Program

Editor for this volume:
Francesc Perera, Universitat Autònoma de Barcelona

 Birkhäuser

Karen R. Strung
Institute of Mathematics
Czech Academy of Sciences
Praha, Czech Republic

ISSN 2297-0304   ISSN 2297-0312   (electronic)
Advanced Courses in Mathematics - CRM Barcelona
ISBN 978-3-030-47464-5   ISBN 978-3-030-47465-2   (eBook)
https://doi.org/10.1007/978-3-030-47465-2

Mathematics Subject Classification (2020): 46L05, 46L06, 46L80, 46L55, 46L35, 19K14, 06F05

This book is published under the imprint Birkhäuser, www.birkhauser-science.com by the registered company
Springer Nature Switzerland AG
The registered company address is: Gewerbestrasse 11, 6330 Cham, Switzerland

Contents

Part II Simple Examples

Part III   Introduction to Classification Theory

Foreword

The study of C*-algebras has attracted a good deal of attention, particularly due to the classification of simple, nuclear C*-algebras and the recent breakthroughs obtained in recent years by many authors. This classification program, initiated by Elliott, bears analogy to Connes' classification of von Neumann factors, but is significantly more complex. The far reaching results that were obtained made particularly blatant the need of good models for the class of C*-algebras considered. A good source of such models can be found in groupoid algebras and dynamical systems.

To cater for this need, an Intensive Research Program was set up at the Centre de Recerca Matemàtica CRM-Barcelona in the spring of 2017. The main focus was Operator Algebras, Dynamical Systems, and the interactions between these. The main goal of this program consisted of bringing together leading experts in the field, as well as talented young researchers in order to foster future collaborations and share their insights. We felt that it was equally important to offer undergraduate students in their final year or recently graduated students the opportunity to have a course that would take them to the forefront of current research starting from scratch.

This course, entitled *An Invitation to C*-algebras*, ran from March 8 to March 21, 2017. It consisted of a series of six lectures delivered by Karen R. Strung (Institute of Mathematics of the Czech Academy of Sciences). This volume contains the set of notes the author used in the lectures, that were expanded later to include further details and additional material. It is therefore not only an excellent introduction to the theory of C*-algebras, but it also takes the reader into the more exciting recent developments, including an overview of the Classification Theorem as well as the Toms–Winter Conjecture.

The volume is structured in three parts, the first of which constitutes an introduction to the basic theory covering the topics of positivity, representations, tensor products, and completely positive maps, among others.

Examples are the main theme of the second part, with special emphasis on simple C*-algebras, that are the focus of the Classification Program. Important examples such as AF algebras, Cuntz algebras, and significant constructions such as group C*-algebras and crossed products are analysed in detail. Quasidiagonality and tracial approximation are also treated in this part, as they provide useful tools towards determining certain structural aspects of simple C*-algebras.

The third and final part of the volume contains the more advanced material. It develops technical tools such as K-Theory, the Cuntz semigroup, and it establishes the now classical result of classification of AF-algebras by means of the K_0-group. The fundamental example of the Jiang–Su algebra and its role in classification are explained here in detail, as well as the notions of strong self-absorption and nuclear dimension. These are key ingredients for stating the Classification

Theorem and the Toms–Winter Conjecture, which bring the volume to an elegant closure.

In order to help the reader gain insight into the explanations, each chapter contains a list of exercises of various degrees of difficulty.

The Scientific Committee for the Intensive Research Program consisted of Nathanial P. Brown (The Pennsylvania State University), Francesc Perera (Universitat Autònoma de Barcelona), Aidan Sims (University of Wollongong), Stuart White (University of Glasgow), and Wilhelm Winter (Münster Universität).

The said research program was made possible not only by the support of the CRM, but also from the Simons Foundation, the National Science Foundation, the Clay Mathematics Institute, and by the Ministerio de Ciencia e Innovación (grant No. MTM2014-53644-P.

It is also a great pleasure to thank Karen as well as the undergraduate and graduate students that participated, without whom this course would have not been possible.

Bellaterra, September 2019   Francesc Perera

Background and context

The study of C*-algebras fits into the broader mathematical framework of operator algebras. A phrase one often hears in relation to operator algebras is that they are the "noncommutative" version of classical objects like topological spaces, measure spaces, groups, and so forth. The study of operator algebras started at the beginning of the 20th century. Of course, some noncommutative mathematics, in particular the study of matrices and abstract linear transformations on finite-dimensional vector spaces, was already known and reasonably developed. Indeed, matrix algebras are themselves (important) examples of operator algebras. However, it is reasonable to say that the subject was brought into being by von Neumann's efforts to make mathematically sound the emerging area of quantum physics. The noncommutative behaviour that popped up in places such as Heisenberg's uncertainty principal led von Neumann to define an abstract Hilbert space (particular Hilbert spaces – though they were not called that – such as L^2 and ℓ^2, were already known) and undertake a study of operators on such spaces. In particular, he introduced self-adjoint subalgebras of bounded operators that were closed in the weak operator topology; these later became known as von Neumann algebras and sometimes as W^*-algebras (though the former is often reserved for those algebras acting on Hilbert space and the latter for abstract *-algebras that are weakly closed) (see for example [125] for more on von Neumann's contributions). Subsequent papers by Murray and von Neumann further developed the subject [84, 85, 86].

C*-algebras, which are the topic of this course, have their origins in the studies of Gelfand and Naimark in the 1940s. While it turns out that von Neumann algebras are a particular type of C*-algebra, by now they are often studied separately; indeed, there will not be much to say about von Neumann algebras in these notes. Given a compact Hausdorff space X, the algebra of continuous functions on that space, $C(X)$, can be given an involution as well as norm which makes it a Banach *-algebra. Moreover, this norm satisfies the C*-equality: $\|f^*f\| = \|f\|^2$ for every $f \in C(X)$. The C*-equality is precisely what is needed to pass from a Banach *-algebra to a C*-algebra. This apparently minor requirement has wide-reaching implications for the structure of C*-algebras; we will see just what this means in the sequel. Further work by Gelfand and Naimark, together with Segal, established a way of constructing a representation of any C*-algebra as a norm-closed self-adjoint subalgebra of bounded operators on a Hilbert space. Gelfand and Naimark showed that in fact there is always a *faithful* representation; thus every C*-algebra is isomorphic to such a subalgebra of operators. Since those early days, the study of C*-algebras has taken off in many directions. It continues to find utility in the study of physics (including quantum gravity, quantum information, statistical mechanics), the study of topological groups as well as the development of quantum groups, dynamical systems, K-theory, and noncommutative geometry.

A main focus of this book is C*-algebraic classification. Classification is a common theme across all areas of mathematics. Finite-dimensional vector spaces are identified, up to isomorphism, by a natural number corresponding to their dimension. Dynkin diagrams classify semisimple Lie algebras. In group theory, there is the remarkable classification theorem of finite simple groups into four broad classes, completed bit-by-bit over many years. The classification program for C*-algebras takes its motivation from the classification of von Neumann algebra factors. Connes made huge advances in operator algebras by providing a detailed analysis of amenable von Neumann algebras and the classification of von Neumann algebra factors. For this he was awarded the Fields medal. The analogous C*-algebras are those that are separable, simple and nuclear. Their classification was initiated by Elliott following his classification of approximately finite C*-algebras. This classification program seeks to classify simple, separable, nuclear C*-algebras by K-theoretic invariants which we now call Elliott invariants.

For C*-algebras, classification is significantly more complicated than it is for von Neumann algebras. Nevertheless, the past few years have seen tremendous breakthroughs via innovations such as nuclear dimension (a noncommutative covering dimension) and successful adaptation of techniques from von Neumann algebras. By now the classification program has proven a resounding success. All simple separable unital nuclear infinite-dimensional C*-algebras with finite nuclear dimension satisfying the UCT ("classifiable" C*-algebras) can be distinguished by their Elliott invariants. The final pieces of the puzzle have only recently been put in place. The Toms–Winter Conjecture, which has been confirmed for all simple separable unital nuclear C*-algebras, under a minor restriction on the tracial state space, tells us that the question of whether a C*-algebra belongs is classifiable is equivalent to the presence of certain regularity properties. This has left us with powerful tools to determine and describe the structure of many C*-algebras.

The first section of the book will concentrate on developing the theory from the beginning, with minimal background requirements save for some knowledge of functional analysis and Hilbert spaces. In the second section, we develop a few examples in more depth and in the final section we deal with the theory of classification of C*-algebras. The first section as well as much of the second section corresponds approximately to an introductory course on C*-algebras, aimed at masters students and students at the beginning of their PhD studies. These sections are based on courses given by the author at the Centre de Recerca Matemàtica in Barcelona during the intensive research program *Operator Algebras: Dynamics and Interactions* during the spring of 2017, as well as at the Noncommutative Geometry semester at the Banach Center at the Institute for Mathematics of the Polish Academy of Sciences in autumn 2016. The decision of what results to include in the first two sections was often made based on what would be required in the final section on the classification program. It also, naturally, reflects my own personal interests and biases. For example, in Part III there is a greater focus on stably finite C*-algebras compared to the attention paid to purely infinite C*-algebras.

For the reader who has some basic knowledge of C*-algebras, most of the first part can be skipped. The material is roughly similar to standard texts such as Murphy's book [83] or Davidson's book [32], with the exception of Chapter 7, which looks at completely positive maps and introduces the completely positive approximation property and will be useful for later sections. I have made an effort to reference results in the book so that if the reader skips the first few chapters but comes across an argument in a proof with which they are unfamiliar, it should not be too difficult to go back and find the required result. Part II includes constructions that may be familiar to those readers who have already studied some C*-algebra theory, with the exception of the final chapter on quasidiagonality and tracial approximation as some of these results are relatively recent.

The final section contains more advanced material, but the book aims to be as self-contained as possible and so the material should be accessible to those readers who work through the first two sections. The aim is to leave the reader able to confidently read and understand the current literature. The final section contains material from a short course given at the spring institute *Noncommutative Geometry and Operator Algebras* at the University of Münster in spring 2018 as well as lectures given at the department of mathematics at Radboud University in Nijmegen.

I would like to thank Francesc Perera for inviting me to give the course at the Centre de Recerca Matemàtica as well as suggesting I prepare these notes. I am also indebted to Tristan Bice, Robin Deeley, Andrey Krutov, Klaas Landsman and Réamonn Ó Buachalla for help with proofreading.

The author was supported by a Radboud Excellence Initiative postdoctoral fellowship, the Sonata 9 NCN grant 2015/17/D/ST1/02529, GAČR project 20-17488 Y, and RVO:67985840.

Part I

Basic Theory

1 Banach algebras and spectral theory

Although the subject of these notes is C*-algebras, we begin with Banach algebras. Many of the basic results were first developed for Banach algebras. Initially working at the level of generality of Banach algebras will furthermore help the reader to later appreciate, by comparison, the manner in which C*-algebras enjoy many structural properties that are absent in the more general setting of Banach algebras.

In this chapter, we first define Banach algebras and look at some examples. In the second section we define and look at properties of the *spectrum* of an element in a Banach algebra: an object which simultaneously generalises the set of eigenvalues of a matrix and the set of values in the range of continuous function. We will see that, unlike in an arbitrary algebra, the spectrum of an element in a Banach algebra is always nonempty. In Section 1.3 we look at ideals and quotients in Banach algebras. The final two sections are concerned with unital commutative Banach algebras. After proving a number of results about their *characters*, we end the chapter with the Gelfand representation, which says that we can represent any commutative Banach algebra A as a Banach algebra of functions on a locally compact space. This theorem will be of fundamental importance when we move on to C*-algebras in the next chapter.

All Banach algebras we will consider will be $\mathbb{C}$-algebras, though they can also be defined over other fields.

1.1 Banach algebras

A *Banach algebra* is an algebra A equipped with a submultiplicative norm

$$\| \cdot \| : A \to [0, \infty),$$

that is complete with respect to its norm.

© The Author(s), under exclusive license to Springer Nature Switzerland AG 2021
K. R. Strung, *An Introduction to C*-Algebras and the Classification Program*, Advanced Courses in Mathematics - CRM Barcelona, https://doi.org/10.1007/978-3-030-47465-2_1

Note that A is not necessarily unital. If A has a unit 1_A then we require that $\|1_A\| = 1$ and in this case we call A a *unital* Banach algebra.

If $B \subset A$ is a subalgebra, then its closure with respect to the norm of A is also a Banach algebra.

1.1.1. Examples. (a) Let X be a topological space and let

$$C_b(X) := \{f : X \to \mathbb{C} \mid f \text{ continuous, bounded}\}.$$

Then $C_b(X)$ is a Banach algebra when equipped with pointwise operations and supremum norm

$$\|f\|_\infty := \sup\{|f(x)| \mid x \in X\}.$$

(b) Let X be a Banach space. Equip the set of linear operators

$$\mathcal{L}(X) = \{T : X \to X \mid T \text{ linear, continuous}\}$$

with pointwise addition, composition for multiplication, and the operator norm

$$\|T\| := \sup\{\|T(x)\| \mid x \in X, \|x\| \le 1\}.$$

Then $\mathcal{L}(X)$ is a Banach algebra.

(c) Let (X, Σ, μ) be a measure space. Let

$$L^\infty(X, \Sigma, \mu) := \{f : X \to \mathbb{C} \mid f \text{ measurable, } \exists K > 0 \text{ s.t. } \mu(\{x \mid |f(x)| > K\}) = 0\}.$$

Define a norm on $L^\infty(X, \Sigma, \mu)$ by

$$\|f\| := \inf_{f = g \text{ a.e. } \mu} \ \sup_{x \in X} |g(x)|.$$

Then $L^\infty(X, \Sigma, \mu)$ is a Banach algebra.

1.2   Spectrum

Let A be a unital Banach algebra. An element $a \in A$ is *invertible* if there is a $b \in A$ such that $ab = ba = 1_A$. In this case we write $b = a^{-1}$. (This makes sense because if such a b exists, it must be unique.) We denote the set of invertible elements in A by

$$\text{Inv}(A) := \{a \in A \mid \text{ there is } b \in A \text{ such that } ba = ab = 1_A\}.$$

1.2.1. Definition. The *spectrum* of an element a in the unital algebra A is defined to be

$$\text{sp}(a) := \{\lambda \in \mathbb{C} \mid \lambda \cdot 1_A - a \notin \text{Inv}(A)\}.$$

1.2.2. Suppose that $1_A - ab$ is invertible with inverse c. Then one can check that $1_A - ba$ is also invertible with inverse given by $1_A + bca$. As a result, for any a, b in a unital Banach algebra A, we have

$$\text{sp}(ab) \setminus \{0\} = \text{sp}(ba) \setminus \{0\}.$$

Let $p(z) = \lambda_0 + \lambda_1 z + \cdots + \lambda_n z^n$, $\lambda_i \in \mathbb{C}$, be a polynomial in the algebra of polynomials in one indeterminate, which we denote by $\mathbb{C}[z]$. For $a \in A$ denote

$$p(a) := \lambda_0 1_A + \lambda_1 a + \cdots \lambda_n a^n.$$

We have the following spectral mapping property for polynomials.

1.2.3. Theorem. *Let A be a unital algebra, $a \in A$ and p a polynomial in $\mathbb{C}[z]$. Suppose that $\mathrm{sp}(a) \neq \emptyset$. Then $\mathrm{sp}(p(a)) = p(\mathrm{sp}(a))$.*

Proof. If p is constant, the result is obvious, so we may assume otherwise. Let $\mu \in \mathbb{C}$ and consider the polynomial $p - \mu$. Since every polynomial over $\mathbb{C}$ splits, we can write

$$p(z) - \mu = \lambda_0 (\lambda_1 - z) \cdots (\lambda_n - z)$$

for some $n \in \mathbb{N} \setminus \{0\}$, $\lambda_0, \ldots, \lambda_n \in \mathbb{C}$ and $\lambda_0 \neq 0$.

If $\mu \notin \mathrm{sp}(p(a))$ then by definition $p(a) - \mu \cdot 1_A$ is invertible, and hence each $\lambda_i \cdot 1_A - a$, $i = 1, \ldots, n$, is invertible. Conversely, it is clear that if each $\lambda_i \cdot 1_A - a$ is invertible, so is $p(a) - \mu \cdot 1_A$. Thus $\mu \in \mathrm{sp}(p(a))$ if and only if there is some $1 \leq i \leq n$ with $\lambda_i \in \mathrm{sp}(a)$ and we have $\mathrm{sp}(p(a)) \subseteq p(\mathrm{sp}(a))$. Now if $\lambda \in \mathrm{sp}(a)$, then $p(a) - p(\lambda) = (\lambda \cdot 1_A - a)b$ for some $b \in A$ and hence is not invertible. Thus $\mathrm{sp}(p(a)) = p(\mathrm{sp}(a))$. $\qquad\square$

For a general unital algebra, it is possible for an element to have empty spectrum, as is shown in Exercise 1.6.9. In a unital Banach algebra, however, this is not the case. To prove this we require a few preliminary results. The first is that in a unital Banach algebra – where unlike in an arbitrary algebra we have a notion of convergence – we can often use a geometric series argument to calculate inverses.

1.2.4. Theorem. *Let A be a unital Banach algebra and $a \in A$ such that $\|a\| < 1$. Then $1_A - a$ is invertible and*

$$(1_A - a)^{-1} = \sum_{n=0}^{\infty} a^n.$$

Proof. First note that by submultiplicativity of the norm, we have

$$\left\| \sum_{n=0}^{\infty} a^n \right\| \leq \sum_{n=0}^{\infty} \|a\|^n,$$

which is finite by the usual convergence of a geometric series in $\mathbb{R}$. Since A is complete, this means that $\sum_{n=0}^{\infty} a^n$ converges to some $b \in A$. Now, we compute that

$$(1_A - a) \sum_{n=0}^{N} a^n = \sum_{n=0}^{N} a^n - \sum_{m=1}^{N+1} a^m = 1_A - a^{N+1},$$

and since $1_A - a^{N+1} \to 1_A$ as $N \to \infty$, we must have $b = (1_A - a)^{-1}$, as claimed. $\quad\square$

1.2.5. Lemma. *Let A be a unital Banach algebra. Then $\mathrm{Inv}(A)$ is open in A and the map*

$$\mathrm{Inv}(A) \to \mathrm{Inv}(A), \quad a \mapsto a^{-1}$$

is differentiable.

Proof. Let $a \in \mathrm{Inv}(A)$. We will show that any b sufficiently close to a is also invertible, which will show the first part of the lemma. Let $b \in A$ such that $\|a - b\|\|a^{-1}\| < 1$. Then

$$\|1_A - ba^{-1}\| \le \|a^{-1}\|\|a - b\| < 1,$$

so, by the previous theorem we have that ba^{-1} is invertible. Similarly, $a^{-1}b$ is invertible. Thus we also have that $b(a^{-1}(ba^{-1})^{-1}) = 1_A = (a^{-1}b)^{-1}a^{-1})b$, so b is invertible.

To show that $a \mapsto a^{-1}$ is differentiable, we need to find a linear map $L : A \to A$ such that, for $a \in \mathrm{Inv}(A)$,

$$\lim_{h \to 0} \frac{\|(a + h)^{-1} - a^{-1} - L(h)\|}{\|h\|} = 0.$$

For $b \in A$ define $L(b) = -a^{-1}ba^{-1}$.

Let $a \in A$ be invertible and let h be small enough that

$$\|h\|\|a^{-1}\| < 1/2.$$

Then $\|a^{-1}h\| < 1/2$ so, by Theorem 1.2.4, the element $1_A + a^{-1}h$ is invertible and

$$
\begin{aligned}
\|(1_A + a^{-1}h)^{-1} - 1_A + a^{-1}h\| &= \|\textstyle\sum_{n=0}^{\infty}(-1)^n(a^{-1}h)^n - 1_A + a^{-1}h\| \\
&= \|\textstyle\sum_{n=2}^{\infty}(-1)^n(a^{-1}h)^n\| \\
&\le \textstyle\sum_{n=2}^{\infty}\|(a^{-1}h)\|^n \\
&\le \|a^{-1}h\|^2/(1_A - \|a^{-1}h\|)^{-1} \\
&\le 2\|a^{-1}h\|^2.
\end{aligned}
$$

Thus

$$
\begin{aligned}
\frac{\|(a + h)^{-1} - a^{-1} - L(h)\|}{\|h\|} &= \frac{\|(a + h)^{-1} - a^{-1} + a^{-1}ha^{-1}\|}{\|h\|} \\
&= \frac{\|(a(a^{-1}a + a^{-1}h))^{-1} - (1_A - a^{-1}h)a^{-1}\|}{\|h\|} \\
&\le \frac{\|(1_A + a^{-1}h)^{-1} - 1_A + a^{-1}h\|\|a^{-1}\|}{\|h\|} \\
&\le \frac{2\|a^{-1}\|^2\|h\|^2}{\|h\|}, \\
&= 2\|a^{-1}\|^{-2}\|h\|,
\end{aligned}
$$

which goes to zero as h goes to zero. $\square$

In a metric space X, we denote the open ball of radius $r > 0$ about a point $x \in X$ by $B(x, r)$. Its closure is denoted by $\overline{B(x, r)}$.

1.2.6. Lemma. *Let A be a unital Banach algebra. Then for any $a \in A$, the spectrum of a is a closed subset of $\overline{B(0, \|a\|)} \subset \mathbb{C}$ and the map*

$$\mathbb{C} \setminus sp(a) \to A, \quad \lambda \mapsto (a - \lambda 1_A)^{-1}$$

is differentiable.

Proof. Once we show that $sp(a) \subset \overline{B(0, \|a\|)}$, the rest follows from the previous lemma. If $|\lambda| > \|a\|$ then $\lambda \cdot 1_A - a$ is invertible, so $sp(a) \subset \overline{B(0, \|a\|)}$. The details are left as an exercise (for a hint, see Exercise 1.6.10). $\square$

1.2.7. Now we are able to prove that every element in a unital Banach algebra has nonempty spectrum. In what follows, A^* denotes the continuous dual space of A, that is,
$$A^* = \{f : A \to \mathbb{C} \mid f \text{ continuous and linear}\}.$$

Endow A^* with the *weak-*-topology*, which is the topology generated by seminorms of the form $p_a(\tau) = |\tau(a)|$ ranging over all $a \in A$. A sequence $(\phi_n)_{n \in \mathbb{N}}$ converges to $\phi \in A^*$ if $\phi_n(a) \to \phi(a), n \to \infty$ for every $a \in A$ (pointwise convergence). Recall that A^* separates points. For further details see, for example, [83, Appendix].

1.2.8. Theorem. *Let A be a unital Banach algebra. Then for every $a \in A$ we have*

$$sp(a) \neq \emptyset.$$

Proof. First of all, we may assume that a is nonzero since $0 \in sp(0)$. So let $a \neq 0$ and assume, for contradiction, that $sp(a) = \emptyset$. We leave it as an exercise to show that the map
$$\mathbb{C} \to Inv(A), \quad \lambda \mapsto (a - \lambda 1_A)^{-1}$$

is bounded on the compact disc of radius $2\|a\|$. Once this has been shown it follows that for any $\phi \in A^*$ the map

$$\lambda \mapsto \phi((a - \lambda 1_A)^{-1})$$

is also bounded. From the previous theorem, this map is also entire, which, by Liouville's theorem, implies it must be constant. Thus $\phi(a^{-1}) = \phi((a - 1_A)^{-1})$ for every $\phi \in A^*$ leading to the contradiction that $a^{-1} = (a - 1_A)^{-1}$. $\square$

1.2.9. Theorem. *Let A be a unital Banach algebra with $Inv(A) = A \setminus \{0\}$. Then $A \cong \mathbb{C}$.*

Proof. Let $a \in A \setminus \{0\}$ and suppose that $\lambda \in sp(a)$. Then $\lambda 1_A - a$ is not invertible, so $\lambda 1_A - a = 0$. It follows that each nonzero element is a scalar multiple of the identity and the map $a - \lambda 1_A \mapsto \lambda$ gives us the required isomorphism. $\square$

1.2.10. The *spectral radius* of an element a in a unital Banach algebra A is defined to be

$$r(a) := \sup_{\lambda \in \mathrm{sp}(a)} |\lambda|.$$

We have the following characterisation of the spectral radius, relating it to the norm of the element a.

1.2.11. Theorem. *For any $a \in A$ we have*

$$r(a) = \inf_{n \geq 1} \|a^n\|^{1/n} = \lim_{n \to \infty} \|a^n\|^{1/n}.$$

Proof. Since $\lambda \in \mathrm{sp}(a)$ implies $\lambda^n \in \mathrm{sp}(a^n)$, we have $|\lambda^n| \leq \|a^n\|$. Therefore $|\lambda| = |\lambda^n|^{1/n} \leq \|a^n\|^{1/n}$ for every $\lambda \in \mathrm{sp}(a)$ and every $n \geq 1$. Thus

$$r(a) = \sup_{\lambda \in \mathrm{sp}(a)} |\lambda| \leq \inf_{n \geq 1} \|a^n\|^{1/n}.$$

By definition we have that $\inf_{n \geq 1} \|a^n\|^{1/n} \leq \liminf_{n \to \infty} \|a^n\|^{1/n}$, so we are finished if we show that $r(a) \geq \limsup_{n \to \infty} \|a^n\|^{1/n}$.

Let $D = B(0, 1/r(a))$ if $r(a) \neq 0$ and $D = \mathbb{C}$ otherwise. If $\lambda \in D$ then $1_A - \lambda a$ is invertible by Theorem 1.2.4. It follows from Lemma 1.2.6 that, for every $\phi \in A^*$, the map

$$f : D \to \mathbb{C}, \quad \lambda \mapsto \phi((1_A - \lambda a)^{-1})$$

is analytic. Thus there are unique complex numbers $(c_n)_{n \in \mathbb{N}}$ such that

$$f(\lambda) = \sum_{n=0}^{\infty} c_n \lambda^n,$$

whenever $\lambda \in D$.

Again, by applying Theorem 1.2.4, for $\lambda < 1/\|a\| \leq 1/r(a)$ we have

$$(1_A - \lambda a)^{-1} = \sum_{n=0}^{\infty} \lambda^n a^n.$$

It follows that $f(\lambda) = \sum_{n=0}^{\infty} \lambda^n \phi(a^n)$, so that $\phi(a^n) = c_n$ for every $n \in \mathbb{N}$. Thus $\phi(a^n) \to 0$ as $n \to \infty$ and therefore the sequence $(\phi(a^n))_{n \in \mathbb{N}}$ is bounded. This is true for every $\phi \in A^*$, so in fact $(\|\lambda^n a^n\|)_{n \in \mathbb{N}}$ is also bounded by some $M_\lambda > 0$. Thus

$$\|a^n\|^{1/n} \leq M_\lambda^{1/n}/|\lambda|,$$

so

$$\limsup_{n \to \infty} \|a^n\|^{1/n} \leq 1/|\lambda|,$$

for every $\lambda \in D$. It follows that

$$\limsup_{n \to \infty} \|a^n\|^{1/n} \leq r(a),$$

as required. $\square$

1.2.12. Let A be a (not necessarily unital) Banach algebra. Let $\tilde{A} := A \oplus \mathbb{C}$ as a vector space. Define a multiplication on $\tilde{A}$ by

$$(a, \lambda) \cdot (b, \mu) = (ab + \lambda b + \mu a, \lambda \mu),$$

and a norm by

$$\|(a, \lambda)\| = \|a\| + |\lambda|.$$

This turns $\tilde{A}$ into a unital Banach algebra (exercise). When A is nonunital, $\tilde{A}$ is called the *unitisation* of A. When we consider C*-algebras in Chapter 2, we will have to be a little bit more careful in defining the norm. The unitisation allows us to make sense of the spectrum of an element in a nonunital Banach algebra.

1.2.13. Let A be a nonunital Banach algebra and let $a \in A$. The spectrum of a is defined to be

$$\mathrm{sp}(a) := \{\lambda \in \mathbb{C} \mid \lambda \cdot 1_{\tilde{A}} - a \notin \mathrm{Inv}(A)\},$$

where $1_{\tilde{A}}$ denotes the unit in the unitisation of A.

1.3 Ideals

As with almost any mathematical object, we are interested in the subobjects of a Banach algebra. We first consider ideals. For Banach algebras, unless otherwise stated, an ideal is norm-closed, and hence a Banach algebra in its own right.

1.3.1. Let A be an algebra. A subalgebra $I \subset A$ is a right (left) ideal if $a \in A$ and $b \in I$ then $ab \in B$ ($ba \in I$). We call $I \subset A$ an algebraic ideal if it is both a right and a left ideal. When I is an algebraic ideal, then A/I is also an algebra with the obvious definitions for multiplication and addition.

The quotient A/I is a unital algebra exactly when I is a *modular ideal*: there exists an element $u \in A$ such that $a - ua \in I$ and $a - au \in I$ for every $a \in A$. (What is $1_{A/I}$?) Note that this implies that every algebraic ideal in a unital algebra is modular.

When A is a Banach algebra, we call $I \subset A$ an *ideal* if I is a norm-closed algebraic ideal. In this case A/I can be given the quotient norm

$$\|a + I\| = \inf_{b \in I} \|a + b\|, \quad a \in A;$$

which makes A/I into a Banach algebra.

1.3.2. We also have the usual notions of *trivial ideals* ($I = 0, A$) and ideals generated by a set $J \subset A$ (= smallest ideal containing J). A *proper (algebraic) ideal* is one which is not equal to A (but may be zero) and a *maximal (algebraic) ideal* is a proper (algebraic) ideal not contained in any other proper (algebraic) ideal. One can use the Kuratowski–Zorn Lemma to show that every proper modular ideal is contained in a maximal modular ideal. In particular, if A is unital then every proper ideal of A is contained in a maximal ideal.

1.3.3. Proposition. *Let A be a Banach algebra and $I \subset A$ an algebraic ideal. If I is proper and modular, then $\overline{I}$ is also proper.*

Proof. Since I is modular, there is an element $u \in A$ such that $a - ua \in I$ and $a - au \in I$ for every $a \in A$. Let $b \in I$ with $\|u - b\| < 1$. Then $1_{\tilde{A}} - u + b$ is invertible as an element of $\tilde{A}$, the unitisation of A (1.2.12). Let c denote its inverse. Then

$$1_{\tilde{A}} = c(1_{\tilde{A}} - u + b) = c - cu + cb \in I,$$

contradicting the assumption that I is proper. Thus any $b \in I$ must satisfy $\|u - b\| \geq 1$. In particular, $u \in A \setminus \overline{I}$, so $\overline{I}$ is proper.   □

1.3.4. Corollary. *If I is a maximal modular ideal in a Banach algebra, then it is closed.*

1.3.5. Proposition. *Let A be a commutative algebra and $I \subset A$ a modular ideal. If A is maximal then A/I is a field.*

Proof. Exercise.   □

1.3.6. If A and B are Banach algebras, a map $\varphi : A \to B$ is called a *homomorphism* if it is an algebra homomorphism that is continuous with respect to the norms of A and B. If A and B are unital and $\varphi(1_A) = 1_B$ then we call φ a unital homomorphism. The norm of a given homomorphism $\varphi : A \to B$ is defined to be

$$\|\varphi\| := \sup\{\|\phi(a)\|_B \mid a \in A, \|a\|_A \leq 1\}.$$

1.3.7. Proposition. *If $\varphi : A \to B$ is a homomorphism of Banach algebras A and B, then $\ker(\varphi)$ is an ideal in A.*

Proof. Exercise.   □

1.3.8. Recall from 1.2.12 that $\tilde{A}$ denotes the unitisation of nonunital Banach algebra A. The map $\iota : A \to \tilde{A}$ given by $\iota(a) = (a, 0)$ is an injective homomorphism, so we may identify A as a subalgebra in $\tilde{A}$. We also have a canonical projection homomorphism $\pi : \tilde{A} \to \mathbb{C}$ given by $\pi((a, \lambda)) = \lambda$. Its kernel is clearly A, so A is in fact an ideal in $\tilde{A}$.

1.4   Characters

Let A be a Banach algebra. We have just seen that the kernel of a homomorphism $\varphi : A \to B$ to some Banach algebra B gives rise to a closed ideal in A. Intuitively the "smaller" the image of φ, the "larger" we expect the corresponding ideal to be, and vice versa. Here, we make this more precise in the case that A is commutative: if the image of a nonzero homomorphism is as small as possible, the kernel is a maximal ideal.

1.4.1. Definition. Let A be a Banach algebra. A *character on A* is a nonzero algebra homomorphism $\tau : A \to \mathbb{C}$. Let

$$\Omega(A) := \{\tau : A \to \mathbb{C} \mid \tau \text{ a character on } A\}.$$

We call $\Omega(A)$ the *character space* of A, or based on what we will see below, the spectrum of A.

For commutative Banach algebras there is an important relation between characters, maximal ideals, and Banach algebras of the form $C_0(X)$ for some locally compact Hausdorff space X.

1.4.2. Proposition. *Let A be a unital commutative Banach algebra. Then*

(i) $\tau(a) \in \mathrm{sp}(a)$ *for every $\tau \in \Omega(A)$ and every $a \in A$;*

(ii) $\|\tau\| = 1$;

(iii) $\Omega(A) \neq \emptyset$ *and if A is not isomorphic to $\mathbb{C}$ then $\tau \mapsto \ker \tau$ is a bijection from $\Omega(A)$ to the set of maximal ideals of A.*

Proof. For (i), suppose that $\tau(a)$ is not in the spectrum of a. Then there exists $b \in A$ such that $(\tau(a)1_A - a)b = 1_A$. Since τ is a homomorphism, we get

$$1 = \tau(1_A) = (\tau(a)\tau(1_A) - \tau(a))\tau(b) = (\tau(a) - \tau(a))\tau(b) = 0,$$

a contradiction. So $\tau(a) \in \mathrm{sp}(a)$.

For (ii), let $a \in A$ be an element with $\|a\| \leq 1$. By Lemma 1.2.6 we have $|\tau(a)| \leq \|a\| \leq 1$, since $\tau(a) \in \mathrm{sp}(a)$ by (i). Thus

$$\|\tau\| = \sup\{|\tau(a)| \mid a \in A, \|a\| \leq 1\} \leq 1.$$

Since $\tau(1_A) = 1$, the result follows.

Now we prove (iii). Suppose that there is an ideal $I \subset A$. Then I must be contained in some maximal ideal J, and by Theorem 1.3.5 we have $A/J \cong \mathbb{C}$. Thus the quotient map $\pi : A \to A/J \cong \mathbb{C}$ is a character. Suppose now that A does not contain any ideals. Let $a \in A \setminus \{0\}$. Then the closure of Aa, the algebraic ideal generated by a, must be all of A. In particular, there exists some $b \in A$ such that $\|ba - 1_A\| < 1$. Then by Theorem 1.2.4, ba is invertible and hence, since A is commutative, $((ba)^{-1}b)a = 1_A = a((ba)^{-1}b)$, which is to say, a itself is invertible. Since a was arbitrary, $A \setminus \{0\} = \mathrm{Inv}(A)$ so by Theorem 1.2.9 $A \cong \mathbb{C}$, and the isomorphism gives us a character on A.

For the final statement, we have already shown that any maximal ideal J is the kernel of a character, namely the kernel of the quotient map $\pi : A \to A/J \cong \mathbb{C}$. Otherwise, if $\tau : A \to \mathbb{C}$ is a character, then $\ker(\tau)$ is a closed ideal and hence contained in some maximal ideal J. Then $\mathbb{C} \cong A/J \subset A/\ker(\tau) \cong \mathbb{C}$. So $A/J = A/\ker(\tau)$ and it follows that $\ker(\tau) = J$ is a maximal ideal. $\qquad\square$

Note that (ii) above says that $\Omega(A)$ is contained in the closed unit ball of the dual space A^*. Thus we may endow $\Omega(A)$ with the weak-$*$ topology inherited from A^*.

1.4.3. Theorem. *Let A be a unital commutative Banach algebra. Then, for any $a \in A$ we have*

$$\mathrm{sp}(a) = \{\tau(a) \mid \tau \in \Omega(A)\}.$$

Proof. Suppose that $\lambda \in \mathrm{sp}(a)$. The ideal generated by $(a - \lambda 1_A)$ is proper since it cannot contain 1_A (note that this uses commutativity and the fact that $\mathrm{Inv}(A)$ is closed). It is therefore contained in some maximal ideal which is of the form $\ker(\tau)$ for some $\tau \in \Omega(A)$, in which case $\tau(a) = \lambda$. The converse is given by Proposition 1.4.2 (i). $\qquad\qquad\square$

The proof of the next corollary is an easy exercise:

1.4.4. Corollary. *Let A be a nonunital commutative Banach algebra. Then, for any $a \in A$,*

$$\mathrm{sp}(a) = \{\tau(a) \mid \tau \in \Omega(A)\} \cup \{0\}.$$

1.4.5. Theorem. *Let A be a commutative Banach algebra. Then $\Omega(A)$ is a locally compact Hausdorff space. If A is unital, then $\Omega(A)$ is compact.*

Proof. It follows from Proposition 1.4.2 that $\Omega(A) \setminus \{0\}$ is a weak-$*$ subset of the closed unit ball A^*. Thus by the Banach–Alaoglu Theorem (see for example [105, Theorem 3.15]), it is compact. Hence $\Omega(A)$ is locally compact. If A is unital then one checks that in fact $\Omega(A)$ itself is closed, hence compact. $\qquad\qquad\square$

1.5   The Gelfand representation

We will now show the existence of the Gelfand representation, which says that we can represent any commutative Banach algebra A as a Banach algebra of functions on a locally compact Hausdorff space that is homeomorphic to $\Omega(A)$. When we move on to the next chapter, we will see that this has important consequences for C*-algebras, in particular, it will give us a continuous functional calculus – an indispensable tool to the theory. For now, we remain in the more general world of Banach algebras.

1.5.1. Let $a \in A$ and define $\hat{a} : A^* \to \mathbb{C}$ by $\hat{a}(\tau) = \tau(a)$. Then $\hat{a} \in C_0(\Omega(A))$ (indeed, the weak-$*$ topology is the coarsest topology making every $\hat{a}$, $a \in A$ continuous; this can be taken as its definition).

The map $a \mapsto \hat{a}$ is called the *Gelfand transform* and $\hat{a}$ is the *Gelfand transform* of a.

1.5.2. Theorem. *Let A be a commutative Banach algebra with $\Omega(A) \neq \emptyset$. Then*

$$A \to C_0(\Omega(A)), \quad a \mapsto \hat{a}$$

is a norm-decreasing homomorphism and, moreover, $r(a) = \|\hat{a}\|$.

If A is unital and $a \in A$ then $\mathrm{sp}(a) = \hat{a}(\Omega(A))$. When A is nonunital and $a \in A$, then $\mathrm{sp}(a) = \hat{a}(\Omega(A)) \cup \{0\}$.

Proof. By Theorem 1.4.3 and Corollary 1.4.4 we have $r(a) = \|\hat{a}\|$. Since $r(a) \leq \|a\|$, the map is norm-decreasing. It is easy to check that it is also a homomorphism. The nonunital case is left as an exercise. $\qquad\square$

1.5.3. Theorem. *Let A be a unital Banach algebra and let $a \in A$. Let $B \subset A$ be the Banach algebra generated by a and 1_A. Then B is commutative and the map*

$$\hat{a} : \Omega(B) \to \mathrm{sp}(a)$$

defined by

$$\hat{a}(\phi) = \phi(a).$$

is a homeomorphism.

Proof. It is clear that B is commutative since it is the norm closure of the space of polynomials in a. Furthermore, $\hat{a}$ is a continuous bijection and $\Omega(B)$ is compact, so it is a homeomorphism. $\qquad\square$

1.6   Exercises

1.6.1. Let X be a compact metric space and

$$C(X) := \{f : X \to \mathbb{C} \mid f \text{ is continuous}\}.$$

For $f, g \in C(X)$, define addition and multiplication, respectively, by

$$(f + g)(x) = f(x) + g(x), \quad (fg)(x) = f(x)g(x), \quad x \in X.$$

Define $\| \cdot \| : C(X) \to \mathbb{R}$ by

$$\|f\| = \sup_{x \in X} |f(x)|.$$

  (i) Show that $C(X)$ is closed under addition and multiplication.
 (ii) Show that $\| \cdot \|$ is a norm.
(iii) Show that $\| \cdot \|$ is submultiplicative.
 (iv) Suppose $(f_n)_{n \in \mathbb{N}} \subset C(X)$ is a Cauchy sequence. Show that there exists $f \in C(X)$ such that $\lim_{n \to \infty} f_n = f$.
  (v) Conclude that $C(X)$ is a Banach algebra.

1.6.2. Check that Examples 1.1.1 (a), (b) and (c) are Banach algebras.

1.6.3. Let $\mathbb{C}[z]$ denote the single-variable $\mathbb{C}$-valued polynomials, equipped with pointwise operations and norm $\|p\| = \sup_{|z|=1} |p(z)|$. Is this a Banach algebra?

1.6.4. Let A be a Banach algebra. Show that multiplication $m : A \times A, (a, b) \mapsto ab$, is continuous.

1.6.5. Let A be a unital algebra and show that the set of invertible elements is a group under multiplication. Show that $\mathrm{Inv}(A)$ is closed.

1.6.6. Let $M_n := M_n(\mathbb{C})$ denote the $n \times n$ matrices over $\mathbb{C}$. What is $\mathrm{sp}(a)$ for $a \in M_n$? Let X be a compact Hausdorff space and equip $C(X)$ with the supremum norm. What is $\mathrm{sp}(f)$ for $f \in C(X)$?

1.6.7. Let X be a compact space and A a unital Banach algebra. Show that

$$C(X, A) := \{f : X \to A \mid f \text{ continuous}\}$$

can be given the structure of a Banach algebra. In the case that $A = M_n$ we have that $C(X, M_n) \cong M_n(C(X))$.

1.6.8. Let H be a Hilbert space with orthonormal basis given by $(e_i)_{i \in I}$. An operator $T \in \mathcal{B}(H)$ is a *Hilbert–Schmidt operator* if $\sum_{i \in I} \|Te_i\|^2$ is finite. The Hilbert–Schmidt norm $\|T\| = (\sum_{i \in I} \|Te_i\|^2)^{1/2}$ can be defined on the set of all Hilbert–Schmidt operators. Equipped with the usual operations for operators on a Hilbert space, is the set of Hilbert–Schmidt operators a Banach algebra?

1.6.9. Let $\mathbb{C}(z)$ denote the field of fractions of $\mathbb{C}[z]$. Show that there is an element in $\mathbb{C}(z)$ which has empty spectrum.

1.6.10. Let A be a unital Banach algebra and $a \in A$. Show that $\mathrm{sp}(a) \subset \overline{B(0, \|a\|)}$. (Hint: show that if $|\lambda| > \|a\|$ then $\lambda - a$ is invertible.)

1.6.11. Let $H = L^2([0, 1]) = \{f : [0, 1] \to \mathbb{C} \mid \int f^2 < \infty\}$, and consider the Banach algebra $\mathcal{B}(H)$. Let $T \in \mathcal{B}(H)$ be defined as

$$T(f)(t) = \int_0^t f(x)dx.$$

Compute the spectral radius of T. What is $\mathrm{sp}(T)$?

1.6.12. Show that the map in Theorem 1.2.8,

$$\mathbb{C} \to \mathrm{Inv}(A), \quad \lambda \mapsto (a - \lambda 1_A)^{-1}$$

is bounded on the compact disc of radius $2\|a\|$.

1.6.13. Let A be a (not necessarily unital) Banach algebra. Show that the unitisation $\tilde{A}$ is a unital Banach algebra.

1.6.14. Let A be a nonunital Banach algebra. Without using Theorem 1.2.8, give a one-line proof of the nonunital version of Theorem 1.2.8.

1.6.15. Let A be a unital Banach algebra and $B \subset A$ with $1_A \in B$.

(i) Show that $\mathrm{Inv}(B)$ is a clopen subset of $\mathrm{Inv}(A) \cap B$.

(ii) Let $b \in B$. Show that $\mathrm{sp}_A(b) \subset \mathrm{sp}_B(b)$ and $\partial \mathrm{sp}_B(b) \subset \partial \mathrm{sp}_A(b)$, where $\partial \mathrm{sp}(\cdot)$ denotes the boundary of $\mathrm{sp}(\cdot)$. Show that if $\mathbb{C} \setminus \mathrm{sp}_A(b)$ has exactly one bounded component ($\mathrm{sp}_A(b)$ has no holes), then $\mathrm{sp}_A(b) = \mathrm{sp}_B(b)$.

1.6.16. Let A be a unital Banach algebra.

(i) Let $a \in \text{Inv}(A)$. Show that $\text{sp}(a^{-1}) = \{\lambda^{-1} \mid \lambda \in \text{sp}(a)\}$.

(ii) If $a \in A$ show that $r(a^n) = (r(a))^n$.

(iii) If A is commutative and $a, b \in A$, show that $\text{sp}(a + b) \subset \text{sp}(a) + \text{sp}(b)$.

(iv) If A is noncommutative show that (iii) need not hold.

1.6.17. Let $\varphi : A \to B$ be a homomorphism of Banach algebras A and B. Show that $\ker(\varphi)$ is an ideal in A.

1.6.18. Let I be an algebraic ideal in a unital Banach algebra A. Show that if I is maximal, then $I = \overline{I}$.

1.6.19. Let $\mathbb{D}$ denote the closed unit disc in the plane. Let $A \subset C(\mathbb{D})$ denote the subalgebra of functions which are analytic on the interior of $\mathbb{D}$. The algebra A is called the *disc algebra*.

(i) Show that A is a unital abelian Banach algebra.

(ii) Let $B := \{f \in A \mid f(0) = 0\}$. Show that B is a closed subalgebra of A

(iii) Show that B contains an ideal which is maximal but not modular.

1.6.20. Let A be a Banach algebra.

(a) Show that every proper modular ideal is contained in a maximal modular ideal.

(b) Let I be a maximal modular ideal. Show that A/I is a field.

1.6.21. Prove Corollary 1.4.4: If A is a nonunital commutative Banach algebra and $a \in A$, then
$$\text{sp}(a) = \{\tau(a) \mid \tau \in \Omega(A)\} \cup \{0\}.$$

1.6.22. Let A be a nonunital commutative Banach algebra with $\Omega(A) \neq \emptyset$. Prove the second part of Theorem 1.5.2: for any $a \in A$ we have $\text{sp}(a) = \hat{a}(\Omega(A)) \cup \{0\}$.

2 C*-algebra basics

In this chapter we introduce the subject of the book, C*-algebras. We will see that equipping a Banach algebra with an involution and asking that the norm satisfy a seemingly mild condition – the C*-equality – will have a significant effect on the structure of what we call C*-algebras. They are in general much more tractable than arbitrary Banach algebras. We will already meet some of the implications of these minor conditions in this chapter. For example, we will see that the structure-preserving maps, the so-called *-homomorphisms, are automatically continuous. We will also revisit the Gelfand transformation of the last chapter within the context of C*-algebras and prove one of the most fundamental results in C*-algebras: that a commutative C*-algebra is *always*, up to isometric *-isomorphism, of the form $C(X)$ for some locally compact Hausdorff space X.

After covering basic definitions and examples of C*-algebras in the first section, we will see how one can adjoin a unit to a nonunital C*-algebra; the situation is a little bit different from what we saw for an arbitrary Banach algebra, and we can also define a maximal unitisation in the form of the *multiplier algebra*. In Section 2.3 we revisit the Gelfand transform and prove the Gelfand–Naimark Theorem which characterises all commutative C*-algebras. In the final section, we apply the Gelfand–Naimark Theorem to define a *continuous functional calculus* for normal elements, one of the most important tools in C*-algebraic theory.

2.1 Basic definitions and examples

A *-algebra is a $\mathbb{C}$-algebra A together with an *involution*, which is to say, a map
$* : A \to A$, $a \mapsto a^*$ satisfying $(a^*)^* = a$ and $(ab)^* = b^*a^*$ for every $a, b \in A$.

2.1.1. Given an element a in a *-algebra A, we call a^* the *adjoint* of a. An element in A is called *self-adjoint* if $a = a^*$. An element $p \in A$ is called a *projection* if it is self-adjoint and $p^2 = p$. An element $a \in A$ that commutes with its adjoint, $a^*a = aa^*$, is called *normal*. When A is unital and $u \in A$ is a normal element such that $u^*u = 1$ then we call u a *unitary*.

We will denote the set of self-adjoint elements in A by A_{sa} and the unitaries by $\mathcal{U}(A)$.

2.1.2. If A is a Banach algebra with involution *, then we call A a Banach *-algebra if $\|a^*\| = \|a\|$ for every $a \in A$. Note that by submultiplicativity this implies that $\|a^*a\| \le \|a\|^2$ for every $a \in A$.

2.1.3. Definition. An *abstract* C*-*algebra* is a complete normed *-algebra A satisfying the C*-*equality*:

$$\|a^*a\| = \|a\|^2 \text{ for every } a \in A.$$

We call a norm satisfying the C*-equality a C*-*norm*.

© The Author(s), under exclusive license to Springer Nature Switzerland AG 2021
K. R. Strung, *An Introduction to C*-Algebras and the Classification Program*, Advanced
Courses in Mathematics - CRM Barcelona, https://doi.org/10.1007/978-3-030-47465-2_2

2.1.4. Note that a C*-algebra is always a Banach *-algebra, but the converse need not hold. Consider, for example $\ell^1(\mathbb{Z})$, the space of sequences indexed by $\mathbb{Z}$ such that

$$\|x\| := \sum_{n=-\infty}^{\infty} |x_n| < \infty.$$

Equip $\ell^1(\mathbb{Z})$ with the convolution product as multiplication, that is, if $x, y \in \ell^1(\mathbb{Z})$, we define

$$(xy)_n = \sum_{j=-\infty}^{\infty} x_j y_{n-j},$$

and an involution defined by

$$(x^*)_n = \overline{x}_{-n}.$$

Note that the norm is submultiplicative with respect to this multiplication and that $\|x^*\| = \|x\|$ for every $x \in \ell^1(\mathbb{Z})$. However, the norm does not satisfy the C*-equality. For example, take $x \in \ell^1(\mathbb{Z})$ where $x_0 = 1$, $x_1 = x_2 = -1$, and $x_n = 0$ for every other $n \in \mathbb{Z}$. Then $\|x^*x\| = 5$ while $\|x\|^2 = 9$.

What may appear to be only a minor requirement for the norm in fact gives a C*-algebra many nice structural properties that we don't see in an arbitrary Banach algebra, or even a Banach *-algebra.

2.1.5. Definition. Let A and B be C*-algebras. A *-homomorphism* $\varphi : A \to B$ is an algebra homomorphism that is involution-preserving, that is, $\varphi(a^*) = \varphi(a)^*$ for every $a \in A$. If A and B are unital, then we say that φ is a *unital *-homomorphism* if $\varphi(1_A) = 1_B$.

Observe that in the above definition, unlike the definition of a Banach algebra homomorphism, we *do not* require that a *-homomorphism between C*-algebras is continuous. Rather, continuity turns out to be automatic, as we will see in Proposition 2.2.5.

2.1.6. Let A be a C*-algebra. Unless otherwise specified, by an *ideal* $I \subset A$ we will mean a two-sided ideal $I \subset A$ that satisfies $I = \overline{I}$ and $I^* = I$, that is to say, is closed and self-adjoint. If A has no nontrivial ideals, then A is called *simple*. Simple C*-algebras will be the main focus of later chapters.

2.1.7. Example. Let $n \in \mathbb{N} \setminus \{0\}$. We denote by $M_n := M_n(\mathbb{C})$ the set of $n \times n$ matrices with complex entries. Equip M_n with the operator norm,

$$\|A\| = \sup_{x \in \mathbb{C}^n, \|x\|=1} \|A(x)\|.$$

Then M_n is a C*-algebra under the usual matrix multiplication and addition and with involution given by taking adjoints (where the adjoint means the conjugate transpose).

More generally, if H is a Hilbert space then $\mathcal{B}(H)$, equipped with the operator norm, is a C*-algebra; the proof is left as an exercise.

2.1.8. Example. Let H be a Hilbert space and let $\mathcal{K}(H)$ denote the subalgebra of compact operators. Then $\mathcal{K}(H)$ is also a C*-algebra with the inherited structure from $\mathcal{B}(H)$.

More generally, if A is any closed self-adjoint subalgebra of $\mathcal{B}(H)$, then A is also a C*-algebra with the inherited structure. Such a C*-algebra is called a *concrete* C*-algebra.

2.1.9. Example. Let X be a locally compact Hausdorff space. We say that a function $f : X \to \mathbb{C}$ *vanishes at infinity* if, for every $\epsilon > 0$, there is a compact set $K \subset X$ such that $|f(x)| < \epsilon$ for every $x \in X \setminus K$. Let

$$C_0(X) := \{f : X \to \mathbb{C} \mid f \text{ is continuous and vanishes at infinity}\},$$

and equip $C_0(X)$ with pointwise operations, supremum norm and for $f \in C_0(X)$, define $f^*(x) := \overline{f(x)}$. Then $C_0(X)$ is a C*-algebra. It is unital if and only if X is compact, in which case we denote $C_0(X)$ by $C(X)$.

2.1.10. We can already observe some structural properties that we obtain from simply having a C*-norm. For example, it is automatic in a unital C*-algebra A that we have $\|1_A\| = 1$. More generally, if u is a unitary, $\|u\| = 1$ and also if p is a projection then $\|p\| = 1$. This gives us information about the spectrum of a unitary u: Using Lemma 1.2.6, if $\lambda \in \mathrm{sp}(u)$ then $|\lambda| \leq \|u\| = 1$. Since u is invertible, we must also have $\lambda^{-1} \in \mathrm{sp}(u^{-1}) = \mathrm{sp}(u^*) \leq \|u^*\| = 1$. Thus $|\lambda| = 1$, so $\mathrm{sp}(u)$ is a closed subset of $\mathbb{T}$.

2.1.11. With a bit more work we can also show that for any $a \in A_{sa}$ we have $\mathrm{sp}(a) \subset \mathbb{R}$. First, note that for any element a in a unital Banach algebra

$$\sum_{n=0}^{\infty} \left\| \frac{a^n}{n!} \right\| \leq \sum_{n=0}^{\infty} \frac{\|a\|^n}{n!},$$

and so $\sum_{n=0}^{\infty} \frac{a^n}{n!}$ converges. We set

$$e^a := \sum_{n=0}^{\infty} \frac{a^n}{n!}.$$

For any $a \in A$, one can check that the map

$$\phi : \mathbb{R} \to A, \quad t \mapsto e^{ta}$$

is differentiable at every $t \in \mathbb{R}$ with derivative $a\phi(t)$. It also satisfies $\phi(0) = e^0 = 1$. These properties completely characterise the function $t \mapsto e^{ta}$ in the sense that if ψ is another function with these properties then necessarily $\phi = \psi$. (The details are left as an exercise; recall how this is done in the case that $A = \mathbb{R}$.)

Using this characterisation, it follows that $e^{a+b} = e^a e^b$ for any $a, b \in A$ with $ab = ba$. In particular, e^a is always invertible, with inverse e^{-a}.

2.1.12. Now let A be a unital C*-algebra. If a is self-adjoint, then e^{ia} is invertible of norm 1, hence it is a unitary (Exercise 2.5.4). As we saw in 2.1.10, this implies $\mathrm{sp}(e^{ia}) \subset \mathbb{T}$. Let $\lambda \in \mathrm{sp}(a)$. Let

$$b := \sum_{n=2}^{\infty} i^n \frac{(a-\lambda)^{n-1}}{n!}.$$

Note that b commutes with a. We have

$$e^{ia} - e^{i\lambda} = (e^{i(a-\lambda)} - 1_A)e^{i\lambda} = (a-\lambda)be^{i\lambda}.$$

Since b commutes with a and hence commutes with $(a-\lambda)$, and since $(a-\lambda)$ is not invertible, we see that $e^{ia} - e^{i\lambda}$ is not invertible. Thus $e^{i\lambda} \in \mathrm{sp}(e^{ia}) \subset \mathbb{T}$ so we must have $\lambda \in \mathbb{R}$.

2.1.13. Theorem. *Let A be a C*-algebra and let $a \in A_{sa}$. Then $r(a) = \|a\|$.*

Proof. Exercise. $\qquad \square$

This means that the norm of an element a in a C*-algebra A depends *only* on spectral data: if $a \in A$ then a^*a is self-adjoint and

$$\|a\| = \|a^*a\|^{1/2} = (r(a))^{1/2} = \left(\sup_{\lambda \in \mathrm{sp}(a)} |\lambda| \right)^{1/2}.$$

This gives us the next theorem.

2.1.14. Theorem. *If A is a *-algebra admitting a norm which makes A into a C*-algebra then that norm is the unique C*-norm on A.*

Proof. Exercise. $\qquad \square$

2.2 Minimal unitisations and multiplier algebras

We saw in 1.2.12 that a nonunital Banach algebra A can be embedded into a unital Banach algebra $\tilde{A}$. There is also a way of defining a unitisation (in fact, more than one) of a nonunital C*-algebra. Unfortunately, we cannot simply take $A \oplus \mathbb{C}$ with multiplication and norm as given in Exercise 1.2.12. The reason is that the norm there is not a C*-norm (check!). So we have to be a bit more careful in how we adjoin a unit to a nonunital C*-algebra.

2.2.1. Let A and B be C*-algebras. If $T : A \to B$ is a linear operator, we equip it with the operator norm:

$$\|T\| := \sup_{a \in A, \|a\| \leq 1} \|T(a)\|_B,$$

which is just the usual norm for a linear operator if we regard A and B as Banach spaces.

2.2.2. Let $\tilde{A} = A \oplus \mathbb{C}$ as a vector space. We endow $\tilde{A}$ with the same multiplication as for the Banach algebra unitisation of A. Define an involution $* : \tilde{A} \to \tilde{A}$ by $(a, \lambda)^* = (a^*, \bar{\lambda})$. Now, to make A into a C*-algebra, we view the elements of $\tilde{A}$ as left multiplication operators on A,

$$\|(a, \lambda)\|_{\tilde{A}} := \sup\{\|ab + \lambda b\|_A \mid b \in A, \|b\| \leq 1\}.$$

One then checks that this makes $\tilde{A}$ into a unital C*-algebra. Moreover, the *-homomorphism $a \mapsto (a, 0)$ identifies A as an ideal in $\tilde{A}$. We call $\tilde{A}$ the *minimal unitisation of A*, or sometimes simply the *unitisation*. This norm is the unique norm making $\tilde{A}$ a C*-algebra and unless otherwise specified, this is the norm we use (rather than the one defined in Exercise 1.2.12) for $\tilde{A}$.

2.2.3. Example. Let $A = C_0(X)$ where X is a locally compact Hausdorff space. Then $\tilde{A} = C(X \cup \{\infty\})$ where $X \cup \{\infty\}$ is just the one point compactification of X. In this way, we think of adjoining a unit as the noncommutative version of one-point compactification.

The nice thing about the unitisation is that we will now be able to prove many theorems in the (usually easier) unital setting without any loss of generality.

2.2.4. If B is unital and $\varphi : A \to B$ is a *-homomorphism then there exists a *unique* extension $\tilde{\varphi} : \tilde{A} \to B$ such that $\tilde{\varphi}$ is unital (exercise). Observe that this implies that if B is any unitisation of A, the inclusion map $A \hookrightarrow B$ extends to an injective *-homomorphism $\tilde{A} \hookrightarrow B$, so we can think of $\tilde{A}$ as a C*-subalgebra of B. This justifies the term "minimal" unitisation for $\tilde{A}$.

2.2.5. The following proposition is another nice implication of the C*-equality.

Proposition. *A *-homomorphism* $\varphi : A \to B$ *between* C*-*algebras* A *and* B *is always norm-decreasing, that is,* $\|\varphi(a)\|_B \leq \|a\|_A$ *for every* $a \in A$. *In particular, it is always continuous.*

Proof. Let $\varphi : A \to B$ be a *-homomorphism, and let $a \in A$. By replacing A and B by their unitisations if necessary, we can assume that φ, A and B are all unital. Since $\varphi(1_A) = 1_B$, it is easy to see that if a is invertible then $\varphi(a)$ is invertible in B. It follows that $\mathrm{sp}(\varphi(a)) \subset \mathrm{sp}(a)$. The result follows from Theorem 2.1.13. $\square$

2.2.6. Corollary. *Any *-isomorphism of* C*-*algebras is automatically isometric.*

2.2.7. The minimal unitisation $\tilde{A}$ of a C*-algebra A is the smallest unital C*-algebra in which A sits as an ideal. Another frequently used unitisation is called the *multiplier algebra* of A.

A *left multiplier* L of A is a bounded linear operator $L : A \to A$ satisfying $L(ab) = L(a)b$ for every $a, b \in A$. Similarly one defines a *right multiplier* $R : A \to A$ as a bounded linear operator satisfying $R(ab) = aR(b)$ for every $a, b \in A$.

To define the multiplier algebra of A, we consider pairs of left and right multipliers (L, R) with the compatibility condition $aL(b) = R(a)b$ for every $a \in A$. The pair (L, R) is called a *double centraliser*.

We denote the set of such pairs by $M(A)$. We call $M(A)$ the *multiplier algebra* of A. An ideal is called *essential* if it has nonempty intersection with every other ideal in A. As the name suggests, we can give the multiplier algebra the structure of an algebra, in fact a unital C*-algebra, which we show below. The multiplier algebra $M(A)$ is a unital C*-algebra containing A as an essential ideal (Proposition 2.2.10).

2.2.8. Proposition. *Let $(L, R) \in M(A)$. Then $\|L\| = \|R\|$, so we define*

$$\|(L, R)\| := \|L\|.$$

Proof. Note that $\|L(b)\| = \sup\{\|aL(b)\| \mid a \in A, \|a\| \leq 1\}$. From this we have

$$
\begin{aligned}
\|L(b)\| &= \sup_{\|a\| \leq 1} \|aL(b)\| \\
&= \sup_{\|a\| \leq 1} \|R(a)b\| \\
&\leq \sup_{\|a\| \leq 1} \|R(a)\|\|b\| \\
&= \|R\|\|b\|.
\end{aligned}
$$

Thus $\|L\| = \sup\{\|L(b)\| \mid b \in A, \|b\| \leq 1\} \leq \|R\|$. One shows similarly that $\|R\| \leq \|L\|$ from which the result follows. $\qquad\square$

It is easy to check that we can give $M(A)$ the structure of a vector space by viewing it as a closed subspace as $\mathcal{B}(A) \oplus \mathcal{B}(A)$, where $\mathcal{B}(A)$ denotes the space of bounded linear operators $A \to A$. To show that $M(A)$ is a C*-algebra we need to define the multiplication and adjoint and then check that the norm above is indeed a C*-norm.

2.2.9. Let $L : A \to A$ be a left multiplier. Define $L^* : A \to A$ by

$$L^*(a) = L(a^*)^*, \quad a \in A.$$

Similarly, for a right multiplier $R : A \to A$ define $R^* : A \to A$ by

$$R^*(a) = R(a^*)^*, \quad a \in A.$$

The adjoint of $(L, R) \in M(A)$ is then given by $(L, R)^* = (R^*, L^*)$ and for $(L_1, R_1), (L_2, R_2) \in M(A)$, we define the multiplication by

$$(L_1, R_1)(L_2, R_2) = (L_1 L_2, R_2 R_1).$$

It is not hard to check that with these operations and the norm defined above $M(A)$ is a unital C*-algebra with unit $(\mathrm{id}_A, \mathrm{id}_A)$.

2.2.10. Given any $a \in A$, we can define $(L_a, R_a) \in M(A)$ by $L_a(b) = ab$ and $R_a(b) = ba$ for $b \in A$.

Proposition. *Let A be a C^*-algebra. Then*

$$A \to M(A), \quad a \to (L_a, R_a)$$

identifies A an essential ideal inside $M(A)$, which is proper if and only if A is nonunital.

Proof. Since the map is easily seen to be an isometric *-homomorphism, we can identify A with its image. Then the only thing we need to check is that A is an essential ideal of $M(A)$. Let us show that the image of A is closed under multiplication on the left and the right by elements in $M(A)$. For a double centraliser $(L, R) \in M(A)$ set $L(a) = x$. Then $(L, R)(L_a, R_a) = (LL_a, R_a R)$. Let $b \in A$. Then $LL_a(b) = L(ab) = L(a)b = L_x(b)$, and $R_a R(b) = R(b)a = bL(a) = bx = R_x(b)$. Thus $(L, R)(L_a, R_a) = (L_x, R_x) \in A$. Similarly one shows that $(L_a, R_a)(L, R) \in A$. So A is an ideal.

To see that A is essential, let J be any proper ideal in $M(A)$. Then there is some nonzero $(L, R) \in J$, which is to say that there is some $a \in A$ with $L(a) \neq 0$. Then $(L, R)(L_a, R_a) \in J \cap A$ is nonzero. So A is essential.

Finally, if A is nonunital the map is proper since the unit of $M(A)$, $(\mathrm{id}_A, \mathrm{id}_A)$, is not in A. Conversely, if A is unital then $1_A \mapsto (\mathrm{id}_A, \mathrm{id}_A)$ which implies $1_{M(A)} \in A$ so $A = M(A)$. $\qquad\square$

The multiplier algebra is in general much larger than the minimal unitisation, and it can be a bit unwieldy. In fact, for a C^*-algebra A, its multiplier algebra $M(A)$ is the largest unital C^*-algebra in which A sits as an essential ideal (see Exercise 2.5.13). For example, if X is a locally compact Hausdorff space which is not compact, then $M(C_0(X)) \cong C(\beta X)$ where βX is the Stone–Čech compactification of X (see for example [91, Proposition 4.3.18]). The multiplier algebra of a C^*-algebra is in this way thought of as its noncommutative Stone–Čech compactification.

2.3   The Gelfand–Naimark Theorem

Recall that $\Omega(A)$ is the spectrum, or character space, of A (see 1.4.1). If X is locally compact and Hausdorff and $x \in X$ then the $\mathrm{ev}_x(f) = f(x)$ is a character on $C_0(X)$. In fact, all characters on $C_0(X)$ are of this form. We show the result for compact X.

2.3.1. Remark. The proof of the next theorem uses the notion of a net. The reader who is unfamiliar with nets can usually replace "net" with "sequence" and "Hausdorff space" with "metric space" or "metrisable space". Nets will also occur in the sequel when we don't assume our C^*-algebras are separable. If a C^*-algebra is separable, then an occurrence of a net can often be replaced with a sequence.

2.3.2. Theorem. *Let $A = C(X)$ for some compact Hausdorff space X. Then*

$$\Omega : X \to \Omega(A), \quad x \to \mathrm{ev}_x$$

is a homeomorphism.

Proof. Let $(x_\lambda)_\Lambda$ be a net in X with $\lim_\lambda x_\lambda \to x \in X$. Then $\mathrm{ev}_{x_\lambda}(f) = f(x_\lambda) \to f(x) = \mathrm{ev}_x$ for every $f \in C(X)$, so $(\mathrm{ev}_\lambda)_\Lambda$ is weak-* convergent to ev_x. Thus the map is continuous.

Suppose now that $x \neq y \in X$. Then by Urysohn's Lemma, there is $f \in C(X)$ such that $f(x) = 1$ and $f(y) = 0$. Thus $\mathrm{ev}_x \neq \mathrm{ev}_y$ and so we see that the map is injective.

Now let us prove surjectivity. Let τ be a character on $C(X)$. Let $M := \ker(\tau)$; this is a maximal, hence proper, ideal in $C(X)$. We show that M separates points. If $x \neq y$ there is $f \in C(X)$ with $f(x) = 1$ and $f(y) = 0$. Now $f - \tau(f) \in M$ satisfies $f(x) - \tau(f) \neq f(y) - \tau(f)$ so by the Stone–Weierstrass theorem, there is $x \in X$ such that $f(x) = 0$ for every $f \in M$.

Thus $(f - \tau(f))(x) = 0$ and so $f(x) = \tau(f)$ for every $f \in C(X)$. It follows that $\mathrm{ev}_x = \tau$ and so Ω is surjective.

Since any continuous bijective map from a compact space is a homeomorphism, the result follows. $\square$

2.3.3. Let A be a *-algebra. Any element $a \in A$ can be written as $a = a_1 + ia_2$ where a_1, a_2 are self-adjoint (exercise). In this sense, the self-adjoint elements play the role of "real" elements in A, in analogy with real numbers in $\mathbb{C}$. We often call a_1 and a_2 the real and imaginary parts of a, respectively.

2.3.4. Now we come to one of the most important results in C*-algebra theory: for a commutative C*-algebra the Gelfand transform of 1.5.1 is an isometric *-isomorphism. This gives us a complete characterisation of commutative C*-algebras: they are *always*, up to *-isomorphism, of the form $C_0(X)$ for some locally compact Hausdorff space X. The next theorem is usually referred to as the Gelfand–Naimark Theorem or simply the Gelfand Theorem (perhaps to avoid confusion with the second Gelfand–Naimark Theorem 4.2.6)

Theorem (Gelfand–Naimark). *Let A be a commutative* C*-*algebra. Then the Gelfand transform*

$$\Gamma : A \to C_0(\Omega(A)), \quad a \to \hat{a}$$

is an isometric *-*isomorphism.*

Proof. If $\phi \in \Omega(A)$ then $\phi(a) \in \mathbb{R}$ whenever $a \in A_{sa}$. Thus for any $c \in A$, where $c = a + ib$ with $a, b \in A_{sa}$, we have $\phi((a + ib)^*) = \phi(a - ib) = \phi(a) - i\phi(b) = \overline{(\phi(a) + i\phi(b))}$, which is to say, ϕ is a *-homomorphism. It follows that $\hat{a}^*(\phi) = \overline{\hat{a}(\phi)}$ for any $a \in A$ and any $\phi \in C_0(\Omega(A))$ meaning Γ is a *-homomorphism. This moreover implies that $\|\hat{a}\|^2 = \|\hat{a}^*\hat{a}\| = \|\widehat{a^*a}\| = r(a^*a) = \|a^*a\| = \|a\|^2$, where the third equality comes from Theorem 1.5.2 and the fourth from Theorem 2.1.13. Thus the map is isometric. Finally, to see that the map is also surjective, we appeal to the Stone–Weierstrass Theorem: the image of A under Γ separates points and contains functions which do not simultaneously vanish anywhere on $\Omega(A)$. Thus $\Gamma(A)$ is exactly $C_0(\Omega(A))$.   □

2.4   The continuous functional calculus

At first glance, Theorem 2.3.4 applies to the relatively small class of commutative C*-algebras. While it is true that in greater generality we don't have such an explicit characterisation for a class of C*-algebras, what we do get is an extremely useful tool: the continuous functional calculus for normal elements.

Given a C*-algebra A and a set of elements $S \subset A$ the C*-subalgebra of A generated by the set S is denoted by $C^*(S)$. In other words, $C^*(S)$ is the intersection of all C*-subalgebras of A containing the set S.

2.4.1. Theorem. *Let A be a unital C*-algebra and let $a \in A$ be a normal element. Then the map*

$$\gamma : C(\mathrm{sp}(a)) \to A, \quad (z \mapsto z) \mapsto a$$

*is an isometric *-homomorphism and $\gamma(C(\mathrm{sp}(a))) = C^*(a, 1_A)$.*

Proof. Since a is a normal element, $C^*(a, 1_A)$ is commutative. Thus by Theorem 2.3.4 we have a *-isomorphism

$$\Gamma : C^*(a, 1) \to C(\Omega(C^*(a, 1_A))), \quad a \mapsto \hat{a}.$$

By Theorem 1.5.3, $h : \Omega(C^*(a, 1_A)) \to \mathrm{sp}(a)$ is a homeomorphism so we also have an isomorphism

$$\psi : C(\mathrm{sp}(a)) \to C(\Omega(C^*(a, 1_A))), \quad f \mapsto f \circ h.$$

Let $f(z) = z$ for $z \in \mathrm{sp}(a)$. Let $\gamma = \Gamma^{-1} \circ \psi$. Since $C(\mathrm{sp}(a))$ is generated by 1_A and f, γ is the unique unital *-homomorphism with $\gamma(f) = a$. Clearly γ is isometric and its image is $C^*(a, 1_A)$.   □

It is also possible to formulate a nonunital version of Theorem 2.4.1 (exercise). The reason for highlighting the unital version is the following categorical result:

2.4.2. Theorem. *The correspondence between X and $C(X)$ is a categorical equivalence between the category of compact Hausdorff spaces and continuous maps to the category of unital C*-algebras and unital *-homomorphisms.*

2.4.3. This is where we get the nomer "noncommutative topology" for the study of C*-algebras. In general it useful to think of the C*-landscape as having two coasts at opposite ends, one of which consists of bounded operators on Hilbert spaces and matrix algebras, the other consisting of commutative C*-algebras. The interesting part of the theory comes as we move inland, as most C*-algebras lie somewhere in between. Some of our best tools are brought in via either coast and it is often useful to keep these two examples in mind.

With the continuous functional calculus result in hand, we will be able to do a lot more in our C*-algebras. If $p \in \mathbb{C}[z_1, z_2]$ is a polynomial, then since A is an algebra, it is clear that $p(a, a^*) \in C^*(a, 1_A)$. (In the nonunital case we would require that p have no constant terms; then $p(a, a^*) \in C^*(a)$.) Since polynomials p of this form are dense in $C(\mathrm{sp}(a))$, we can then define, using Theorem 2.4.1,

$$f(a) := \gamma(f) \in C^*(a, 1_A),$$

or when A is nonunital, for $f \in C_0(\mathrm{sp}(a))$,

$$f(a) := \gamma(f) \in C^*(a),$$

where γ is as in Theorem 2.4.1.

The following is sometimes called the *Spectral Mapping Theorem*.

2.4.4. Theorem. *Let A be a C*-algebra and let $a \in A$ be a normal element. Then for any $f \in C_0(\mathrm{sp}(a))$ the element $f(a)$ is normal and we have*

$$\mathrm{sp}(f(a)) = f(\mathrm{sp}(a)).$$

Furthermore, if $g \in C_0(\mathrm{sp}(f(a))$ then $g \circ f(a) = g(f(a))$.

Proof. Exercise.   □

2.5 Exercises

2.5.1. Let A be a *-algebra and let $a \in A$. Describe $\mathrm{sp}(a^*)$ in terms of $\mathrm{sp}(a)$.

2.5.2. Let H be a (finite- or infinite-dimensional) Hilbert space. Show that the operator norm satisfies the C*-equality and that $\mathcal{B}(H)$ is a *-algebra.

2.5.3. Let A be a C*-algebra and $p \in A$ a nonzero projection. What is $\mathrm{sp}(p)$?

2.5.4. Let A be a unital C*-algebra and let $u \in \mathrm{Inv}(A)$. Show that u is a unitary if and only if $\|u\| = \|u^{-1}\| = 1$.

2.5.5. Let A be a commutative C*-algebra and suppose that A contains a nonzero projection p. Show that if $p \neq 1_A$ then $\Omega(A)$ is disconnected.

2.5.6. Let $\mathcal{H} = \ell^2(\mathbb{N}) = \{(\lambda_n)_{n \in \mathbb{N}} \mid \sum_{n=0}^{\infty} |\lambda_n|^2 \text{ converges}\}$. Define the unilateral shift operator $S : \ell^2(\mathbb{N}) \to \ell^2(\mathbb{N})$ by

$$S((\lambda_n)_{n \in \mathbb{N}}) = (\mu_n)_{n \in \mathbb{N}},$$

where $\mu_n = \lambda_{n-1}$.

(i) Show that $S \in \mathcal{B}(H)$

(ii) What is S^*? Is S invertible? If so, what is its inverse?

(iii) Show that S has no eigenvalues (that is, for every $\lambda \in \mathbb{C}$ there is no $\xi \in \mathcal{H}$ such that $S\xi = \lambda \cdot \xi$.)

(iv) Show that if $|\lambda| = 1$ then $\lambda \cdot 1_{\mathcal{B}(H)} - S$ is not invertible.

2.5.7. Let $\psi : \mathbb{R} \to A$ be a differentiable map with derivative $a\psi(t)$ and $\psi(0) = 1$. Show that $\psi(t) = e^{ta}$.

2.5.8. Let A be a C*-algebra. Show that any element $a \in A$ can be written as $a = a_1 + ia_2$ where a_1, a_2 are self-adjoint.

2.5.9. Let A be a C*-algebra. Let $a, b \in A$.

(i) Let $f : \mathbb{C} \to \tilde{A}$ be defined by $f(\lambda) = e^{i\lambda a^*} b e^{-i\lambda a^*}$ is differentiable and $f'(0) = i(a^*b - ba^*)$.

(ii) Suppose that a is normal and b commutes with a. By Exercise 2.5.8, we can write $\lambda a^* = c_1(\lambda) + ic_2(\lambda)$ where $c_1(\lambda)$ and $c_2(\lambda)$ are self-adjoint. Show that $f(\lambda) = e^{2ic_1(\lambda)} b e^{-2ic_1(\lambda)}$ and hence $f(\lambda)$ is bounded.

(iii) Suppose that a is normal and b commutes with a. Use (ii) and Louiville's Theorem to conclude that a also commutes with b^*. This result is called *Fuglede's Theorem* as it was first proved by Fuglede in [47].

2.5.10. Below, $\mathcal{K}$ denotes the C*-algebra of compact operators on a separable Hilbert space. Since there is only one separable Hilbert space up to isomorphism we often don't reference the underlying Hilbert space.

(i) Give an example of a finite-dimensional simple C*-algebra. Prove that it is simple.

(ii) Show that the compact operators $\mathcal{K}$ are a simple C*-algebra.

2.5.11. Show that if a unital C*-algebra A is simple, then A contains no nontrivial two-sided *algebraic* ideals.

2.5.12. Let $A = C(X)$ where X is a compact Hausdorff space. Show that if $F \subset X$ is a closed subset then

$$\{f \in A \mid f|_F = 0\}$$

is an ideal. Show that every ideal in A has this form. Describe the simple commutative C*-algebras.

2.5.13. Let A be a nonunital C*-algebra. Suppose that B is a unitisation of A in which A is an essential ideal and let $M(A)$ denote the multiplier algebra of A. Show that there is an injective *-homomorphism $\varphi : B \hookrightarrow M(A)$, the multiplier algebra of A.

2.5.14. Let H be a Hilbert space. Show that the compact operators $\mathcal{K}(H)$ form an ideal in $\mathcal{B}(H)$.

2.5.15. Let A be a nonunital C*-algebra. Show that the unitisation of A as a Banach algebra, as given in 1.2.12, does not define a C*-algebra.

2.5.16. Let $A \subset \mathcal{B}(H)$ be a concrete C*-algebra. Show that $M_n(A)$, the *-algebra of $n \times n$ matrices with entries in A, admits a C*-norm and is a C*-algebra with respect to this norm.

2.5.17. (Uniqueness of the C*-norm) Let A be a C*-algebra and let $a \in A_{sa}$. Show that $r(a) = \|a\|$, and hence if A is a *-algebra admitting a complete C*-norm, then the norm is the unique C*-norm on A.

2.5.18. Let A be a unital C*-algebra. Suppose $a, b \in A$ are normal elements that are unitarily equivalent (that is, there exists a unitary $u \in \mathcal{U}(A)$ such that $u^*au = b$). Show that the C*-subalgebras $C^*(a, 1)$ and $C^*(b, 1)$ are *-isomorphic.

2.5.19. Let A be a unital C*-algebra. Let $a \in A_{sa}$ and $0 < \epsilon < 1/4$. Suppose $\mathrm{sp}(a) \subset [0, \epsilon] \cup [1 - \epsilon, 1]$. Show that there is a projection $p \in A$ with $\|p - a\| \leq \epsilon$.

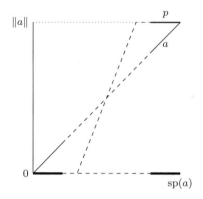

2.5.20. Write down and prove the nonunital version of Theorem 2.4.1.

2.5.21. Let $\iota : (0, 1) \hookrightarrow \mathbb{R}$ be the inclusion map. Then ι is continuous. Show that $\iota_* : C_0(\mathbb{R}) \to C_0((0, 1))$ defined by $\iota_*(f) = f \circ \iota$ is not a *-homomorphism. What goes wrong?

2.5.22. Prove the spectral mapping theorem.

2.5.23. Let $a \in A$ be a normal element. Suppose that $\lambda \notin \mathrm{sp}(a)$. Show that

$$\|(\lambda 1_A - a)^{-1}\| = \frac{1}{\mathrm{dist}(\lambda, \mathrm{sp}(a))}.$$

2.5.24. Show that C*-algebra A is simple if and only if, whenever B is another nonzero C*-algebra and $\psi : A \to B$ is a surjective *-homomorphism, then φ is injective.

2.5.25. Let A be a unital C*-algebra and $a, b \in A$ with b normal. Show that if $f \in C(\mathrm{sp}(b))$ and a commutes with b, then a commutes with $f(b)$.

3 Positive elements

Positive elements play a very important role in the theory of C*-algebras. They allow us to define a partial order on the self-adjoint elements which is a key part of the structure of any C*-algebra. We will see, for example, that the positive elements provide nonunital C*-algebras (so for example, any ideal in a C*-algebra) with *approximate units*, useful objects that need not exist in arbitrary Banach algebras. Positive elements also generate *hereditary* C*-subalgebras, which are C*-subalgebras that retain much of the structure – for example, simplicity, which we prove below – of the larger C*-algebra.

In the first section, we define positive elements and the partial order on self-adjoints, and show that we can use the functional calculus to determine how various self-adjoint elements relate to one another with respect to this order structure. In the second section, we prove the existence of approximate units, define hereditary C*-subalgebras and see how they relate to closed left ideals. We show that a separable hereditary C*-subalgebra is generated by a single positive element. Finally, in Section 3.3 we show that an operator in $\mathcal{B}(H)$ admits a polar decomposition, and collect similar factorisation results for elements in arbitrary C*-algebras where polar decompositions need not exist.

3.1 Positivity and partial order

We will give the functional calculus a workout in this section. The first thing is to define positive elements in a C*-algebra as well as a partial order on its self-adjoint elements.

3.1.1. Let A be a C*-algebra. An element $a \in A$ is said to be *positive* if $a \in A_{sa}$ and $\mathrm{sp}(a) \subset [0, \infty)$. The set of positive elements is denoted A_+.

3.1.2. Definition. If S is a set and $\leq$ is a binary relation, then $\leq$ is said to be a *partial order* if

(i) $\leq$ is reflexive: for every $x \in S$, $x \leq x$,
(ii) $\leq$ is transitive: for every $x, y, z \in S$, if $x \leq y$ and $y \leq z$ then $x \leq z$,
(iii) $\leq$ is symmetric: for every $x, y \in S$, if $x \leq y$ and $y \leq x$ then $x = y$.

3.1.3. Let A be a C*-algebra. For $a, b \in A_{sa}$, write $a \leq b$ if $b - a \in A_+$. We leave it as an easy exercise to show that this defines a partial order on A_{sa}. (Note in particular that if $a \geq 0$ then a is positive.)

3.1.4. Given an element a of a C*-algebra A, a *square root* of a is an element $b \in A$ satisfying $b^2 = a$.

Lemma. *Let X be a locally compact Hausdorff space and let $f \in C_0(X)$ satisfy $f(x) \geq 0$ for all $x \in X$. Then f has a unique positive square root.*

© The Author(s), under exclusive license to Springer Nature Switzerland AG 2021
K. R. Strung, *An Introduction to C*-Algebras and the Classification Program*, Advanced Courses in Mathematics - CRM Barcelona, https://doi.org/10.1007/978-3-030-47465-2_3

Proof. For every x, the real number $f(x)$ has a unique positive square root, $f(x)^{1/2}$. Define $f^{1/2}(x) := f(x)^{1/2}$. Then $f^{1/2} \in C_0(X)$ and clearly satisfies $(f^{1/2})^2 = f$. It is also clear, using for example Exercise 1.6.6, that $f^{1/2} \geq 0$.

Now suppose that $g \in C_0(X)$ satisfies $g \geq 0$ and $g^2 = f$ but $g \neq f^{1/2}$. Then there is $x \in X$ such that $g(x) - f^{1/2}(x) \neq 0$. Since $g(x)$ and $f^{1/2}(x)$ are both positive and $g(x) \neq f^{1/2}(x)$ then also $g(x) + f^{1/2}(x) > 0$. It follows that $(g(x) - f^{1/2}(x))(g(x) + f^{1/2}(x)) \neq 0$. But $(g(x) - f^{1/2}(x))(g(x) + f^{1/2}(x)) = g^2(x) - f(x) = 0$, a contradiction. Thus we have $g(x) = f^{1/2}(x)$ for every $x \in X$, showing uniqueness. $\qquad\square$

3.1.5. The following fact is a fun use of the functional calculus and is also a result that will come in handy time and again.

Proposition. *Every positive element has a unique positive square root.*

Proof. Let $a \in A_+$. It is clear this holds if $a = 0$, so assume otherwise. Since $\mathrm{sp}(a) \subset [0, \infty)$ there is a function $f \in C_0(\mathrm{sp}(a))$, $f \geq 0$ that satisfies $f(t)^2 = t$ for every $t \in \mathrm{sp}(a)$. Since a is normal, we can set $a^{1/2} := f(a)$, which is a positive square root for $a \in A$. Suppose there is another $b \in A_+$ with $b^2 = a$. Since $b \neq 0$ and b^2 commutes with a, so does b, and by approximating by polynomials, b also commutes with $a^{1/2}$ (Exercise 2.5.25). Thus the C*-algebra B generated by a and b is commutative, and moreover $\mathrm{sp}(a) \subset \Omega(B)$. Thus by uniqueness of the square root in $C_0(\Omega(B))$, $b = a^{1/2}$. $\qquad\square$

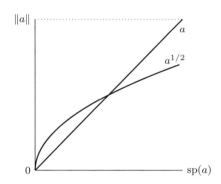

3.1.6. Proposition. *A unital C*-algebra is linearly spanned by its unitaries.*

Proof. We saw that every element can be written as a linear combination of self-adjoint elements, so we need only show that the self-adjoints are spanned by unitaries. Let $a \in A_{sa}$. By scaling if necessary, we may assume that $\|a\| \leq 1$. In this case, $1_A - a^2 \geq 0$ (Exercise 3.4.1) and so has a positive square root, $\sqrt{1_A - a^2}$. Let $u_1 = a - i\sqrt{1_A - a^2}$ and $u_2 = a + i\sqrt{1_A - a^2}$. It is easily checked that these are unitaries and that $a = u_1/2 + u_2/2$. $\qquad\square$

3.1.7. A *convex cone* is a subset of a vector space that is closed under linear combinations with positive coefficients.

Theorem. *Let A be a C^*-algebra. Then A_+ is closed convex cone.*

Proof. Without loss of generality, we may assume that A is unital. Let $a \in A_+$ and $\lambda \geq 0$. Then clearly $\lambda a \geq 0$ so we only need to check that the sum of two positive elements is again positive.

First, we claim that if $f \in C(X)$ where X is a compact subset of $\mathbb{R}$ that f is positive if there is $r \in \mathbb{R}_{\geq 0}$ such that $\|f - r\| \leq r$. If $r = 0$ we must have $f = 0$, which is positive. Suppose that $r > 0$ but f is not positive. Then there is some $t \in X$ such that $f(t) < 0$. Then $|f(t) - r| > r$ so $\|f - r\| = \sup_{t \in X} |f(t) - r| > r$, a contradiction. This proves the claim. Note that furthermore, if f is positive then $\|f - r\| < r$ if $r \leq \|f\|$.

Let $a, b \in A_+$. Since $a + b$ is self-adjoint, we may identify $C^*(a + b, 1_A) \cong C(\mathrm{sp}(a + b))$ and $a + b$ with $f(t) = t$. Thus we need only find an r satisfying $\|a+b-r\| \leq r$. Let $r := \|a\|+\|b\|$. We have that $\|a-\|a\|\| \leq \|a\|$ and $\|b-\|b\|\| \leq \|b\|$ by the above. Thus

$$\|a + b - r\| = \|a + b - \|a\| - \|b\|\| = \|a - \|a\| + b - \|b\|\|$$
$$\leq \|a - \|a\|\| + \|b - \|b\|\| \leq \|a\| + \|b\|.$$

Hence $a + b$ is positive.

To show that it is closed, first note that by the above

$$B := \{a \in A_+ \mid \|a\| \leq 1\} = \{a \mid \|a - 1_A\| \leq 1\} \cap A_{sa}.$$

Both $\{a \mid \|a - 1_A\| \leq 1\}$ and A_{sa} are closed, thus $\mathbb{R}B = \mathbb{R}A_+ = A_+$ is closed. $\square$

3.1.8. We say that two positive elements $a, b \in A$ are *orthogonal* if $ab = ba = 0$.

Let $a \in A_{sa}$. Let $a_+, a_- \in C_0(\mathrm{sp}(a))$ be the functions

$$a_-(t) = \begin{cases} -t & t \leq 0 \\ 0 & \text{otherwise,} \end{cases} \qquad a_+(t) = \begin{cases} t & t \geq 0 \\ 0 & \text{otherwise.} \end{cases}$$

Then $a_+, a_- \in A_+$ and $a = a_+ - a_-$. Note also that $a_+ a_- = a_- a_+ = 0$. The above observation also means that $a_+ - a = a_- \in A_+$ and $a \leq a_+$. Thus we have proved the following proposition.

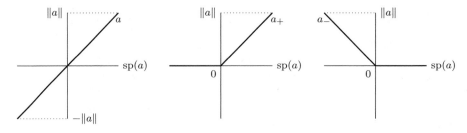

Proposition. *Any $a \in A_{sa}$ has a unique decomposition $a = a_+ - a_-$ into orthogonal positive elements $a_+, a_- \in A$.*

3.1.9. A partially ordered set $(S, \leq)$ is called *upwards directed* if, for every $a, b \in S$, there exists $c \in S$ such that $a \leq c$ and $b \leq c$. The previous proposition and theorem thus gives us the next corollary.

Corollary. $(A_{sa}, \leq)$ *is upwards directed.*

3.1.10. Theorem. *For every $a \in A$, the element a^*a is positive.*

Proof. Clearly $a^*a \in A_{sa}$. Suppose that $-a^*a \in A_+$. Then by 1.2.2, so is $-aa^*$. We have $a = b + ic$ for some $b, c \in A_{sa}$, so $a^*a = b^2 + ibc - icb + c^2$ and $aa^* = b^2 - ibc + icb + c^2$. Hence $a^*a = 2b^2 + 2c^2 - aa^*$ is the sum of positive elements and must also be positive. Thus $\|a^*a\| = 0$, hence $\|a\|^2 = 0$ and therefore $a = 0$.

Suppose that $a \neq 0$. Then by the above, $-a^*a$ is not positive. As we saw in 3.1.8, we can write $a^*a = b - c$ where $b, c \in A_+$ and $bc = 0$. We will use the above to show that $c = 0$. Consider ac. We have $-(ac)^*(ac) = -ca^*ac = -c(b - c)c = c^3 \in A_+$, hence $ac = 0$. Since a is nonzero, this implies $c = 0$. $\qquad\square$

Observe that if $a \geq 0$ then a has a positive square root by Theorem 3.1.5. Hence $a = a^{1/2}a^{1/2} = (a^{1/2})^*a^{1/2}$, which is to say that the result above implies

$$A_+ = \{a^*a \mid a \in A\}.$$

3.1.11. Proposition. *Let A be a C*-algebra. Let $a, b \in A_{sa}$ with $a \leq b$. Then the following hold:*

(i) *for any $c \in A$, we have $c^*ac \leq c^*bc$;*

(ii) $\|a\| \leq \|b\|$;

(iii) *if A is unital and $a, b \in \text{Inv}(A)$ then $b^{-1} \leq a^{-1}$.*

Proof. For (i), since $a \leq b$, the element $b - a \in A_+$ and therefore has a positive square root. Then $c^*bc - c^*ac = c^*(b - a)c = ((a - b)^{1/2}c)^*((a - b)^{1/2}c)$, which is positive by Proposition 3.1.10. Hence $c^*ac \leq c^*bc$.

For (ii), without loss of generality, we may assume that A is unital. Then, using the functional calculus, we have $b \leq \|b\|$, so also $a \leq \|b\|$. Since $\text{C}^*(a, 1_A) \to C(\text{sp}(a))$ is isometric, we have that $\|a\| \leq \|b\|$.

Finally, we show (iii). For any $a \in A$, if $c \in \text{Inv}(A)$ then $\lambda \in \text{sp}(c)$ if and only if $\lambda^{-1} \in \text{sp}(c)$ so $a^{-1}, b^{-1} \in A_{sa}$. By (i), $b^{-1/2}ab^{-1/2} \leq b^{-1/2}bb^{-1/2} = 1_A$. So, using (ii), $\|b^{-1/2}a^{1/2}\|^2 < 1$. Thus $\|a^{1/2}b^{-1}a^{1/2}\| \leq 1_A$ and $a^{1/2}b^{-1}a^{1/2} \leq 1_A$. By (i) again, multiplying $a^{-1/2}$ on either side gives $b^{-1} < a^{-1}$. $\qquad\square$

We also have the next theorem, which we include separately because the proof is a little trickier.

3.1.12. Theorem. *Let A be a C*-algebra and let $a, b \in A_+$ with $a \leq b$. Then $a^n \leq b^n$ for all $0 < n \leq 1$.*

Proof. Let $a, b \in A_+$ with $a \leq b$. Let $\epsilon > 0$ and put $c = b + \epsilon 1_{\tilde{A}}$. Then $a \leq c$ and c is invertible in $\tilde{A}$ since $0 \notin \text{sp}(c)$. Set

$$S := \{n \in (0, \infty) \mid a^n \leq c^n\}.$$

Note that $1 \in S$. Suppose that $(t_n)_{n \in \mathbb{N}}$ is a sequence in S converging to $t \in (0, \infty)$. Then $\lim_{n \to \infty} a^{t_n} - c^{t_n} = a^t - c^t \geq 0$ since A_+ is closed.

Since $c \geq 0$, we also have $c^{-1} \geq 0$, and thus c^{-1} has a positive square root. By Proposition 3.1.11 (i), this implies that $c^{-1/2} a c^{-1/2} \leq 1_{\tilde{A}}$, and using the C*-equality, that $\|a^{1/2} c^{-1/2}\| \leq 1$. We saw earlier (1.2.2) that for any $x, y \in A$ we have $\text{sp}(xy) \setminus \{0\} = \text{sp}(yx) \setminus \{0\}$. Thus

$$\|c^{-1/4} a^{1/2} c^{-1/4}\| = r(c^{-1/4} a^{1/2} c^{-1/4}) = r(a^{1/2} c^{-1/2}) \leq 1,$$

so $c^{-1/4} a^{1/2} c^{-1/4} \leq 1_{\tilde{A}}$ and $a^{1/2} \leq c^{1/2}$. Thus $1/2 \in S$.

One shows similarly that if $s, t \in S$ then $(s + t)/2 = z \in S$. Then,

$$\|c^{-z/2} a^z c^{-z/2}\| = r(c^{-z/2} a^z c^{z/2}) = r(c^{-s/2} a^{s/2} a^{t/2} c^{-t/2})$$
$$= \|(c^{-s/2} a^{s/2})(a^{t/2} c^{-t/2})\| \leq \|c^{-s/2} a^{s/2}\| \|a^{t/2} c^{-t/2}\| \leq 1.$$

Thus $(0, 1] \subset S$. Since $\epsilon > 0$ was arbitrary, we have shown that $a^n \leq b^n$ for every $n \in (0, 1]$. $\qquad\square$

3.1.13. What happens for $n > 1$? If $A = C_0(X)$ is commutative and $f, g \geq 0$ with $f \leq g$, then certainly $f^2 \leq g^2$. In this case, however, an attempt to look to the commutative algebras leads us astray. Let us consider things from the other side, then, in spirit of 2.4.3. In fact, we need only consider M_2 to see that things can go wrong. The conclusion of the last theorem no longer necessarily holds. Let

$$a = \begin{pmatrix} 1 & 0 \\ 0 & 0 \end{pmatrix} \text{ and } b = \begin{pmatrix} 1/2 & 1/2 \\ 1/2 & 1/2 \end{pmatrix}$$

Then $a \leq a + b$ but

$$(a + b)^2 - a^2 = ab + ba + b^2 = a + 2ab + b = \begin{pmatrix} 3/2 & 1 \\ 1 & 1/2 \end{pmatrix}$$

is not positive.

By the way, this above example is also a good illustration of the following: It is often very useful to look to the 2×2 matrices as a test case when deciding whether or not something holds in a C*-algebra.

3.1.14. We end this section with a proposition about the order structure applied to projections. By Exercise 2.5.3, every projection is positive. This is also immediate from Theorem 3.1.10.

Proposition. *Let A be a C*-algebra and $p, q \in A$ projections. Then $p \leq q$ if and only if $pq = qp = p$.*

Proof. Suppose that $p \le q$. Then $0 \le q - p$ and $q - p \le 1_{\tilde{A}} - p$, so by Proposition 3.1.11 (i),

$$0 = p\,0\,p \le p(q - p)p \le p(1_{\tilde{A}} - p)p = p - p = 0.$$

Thus $p(q - p)p = pqp - p = 0$, hence, using the C*-equality,

$$\|qp - p\|^2 = \|(qp - p)^*(qp - p)\| = \|pqp - pqp - pqp + p\| = \|p - pqp\| = 0,$$

so $qp - p = 0$, and also $pq - p = (qp - p)^* = 0$.

Conversely, suppose that $pq = qp = p$. Then

$$(q - p)(q - p) = q^2 - qp - pq + p^2 = q - p - p + p = (q - p),$$

and also $(q - p)^* = q^* - p^* = q - p$. So $q - p$ is a projection. In particular, $q - p \ge 0$, so $p \le q$. $\qquad\square$

3.2 Approximate units and hereditary C*-subalgebras

We have already seen that it is always possible to adjoin a unit to a nonunital C*-algebra. In many cases, however, this is not necessarily useful. For example, if one is investigating simple C*-algebras (no nontrivial ideals), then attaching a unit destroys simplicity. We can, however, always find an *approximate* unit. Let us consider continuous functions on some open interval I in $\mathbb{R}$, for example. Then the C*-algebra $C_0(I)$ is of course nonunital. However, for any finite subset of functions $F \subset C_0(I)$ and any $\epsilon > 0$, we can always find another function, g such that $\|gf - f\| < \epsilon$ for every $f \in F$. Since $C_0(I)$ is separable, one can find a nested increasing sequence of these finite subsets $F_0 \subset F_1 \subset \cdots$ that exhausts the algebra. Then, for every $n \in \mathbb{N}$ find g_n such that $fg_n = f$ for every $f \in F_n$. This gives us a sequence $(g_n)_{n \in \mathbb{N}} \subset C_0(I)$ which itself will not converge, but which still satisfies

$$\lim_{n \to \infty} g_n f = f \text{ for every } f \in C_0(I).$$

In the completely noncommutative case, consider the C*-algebra of compact operators $\mathcal{K} := \mathcal{K}(H)$ on a separable Hilbert space H with orthonormal basis $(e_i)_{i \in \mathbb{N}}$. This is again a nonunital C*-algebra. Let p_n be the projection onto the subspace spanned by the first n basis elements. Then each p_n is a compact operator. As in the previous example, the sequence $(p_n)_{n \in \mathbb{N}}$ which will not converge in $\mathcal{K}$, but, for any $a \in \mathcal{K}$ we have

$$\lim_{n \to \infty} p_n a = \lim_{n \to \infty} a p_n = a.$$

3.2.1. Definition. Let A be a C*-algebra. An *approximate unit* for A is an increasing net $(u_\lambda)_{\lambda \in \Lambda}$ of positive elements such that

$$\lim_\lambda u_\lambda a = \lim_\lambda a u_\lambda = a,$$

for every $a \in A$.

3.2.2. Lemma. *Let A be a C*-algebra and let $A^1_+ := \{a \in A \mid a \in A_+, \|a\| < 1\}$. Then the set of $(A^1_+, \leq)$ is upwards-directed, that is, for every $a, b \in A^1_+$ there exists $c \in A^1_+$ such that $a \leq c$ and $b \leq c$.*

Proof. First, let $a, b \in A_+$. Then $(1_{\tilde{A}} + a), (1_{\tilde{A}} + b) \in \mathrm{Inv}(\tilde{A})$. Suppose that $a \leq b$. Then $1_{\tilde{A}} + a \leq 1_{\tilde{A}} + b$ so $(1_{\tilde{A}} + b)^{-1} \leq (1_{\tilde{A}} + a)^{-1}$ and

$$a(1_{\tilde{A}} + a)^{-1} = 1_{\tilde{A}} - (1_{\tilde{A}} + a)^{-1} \leq 1_{\tilde{A}} - (1_{\tilde{A}} + b)^{-1} = b(1_{\tilde{A}} + b)^{-1}.$$

Note that $a(1_{\tilde{A}} + a)^{-1}$ and $b(1_{\tilde{A}} + b)^{-1}$ are both in A^1_+.

Now suppose that $a, b \in A^1_+$. Let $x = a(1_{\tilde{A}} - a)^{-1}$ and $y = b(1_{\tilde{A}} - b)^{-1}$ and set $c = (x + y)(1_{\tilde{A}} + x + y)^{-1}$. Since $x \leq x + y$ we have $a = x(x + 1_{\tilde{A}})^{-1} \leq c$. Similarly, $b \leq c$. $\square$

3.2.3. Theorem. *Every C*-algebra A has an approximate unit. If A is separable, then A has a countable approximate unit.*

Proof. Let Λ be the upwards directed set $(A^1_+, \leq)$ and put $u_\lambda = \lambda$ for each $\lambda \in \Lambda$. Then $(u_\lambda)_\Lambda$ is an increasing net of positive elements with norm less than 1. We must show that $\lim_\lambda u_\lambda a = \lim_\lambda a u_\lambda = a$ for every $a \in A$. It is enough to show that this holds when $a \in A_+$. Let $0 < \epsilon < 1$. Let $\Gamma : \mathrm{C}^*(a) \to C_0(\mathrm{sp}(a)) =: C_0(X)$ be the Gelfand transform. Let $f = \Gamma(a)$ and let $K = \{x \in X \mid |f(x)| \geq \epsilon\}$.

Let $\delta > 0$ such that $\delta < 1$ and $1 - \delta < \epsilon$. Let $g_\delta(x) = \delta$ if $x \in K$ and vanishing outside K (this is possible since K is compact). Then $g_\delta \in C_0(X)^1_+$ and $\|g_\delta f - f\| < \epsilon$. Since Γ is isometric, $\Gamma^{-1}(g_\delta) = \mu$ for some $\mu \in \Lambda$ and $\|u_\mu a - a\| = \|a u_\mu - a\| < \epsilon$.

Suppose that $\lambda \in \Lambda$ satisfies $\mu \leq \lambda$. Then $1_{\tilde{A}} - u_\lambda \geq 1_{\tilde{A}} - u_\mu$ so

$$a^{1/2}(1_{\tilde{A}} - u_\lambda)a^{1/2} \leq a^{1/2}(1_{\tilde{A}} - u_\mu)a^{1/2},$$

and hence

$$\|a^{1/2}(1_{\tilde{A}} - u_\lambda)a^{1/2}\| \leq \|a^{1/2}(1_{\tilde{A}} - u_\mu)a^{1/2}\|.$$

Applying the C*-equality, $\|a - a u_\lambda\| < \|a - a u_\mu\| < \epsilon$. Hence $\lim_\lambda a u_\lambda = \lim_\lambda u_\lambda a = a$.

The proof that if A is separable then A admits a countable approximate unit is left as an exercise. $\square$

The following is a very useful theorem. We omit the proof for now because it uses some techniques that have not been described, but we will return to it in Chapter 5 (see 5.2.15).

3.2.4. Let A be a C*-algebra and I an ideal in A. If I has an approximate unit $(u_\lambda)_{\lambda \in \Lambda}$ satisfying

$$\lim_\lambda \|u_\lambda a - a u_\lambda\| = 0, \quad \text{for every } a \in A,$$

then we say $(u_\lambda)_{\lambda \in \Lambda}$ is an *approximate unit quasicentral for A*.

Theorem. *Let A be a C*-algebra and I an ideal in A. Then I has an approximate unit $(u_\lambda)_{\lambda \in \Lambda}$ quasicentral for A.*

3.2.5. We have already defined C*-subalgebras as well as ideals. There is another important substructure in C*-algebras that we can define now that we have an order structure.

Definition. A C*-subalgebra $B \subset A$ is called *hereditary* if, whenever $b \in B$ and $a \leq b$, then $a \in B$.

3.2.6. Let A be a C*-algebra and $p \in A$ a projection. Then pAp, introduced in the theorem below, is called a *corner* and is the first example of a hereditary C*-subalgebra.

Theorem. *Let $p \in A$ be a projection. Then $pAp = \{pap \mid a \in A\}$ is a hereditary C*-subalgebra of A.*

Proof. Since p is a projection, it is easy to check that pAp is C*-subalgebra. We will show that if $a \in pAp$ is self-adjoint and $b \in A$ satisfies $b \leq a$, then $b \in pAp$. First, observe that we may assume that $a \in pAp$ is positive since if $b \leq a$ then $b \leq a_+$. Suppose first $b \leq a$ with b also positive or $b = -c$ where c is positive. Then

$$(1_{\tilde{A}} - p)b(1_{\tilde{A}} - p) \leq (1_{\tilde{A}} - p)a(1_{\tilde{A}} - p) = 0.$$

So $\|(1_{\tilde{A}} - p)b(1_{\tilde{A}} - p)\| = \|(1_{\tilde{A}} - p)b^{1/2}\|^2 = 0$, and we see that $pb = pb^{1/2}b^{1/2} = b^{1/2}b^{1/2} = b$. Similarly, $bp = b$, hence $b \in pAp$. Now suppose b is self-adjoint. Let $b = b_+ - b_-$ with b_+, b_- positive and $b_+b_- = b_-b_+ = 0$ (3.1.8). Since $b \leq a$ we have $b_+^2 \leq b_+^{1/2}ab_+^{1/2}$ and

$$\|b\|^{-1}b_+^2 \leq (b_+^{1/2}\|b\|^{-1/2})a(b_+^{1/2}\|b\|^{-1/2}) \leq a.$$

We have $b_+^2/\|b\| \in pAp$, which follows from above, since $b_+^2/\|b\|$ is positive. Hence $b_+ \in pAp$. Then since $-b_- \leq b \leq a \in pAp$, we also have $b_- \in pAp$ and so $b \in pAp$, as required. $\square$

For a positive element, we have the following generalisation:

3.2.7. Theorem. *Let A be a C*-algebra and $a \in A_+$ a positive element. Then $a \in \overline{aAa}$ and $\overline{aAa}$ is the hereditary C*-subalgebra generated by a. If B is a separable hereditary C*-subalgebra, then $B = \overline{aAa}$ for some $a \in A_+$.*

Proof. It is clear that $\overline{aAa}$ is a C*-subalgebra. Let $(u_\lambda)_\Lambda$ be an approximate unit for A. Then $\lim_\lambda au_\lambda a = a^2$ so $a^2 \in \overline{aAa}$. Thus $C^*(a^2) \subset \overline{aAa}$ and by uniqueness of the positive square root, also $a \in \overline{aAa}$.

The proof that $\overline{aAa}$ is hereditary is similar to the case for a corner, so is left as an exercise.

Now suppose that $B \subset A$ is hereditary and separable. Since B is separable, it contains a countable approximate unit, say $(u_n)_{n \in \mathbb{N}}$. Let $a = \sum_{n=1}^\infty 2^{-n}u_n$. Then

$a \in B$ and $a \geq 0$. Thus $\overline{aAa} \subset B$. For each $\mathbb{N} \setminus \{0\}$ we have that $2^{-n}u_n \leq a$ so $u_n \in \overline{aAa}$. Thus, if $b \in B$ we have $b = \lim_{n \to \infty} u_n b u_n$ where each $u_n b u_n \in \overline{aAa}$ hence $b \in \overline{aAa}$ and so we have shown that $B = \overline{aAa}$. $\square$

3.2.8. Lemma. *Let I be a closed left ideal of A. Then I has a left approximate unit, that is, $(u_\lambda)_{\lambda \in \Lambda} \subset I$ with*

$$\lim_\lambda au_\lambda = a,$$

for every $a \in A$.

Proof. Observe that $I \cap I^*$ is a C*-subalgebra of A and thus by Theorem 3.2.3 has an approximate unit $(u_\lambda)_{\lambda \in \Lambda}$. Let $a \in I$. Then $a^*a \in I \cap I^*$. Since

$$\|a - au_\lambda\|^2 = \|a^*a - a^*au_\lambda - u_\lambda a^*a + u_\lambda a^*au_\lambda\|,$$

we have

$$\|a - \lim_\lambda au_\lambda\|^2 = 0.$$

Thus $a = \lim_\lambda au_\lambda$. $\square$

3.2.9. Theorem. *Let A be a C*-algebra. There is a one-to-one correspondence between closed left ideals of A and hereditary C*-subalgebras of A given by*

$$I \mapsto I^* \cap I, \qquad B \mapsto \{a \in A \mid a^*a \in B\}.$$

Proof. Let I be a closed left ideal in A and suppose $b \in (I \cap I^*)_+$. Since I is a closed left ideal, it contains a left approximate unit $(u_\lambda)_\Lambda$. If $a \in A_+$ satisfies $a \leq b$, then $(1_{\tilde{A}} - u_\lambda)a(1_{\tilde{A}} - u_\lambda) \leq (1_{\tilde{A}} - u_\lambda)b(1_{\tilde{A}} - u_\lambda)$. Then

$$\begin{aligned}
\|a^{1/2}(1_{\tilde{A}} - u_\lambda)\|^2 &= \|(1_{\tilde{A}} - u_\lambda)a(1_{\tilde{A}} - u_\lambda)\| \\
&\leq \|(1_{\tilde{A}} - u_\lambda)b(1_{\tilde{A}} - u_\lambda)\| \\
&= \|b^{1/2} - b^{1/2}u_\lambda\|^2 \\
&\to 0.
\end{aligned}$$

Thus $\|a^{1/2}(1_{\tilde{A}} - u_\lambda)\|^2 \to 0$ and so $a^{1/2}$ is a limit of elements in I. It follows that $a^{1/2}$, and hence a, are both contained in I. Thus $I^* \cap I$ is a hereditary C*-subalgebra of A.

Now suppose that B is a hereditary C*-subalgebra. Let $I = \{a \in A \mid a^*a \in B\}$. Let $a \in I$, $b \in A$ and without loss of generality, assume $\|b\| \leq 1$. Then

$$(ab)^*(ab) = b^*a^*ab \leq a^*a,$$

so $ab \in I$ since B is hereditary. It is clear that I is closed since B is; thus I is a closed left ideal.

Finally, it is easy to check that the maps are mutual inverses. $\square$

3.2.10. Corollary. *Every closed ideal is hereditary.*

Proof. Let $I \subset A$ be a closed ideal. Then in particular I is a closed left ideal and we have $I = I \cap I^*$ since if $a \in I$ and $(u_\lambda)_\Lambda$ is a left approximate unit for I then $a^* = \lim_\lambda (au_\lambda)^* = \lim_\lambda u_\lambda a^* \in I$. $\qquad\square$

A hereditary C*-subalgebra often inherits properties of the C*-algebra itself. We will see more examples later, but for now we show the following

3.2.11. Theorem. *If A is a simple C*-algebra, then so is every hereditary C*-subalgebra $B \subset A$.*

Proof. We claim that every ideal $J \subset B$ is of the form $I \cap B$ for some ideal $I \subset A$. Suppose that I is an ideal in A and let $b \in I \cap B$. Clearly $I \cap B$ is closed under addition and scalar multiplication and is norm-closed. Let $c \in B$. Then $cb, bc \in B$ and since I is an ideal we also have that $cb, bc \in I$. Thus $cb, bc \in I \cap B$.

Now suppose that $J \subset B$ is an ideal. Let $(u_\lambda)_\Lambda$ be an approximate unit for J quasicentral for A. Set $I = \{au_\lambda \mid a \in A, \lambda \in \Lambda\}$. Then $J = I \cap B$; the details are easy to check. $\qquad\square$

3.3 Some factorisation results

We have already seen that in a C*-algebra any element can be decomposed into a linear combination of self-adjoint elements, or that in a unital C*-algebra, a linear combination of unitaries. There are other ways in which one may also want to decompose elements. For example, any nonzero complex number λ can be written $\lambda = e^{it}|\lambda|$ for some $t \in \mathbb{R}$, and any $n \times n$ matrix can be decomposed into the product of a unitary matrix and a positive matrix.

3.3.1. For an element a in a C*-algebra A, we define its absolute value to be the positive element $|a| := (a^*a)^{1/2}$. A *partial isometry* in a C*-algebra is an element satisfying $v = vv^*v$ (and hence also $v^* = v^*vv^*$).

3.3.2. Extending the case of $n \times n$ matrices, any $a \in \mathcal{B}(H)$ also admits a *polar decomposition* $a = v|a|$ for some partial isometry $v \in \mathcal{B}(H)$.

Proposition. *For any $a \in \mathcal{B}(H)$ there exists a unique partial isometry $v \in \mathcal{B}(H)$ such that $a = v|a|$, $\ker(v) = \ker(a)$, and $v^*a = |a|$.*

Proof. Define a map

$$|a|(H) \to H, \qquad |a|(\xi) \mapsto a(\xi).$$

Then, for any $\xi \in H$ we have

$$\||a|(\xi)\|^2 = \langle |a|(\xi), |a|(\xi) \rangle = \langle |a|^2(\xi), \xi \rangle$$
$$= \langle a^*a(\xi), \xi \rangle = \langle a(\xi), a(\xi) \rangle = \|a(\xi)\|^2,$$

so the map above is a well-defined isometric map, which is easily seen to be linear. Thus it extends to an isometric linear map $v_0 : \overline{|a|(H)} \to H$. We now define an operator on all of H

$$v(\xi) = \begin{cases} v_0, & \xi \in \overline{|a|(H)} \\ 0, & \xi \in \overline{|a|(H)}^\perp. \end{cases}$$

Then $v|a| = a$. Since $\ker(v) = \overline{|a|(H)}^\perp$, the restriction $v|_{\ker(v)^\perp}$ is isometric. Thus, if $\xi \in \ker(v)^\perp$ we have

$$\langle v^*v(\xi), \xi \rangle = \langle v(\xi), v(\xi) \rangle = \langle \xi, \xi \rangle,$$

while for $\xi \in \ker(v)$ we obviously have $v^*v(\xi) = 0$. It follows that v^*v is the projection onto $\ker(v)^\perp$. Then

$$\langle vv^*v(\xi_1), \xi_2 \rangle = \langle v(\xi_1), \xi_2 \rangle$$

for every $\xi_1, \xi_2 \in H$. So $v = vv^*v$, which is to say, v is a partial isometry. Observe that $\ker(v) = \ker(|a|)$. Let $\xi_1, \xi_2 \in H$. Then

$$\langle v^*a(\xi_1), |a|(\xi_2) \rangle = \langle a(\xi_1), v|a|(\xi_2) \rangle = \langle a(\xi_1), a(\xi_2) \rangle$$
$$= \langle a^*a(\xi_1), \xi_2 \rangle = \langle |a|(\xi_1), |a|(\xi_2) \rangle,$$

so that $v^*a = |a|$, from which it also follows that $\ker(|a|) = \ker(a)$ and hence $\ker(v) = \ker(a)$.

Finally, we show uniqueness. Suppose that w is another partial isometry satisfying $a = w|a|$, $\ker(w) = \ker(a)$ and $w^*a = |a|$. Then $w|_{\overline{|a|(H)}} = v|_{\overline{|a|(H)}}$ and since $\overline{|a|(H)}^\perp = \ker(w) = \ker(a) = \ker(v)$, we also have $w|_{\overline{|a|(H)}^\perp} = v|_{\overline{|a|(H)}^\perp}$. Thus $w = v$. $\square$

3.3.3. In an arbitrary C*-algebra A polar decompositions need not exist: it may be the case that the partial isometry v is not itself contained in A. Proposition 3.3.4 below, however, gives us something close. First, a lemma.

Lemma. *Let A be a C*-algebra and suppose there are $x, y \in A$ and $a \geq 0$ satisfying $x^*x \leq a^{t_1}$ and $yy^* \leq a^{t_2}$ where $t_1 + t_2 \geq 1$. Put*

$$u_n := x(n^{-1}1_{\tilde{A}} + a)^{-1/2}y.$$

Then the sequence $(u_n)_{n \in \mathbb{N}}$ converges to some $u \in A$ with $\|u\| \leq \|a^{(t_1+t_2-1)/2}\|$.

Proof. Define $f_n \in C_0(0, \|a\|)$ by $f_n(t) = (n^{-1}1_{\tilde{A}} + t)^{-1/2}t^{(t_1+t_2)/2}$. Then the sequence $(f_n)_{n \in \mathbb{N}}$ is increasing and converges pointwise to $f(t) = t^{(t_1+t_2-1)/2}$. Hence, by Dini's Theorem, the convergence is uniform, and by the functional calculus we have $f_n(a) \to f(a)$ in A as $n \to \infty$. Let $d_{nm} := (n^{-1}1_{\tilde{A}} + a)^{-1/2} - (m^{-1}1_{\tilde{A}} + a)^{-1/2}$. Note that d_{nm} commutes with a. Now

$$\begin{aligned}
\|u_n - u_m\| = \|x d_{nm} y\|^2 &= \|y^* d_{nm} x^* x d_{nm} y\| \le \|y^* d_{nm} a^{t_1} d_{nm} y\| \\
&= \|y^* d_{nm} a^{t_1/2}\|^2 = \|a^{t_1/2} d_{nm} y y^* d_{nm} a^{t_1/2}\| \\
&\le \|a^{t_1/2} d_{nm} a^{t_2} d_{nm} a^{t_2/2}\| = \|d_{nm} a^{(t_1+t_2)/2}\|^2,
\end{aligned}$$

so by the observation above, $\|u_n - u_m\| \to 0$, and therefore u_n converges in norm to some $u \in A$. We also have

$$\|u_n\| = \|x(n^{-1}1_{\tilde{A}} + a)^{-1/2} y\| \le \|a^{t_1/2}(n^{-1}1_{\tilde{A}} + a)^{-1/2} a^{t_2/2}\| \le \|a^{(t_1+t_2-1)/2}\|,$$

so $\|u\| \le \|a^{(t_1+t_2-1)/2}\|$. $\qquad\qquad\qquad\qquad\qquad\qquad\qquad\qquad\qquad\quad \square$

3.3.4. Proposition. *Let A be a C^*-algebra. If $x, a \in A$ with $a \ge 0$ and $x^* x \le a$, then for every $0 < t < 1/2$ there is $u \in A$ with $\|u\| \le \|a^{1/2-t}\|$ such that $x = u a^t$.*

In particular, for any $0 < t < 1$ there is some $u \in A$ such that we can decompose x as $x = u|x|^t$.

Proof. For $n \in \mathbb{N} \setminus \{0\}$, let $u_n := x(n^{-1}1_{\tilde{A}} + a)^{-1/2} a^{1/2-t}$. Let $t_1 = 1$, $t_2 = 1 - 2t$ and $y = a^{1/2-t}$. Then $t_1 + t_2 \ge 1$, $x^* x \le a^{t_1}$ and $yy^* = a^{1-2t} \le a^{t_2}$. Applying Lemma 3.3.3, the sequence $(u_n)_{n \in \mathbb{N}}$ converges to some $u \in A$ and u satisfies

$$\|u\| \le \|a^{(t_1+t_2-1)/2}\| = \|a^{(1+1-2t-1)/2}\| = \|a^{1/2-t}\|.$$

Also,

$$\begin{aligned}
\|x - u_n a^t\|^2 &= \|x(1_{\tilde{A}} - (n^{-1}1_{\tilde{A}} + a)^{-1/2} a^{1/2})\|^2 \\
&= \|(1_{\tilde{A}} - (n^{-1}1_{\tilde{A}} + a)^{-1/2} a^{1/2})^* x^* x (1_{\tilde{A}} - (n^{-1}1_{\tilde{A}} + a)^{-1/2} a^{1/2})\| \\
&\le \|(1_{\tilde{A}} - (n^{-1}1_{\tilde{A}} + a)^{-1/2} a^{1/2})\|^2 \|x^* x\| \\
&\le \|a^{-1/2} - (n^{-1}1_{\tilde{A}} + a)^{-1/2}\|^2 \|a^{1/2}\|^2 \|a\| \\
&\to \|a^{-1/2} - a^{-1/2}\|^2 \|a^{1/2}\|^2 \|a\| = 0
\end{aligned}$$

as $n \to \infty$. It follows that $x = u a^t$.

For the second statement, let $a = x^* x$. Then trivially $x^* x \le a$, so for any $0 < t < 1$ there is $u \in A$ such that $x = u a^{t/2} = u(x^* x)^{t/2} = u|x|^t$. $\qquad \square$

3.3.5. When considering a partially ordered structure on a monoid, we would like to know how the order structure interacts with addition. Consider, for example, the natural numbers with their usual order structure. If $n \le m_1 + m_2$ then we can always find $n_1 \le m_1$ and $n_2 \le m_2$ such that $n = n_1 + n_2$. This is also possible in any lattice-ordered monoid (by *lattice-ordered* we mean that every pair of elements have least upper bound and greatest lower bound). If E is a partially ordered Banach space, then the property that whenever $x \le y + z$ there are $v \le x$

and $w \leq y$ such that $x = v + w$ is called the *Riesz decomposition property*. This will not hold in an arbitrary C*-algebra A, however we do have a very useful noncommutative version of the Riesz decomposition property for A_{sa}.

Proposition. *Let A be a C*-algebra and for $m \geq 2$ suppose $x, w_1, \ldots, w_m \in A$ are elements satisfying $x^*x \leq \sum_{k=1}^{m} w_k w_k^*$. Then there are $u_1, \ldots, u_m \in A$ such that $u_k^* u_k \leq w_k^* w_k$ for every $1 \leq k \leq m$ and $xx^* = \sum_{k=1}^{m} u_k u_k^*$.*

Proof. Let $a := \sum_{k=1}^{m} w_k w_k^*$ and let $u_k^{(n)} := x(n^{-1}1_{\tilde{A}} + a)^{-1/2} w_k$, $1 \leq k \leq m$, $n \in \mathbb{N}$. By Lemma 3.3.3, the sequences $(u_k^{(n)})_{n \in \mathbb{N}}$ are norm-convergent. For each $1 \leq k \leq m$, set $u_k := \lim_{n \to \infty} u_k^{(n)}$. Then, for $1 \leq k \leq m$ we have

$$(u_k^{(n)})^*(u_k^{(n)}) = w_k^*(n^{-1}1_{\tilde{A}} + a)^{-1/2} x^* x (n^{-1}1_{\tilde{A}} + a)^{-1/2} w_k$$
$$\leq w_k^*(n^{-1}1_{\tilde{A}} + a)^{-1/2} a (n^{-1}1_{\tilde{A}} + a)^{-1/2} w_k.$$

Since $w_k^*(n^{-1}1_{\tilde{A}} + a)^{-1/2} a (n^{-1}1_{\tilde{A}} + a)^{-1/2} w_k \to w_k^* w_k$ as $n \to \infty$ and A_+ is closed (Theorem 3.1.7), we have $u_k^* u_k \leq w_k^* w_k$. Also, for every $n \in \mathbb{N}$ we have

$$\left\| \sum_{k=1}^{m} u_k^{(n)}(u_k^{(n)})^* - xx^* \right\| = \left\| \sum_{k=1}^{m} x(n^{-1}1_{\tilde{A}} + a)^{-1/2} w_k w_k^* (n^{-1}1_{\tilde{A}} + a)^{-1/2} x^* - xx^* \right\|$$
$$= \| x(n^{-1}1_{\tilde{A}} + a)^{-1/2} a (n^{-1}1_{\tilde{A}} + a)^{-1/2} x^* - xx^* \|$$
$$= \| x((n^{-1}1_{\tilde{A}} + a)^{-1} a - 1_{\tilde{A}}) x^* \|$$
$$\leq \|a\|/n \to 0$$

as $n \to \infty$. Thus $xx^* = \sum_{k=1}^{m} u_k u_k^*$, as desired. $\square$

3.4   Exercises

3.4.1. Let A be a unital C*-algebra and let $a \in A_{sa}$. Suppose that $\|a\| \leq 1$. Show that $1 - a^2 \geq 0$.

3.4.2. Let A be a C*-algebra. Show that any element in a C*-algebra can be written as the linear combination of four positive elements.

3.4.3. Let $v \in A$ be a partial isometry (see 3.3.1).

  (i) Show that if v is a partial isometry then v^*v and vv^* are projections.

 (ii) Suppose that $v \in A$ and v^*v is a projection. Show that vv^* is also a projection and that v is a partial isometry.

(iii) Let $A = M_n$ and let p, q be projections. If $\mathrm{tr}_n(p) \leq \mathrm{tr}_n(q)$ show that there is a partial isometry $v \in M_n$ such that $v^*v = p$ and $vv^* \leq q$. (Here tr_n denotes the normalised trace on M_n, that is, $\mathrm{tr}_n((a_{ij})_{ij}) = \frac{1}{n} \sum_{i=1}^{n} a_{ii}$.)

(iv) Projections p and q are *Murray–von Neumann equivalent* if there is a partial isometry $v \in A$ with $v^*v = p$ and $vv^* = q$. Check that this is an equivalence relation. If $A = M_n$ describe the equivalence classes.

3.4.4. Let A be a separable C*-algebra. For two positive elements a, b in A we say that a is *Cuntz subequivalent* to b and write $a \precsim b$ if there are $(v_n)_{n \in \mathbb{N}} \subset A$ such that $\lim_{n \to \infty} \|v_n b v_n^* - a\| = 0$. We write $a \sim b$ and say a and b are *Cuntz equivalent* if $a \precsim b$ and $b \precsim a$.

(i) Show that if $a \precsim b$ and $b \precsim c$ then $a \precsim c$.

(ii) Show that $\sim$ is an equivalence relation on the positive elements.

(iii) Let $f, g \in C_0(X)_+$ where X is a locally compact metric space. Show that if $\operatorname{supp}(f) \subset \operatorname{supp}(g)$ then for any $\epsilon > 0$ there is a positive function $e \in C(X)$ such that $\|f - ege\| < \epsilon$. Now show that $f \precsim g$ if and only if $\operatorname{supp}(f) \subset \operatorname{supp}(g)$.

(iv) Let $a \in A_+$ where a is a separable C*-algebra. Show that $a \precsim a^n$ for every $n \in \mathbb{N}$.

3.4.5. Let A and B be concrete C*-algebras. We saw in Exercise 2.5.16 that $M_n(A)$ is a C*-algebra. A linear map $\varphi : A \to B$ is *positive* if $\varphi(A_+) \subset B_+$. For any map $\varphi : A \to B$ and $n \in \mathbb{N}$ we can define $\varphi^{(n)} : M_n(A) \to M_n(B)$ by applying φ entry-wise. If $\varphi^{(n)} : M_n(A) \to M_n(B)$ is positive for every $n \in \mathbb{N}$ then we say φ is *completely positive*.

(i) Show that a *-homomorphism $\varphi : A \to B$ is completely positive.

(ii) Let $A = B = M_2$ and let $\tau : M_2 \to M_2$ be the map taking a matrix to its transpose. Show that τ is positive but not completely positive.

(iii) Let $\varphi : A \to B$ be a *-homomorphism and let $v \in B$. Show that the map

$$v^*\varphi v : A \to B, \qquad a \mapsto v^*\varphi(a)v$$

is completely positive.

Remark. In fact, everything in this exercise makes sense when A and B are abstract C*-algebras. However, we haven not yet shown that $M_n(A)$ is actually a C*-algebra when A is abstract. That $M_n(A)$ is a C*-algebra will follow from results in the next chapter.

3.4.6. Let A be a separable unital C*-algebra and $a \in A_+$. Let $B := \mathrm{C}^*(a, 1_A)$. Show that for any finite subset $\mathcal{F} \subset B$ and any $\epsilon > 0$ there are a finite-dimensional C*-algebra F, a *-homomorphism $\psi : B \to F$, and a completely positive contractive map $\varphi : F \to B$ such that $\|\varphi \circ \psi(f) - f\| < \epsilon$ for every $f \in \mathcal{F}$.

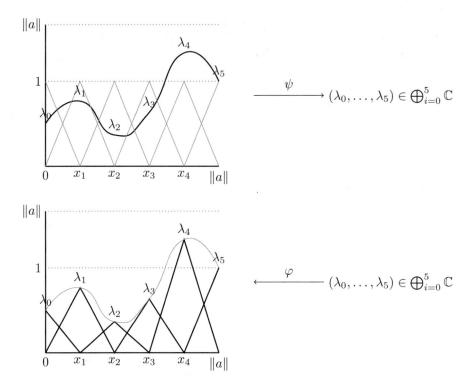

(Hint: Use the Gelfand–Naimark Theorem to identify B with $C(X)$ for some compact metric space X and use a partition of unity argument. See the picture above, where we approximate a finite subset $\mathcal{F}$ consisting of a single element by a relatively large ϵ.)

3.4.7. Prove that every separable C*-algebra admits a countable approximate unit.

3.4.8. Let A be a C*-algebra. Show that if $a \in A_+$ then $\overline{aAa}$ is a hereditary C*-subalgebra.

3.4.9. Let A be a unital C*-algebra, $a \in \mathrm{Inv}(A)$ and $p \in A$ a projection. Show that if a commutes with p then a is invertible in the corner pAp. If $a \in A$ is invertible in pAp, is $a \in \mathrm{Inv}(A)$?

3.4.10. A C*-algebra A is called *elementary* if there is a Hilbert space H such that A is isomorphic to $\mathcal{K}(H)$, the compact of operators on H.

  (i) Show that if A is elementary, then A contains hereditary C*-subalgebras in which every element has finite spectrum.

 (ii) Let A be a simple, nonelementary C*-algebra. Show that every hereditary C*-subalgebra $B \subset A$ contains a positive element b such that $\|b\| = 1$ and $\mathrm{sp}(b)$ contains infinitely many points.

3.4.11. Let A be a C*-algebra and $I \subset A$ an ideal. Then A/I is a Banach algebra (1.3.1). In this exercise we will show it is moreover a C*-algebra.

(i) Let $(u_\lambda)_{\lambda \in \Lambda}$ be an approximate unit for I (not necessarily quasidiagonal). Show that for every $a \in A$ we have

$$\|a + I\| = \lim_\lambda \|a - u_\lambda a\| = \lim_\lambda \|a - au_\lambda\|.$$

(ii) Show that the quotient norm is a C*-norm on A/I, and that A/I is complete, so A/I is a C*-algebra.

(iii) Let B be another C*-algebra and $\varphi : A \to B$ be a *-homomorphism. Show that $\varphi(A)$ is a C*-subalgebra of B.

(iv) Suppose C is a C*-subalgebra of A. Show that $B + I$ is also a C*-subalgebra of A.

3.4.12. Let A be a C*-algebra and $I \subset A$ an ideal. Suppose that J is an ideal in I. Show that J is also an ideal of A. (Hint: Use Exercise 3.4.2.)

4 Positive linear functionals and representations

In this section we will see that every abstract C*-algebra is *-isomorphic to a self-adjoint closed subalgebra of $\mathcal{B}(H)$, for some Hilbert space H. That is to say, the class of abstract C*-algebras and the class of concrete C*-algebras coincide. We say that a C*-algebra A can be *represented* on a Hilbert space H if there is a *-homomorphism from A to $\mathcal{B}(H)$. C*-algebra representations are intimately connected to linear functionals. Every linear functional on a C*-algebra A leads to a representation of A, and because we have "enough" functionals, we are able to add all these representations together to arrive at a *-isomorphism mapping A to a concrete C*-algebra (Theorem 4.2.6).

Knowing that an abstract C*-algebra can be replaced with a *-isomorphic copy acting on some Hilbert space gives us many more tools with which to work. For example, as we will see in the next chapter, we can always put a C*-algebra A into a *von Neumann algebra*. This often allows us to use von Neumann algebra methods to determine structural properties of A. We will return to this later.

In this chapter, we begin by defining positive linear functionals, states, and tracial states. In Section 4.2 we use positive linear functionals to construct representations of C*-algebras, and most importantly, show that every abstract C*-algebra is *-isomorphic to a concrete algebra by constructing a faithful representation. In the final section, we consider *cyclic* representations and show that they are the building blocks of any nondegenerate representation in the sense that any nondegenerate representation is the direct sum of cyclic representations.

4.1 Functionals

A linear map $\phi : A \to B$ between C*-algebras is called *positive* if $\phi(A_+) \subset B_+$. A *-homomorphism is always positive. If $\phi : A \to \mathbb{C}$ is positive, it is called a *positive linear functional*. Any positive linear functional ϕ also satisfies $\phi(A_{sa}) \subset \mathbb{R}$ and hence $\phi(a^*) = \overline{\phi(a)}$ for every $a \in A$ (Exercise 4.4.2).

4.1.1. If ϕ is a positive linear functional with $\|\phi\| = 1$, then ϕ is called a *state*. If in addition it satisfies $\phi(ab) = \phi(ba)$ for every $a, b \in A$ then it is called a *tracial state*. The set of states, respectively tracial states, on A will be denoted by $S(A)$, respectively $T(A)$. A state ϕ is *faithful* if $\phi(a^*a) = 0$ implies $a = 0$.

4.1.2. Example. Let $A = M_n$ and let $a = (a_{ij})_{ij} \in A$ be an $n \times n$ matrix. Then the usual normalised trace, $\mathrm{tr}_n : A \to \mathbb{C}$ given by

$$\mathrm{tr}_n(a) = \frac{1}{n} \sum_{i=1}^{n} a_{ii}$$

is a faithful tracial state.

© The Author(s), under exclusive license to Springer Nature Switzerland AG 2021
K. R. Strung, *An Introduction to C*-Algebras and the Classification Program*, Advanced Courses in Mathematics - CRM Barcelona, https://doi.org/10.1007/978-3-030-47465-2_4

4.1.3. Example. Let $A = C(X)$ be a commutative C*-algebra. Then any character is a tracial state, but not every tracial state is of this form. Suppose that μ is a Borel probability measure on X. Then the map $\tau : A \to \mathbb{C}$ given by

$$\tau(f) = \int f d\mu$$

is a tracial state. One often thinks of a tracial state as a noncommutative probability measure.

4.1.4. Example. Let H be a Hilbert space and $\xi \in H$ a nonzero vector. Then

$$\phi(a) = \langle a\xi, \xi \rangle$$

is a positive linear functional on $\mathcal{B}(H)$, but is not tracial unless $H = \mathbb{C}$.

The next theorem is another example of how C*-algebras are in general better behaved than arbitrary Banach algebras.

4.1.5. Theorem. *Any positive linear functional on a C*-algebra is bounded.*

Proof. Suppose not. Let A be a C*-algebra admitting an unbounded linear functional $\phi : A \to \mathbb{C}$. Since ϕ is unbounded, we can find a sequence $(a_n)_{n \in \mathbb{N}}$ of elements in the unit ball of A such that $|\phi(a_n)| \to \infty$ as $n \to \infty$. Without loss of generality, we may assume that each $a_n \in A_+$, for if ϕ was bounded on every $a_n \in A_+$ then ϕ would be bounded everywhere. Passing to a subsequence if necessary, we may further assume that for every $n \in \mathbb{N}$ we have $\phi(a_n) \geq 2^n$. Let $a := \sum_{n \in \mathbb{N}} 2^{-n} a_n \in A_+$. Then, for every $N \in \mathbb{N}$,

$$\phi(a) = \sum_{n \in \mathbb{N}} 2^{-n} \phi(a_n) \geq \sum_{n \in \mathbb{N}} 1 > N,$$

which is impossible. $\qquad\qquad\qquad\qquad\qquad\qquad\qquad\qquad\qquad\qquad\qquad\square$

Positive linear functionals admit the following Cauchy–Schwarz inequality.

4.1.6. Proposition. *Let A be a C*-algebra and suppose that $\phi : A \to \mathbb{C}$ is a positive linear functional. Then*

$$|\phi(a^*b)|^2 \leq \phi(a^*a)\phi(b^*b)$$

for every $a, b \in A$.

Proof. We may assume that $\phi(a^*b) \neq 0$. Since ϕ is positive, for any $\lambda \in \mathbb{C}$ we have $\phi((\lambda a + b)^*(\lambda a + b)) \geq 0$. In particular, this holds for

$$\lambda := t\frac{|\phi(a^*b)|}{\phi(b^*a)},$$

for any $t \in \mathbb{R}$, giving

$$t^2\phi(a^*a) - 2t|\phi(a^*b)| + \phi(b^*b) \geq 0.$$

If we have $\phi(a^*a) = 0$ then $\phi(b^*b) \geq 2t|\phi(a^*b)|$ for every $t \in \mathbb{R}$, which is impossible unless $|\phi(a^*b)|$ is also zero; in this case the inequality holds. If $\phi(a^*a) \neq 0$ then let

$$t := -\frac{|\phi(a^*b)|}{\phi(a^*a)}.$$

Then

$$\frac{|\phi(a^*b)|^2}{\phi(a^*a)} - 2\frac{|\phi(a^*b)|^2}{\phi(a^*a)} + \phi(b^*b) \geq 0,$$

from which the result follows.   □

4.1.7. Proposition. *Let A be a C*-algebra and let $(u_\lambda)_{\lambda \in \Lambda}$ be an approximate unit. Let $\phi \in A^*$. Then ϕ is positive if and only if $\lim_\lambda \phi(u_\lambda) = \|\phi\|$. In particular if A is unital then $\phi(1_A) = \|\phi\|$.*

Proof. Suppose that ϕ is positive. Then, since $(u_\lambda)_\Lambda$ is an increasing net of positive elements, $(\phi(u_\lambda))_\Lambda$ is increasing in $\mathbb{R}_+$. It is moreover bounded, so converges to some $r \in \mathbb{R}_+$. Since each u_λ has norm less than one, $r \leq \|\phi\|$. Now if $a \geq 0$ with $\|a\| \leq 1$, then, using the Cauchy–Schwarz inequality,

$$|\phi(au_\lambda)|^2 \leq \phi(a^*a)\phi(u_\lambda^2) \leq \phi(a^*a)\phi(u_\lambda) \leq r\phi(a^*a).$$

Since ϕ is continuous and $au_\lambda \to a$, it follows that

$$|\phi(a)|^2 \leq r\phi(a^*a) \leq r\|\phi\|,$$

hence $\|\phi\|^2 \leq r\|\phi\|$, which is to say $\lim_\lambda \phi(u_\lambda) = \|\phi\|$.

For the converse, first let $a \in A_{sa}$. We will show that $\phi(a) \in \mathbb{R}$. Let α and β be real numbers such that $\phi(a) = \alpha + i\beta$. Without loss of generality, assume that $\beta \geq 0$. For $n \in \mathbb{N}$ let λ be sufficiently large that $\|u_\lambda a - au_\lambda\| < 1/n$. Then

$$\|nu_\lambda - ia\|^2 = \|(nu_\lambda + ia)(nu_\lambda - ia)\|$$
$$= \|n^2 u_\lambda^2 + a^2 - in(u_\lambda a - au_\lambda)\|$$
$$\leq n^2 + \|a\| + 1 \leq n^2 + 2.$$

We also have

$$\lim_\lambda |\phi(nu_\lambda - ia)|^2 = (n\|\phi\| + \beta - i\alpha)(n\|\phi\| + \beta + i\alpha)$$
$$= (n\|\phi\| + \beta)^2 + \alpha^2.$$

Thus

$$(n\|\phi\| + \beta)^2 + \alpha^2 = \lim_\lambda |\phi(nu_\lambda - ia)|^2 \leq \|\phi\|^2\|nu_\lambda - ia\|^2$$
$$\leq \|\phi\|^2(n^2 + 2),$$

and then

$$n^2\|\phi\|^2 + 2n\|\phi\|\beta + \beta^2 + \alpha^2 \leq n^2\|\phi\|^2 + 2,$$

for every $n \in \mathbb{N}$. Since $\beta^2 + \alpha^2 \geq 0$ and we assumed that $\beta \geq 0$, for large enough n this can only hold with $\beta = 0$. Thus $\phi(a) \in \mathbb{R}$.

Now if $a \in A_+$ and $\|a\| \leq 1$, then $u_\lambda - a \in A_{sa}$ and $u_\lambda - a \leq u_\lambda$, so

$$\lim_\lambda \phi(u_\lambda - a) \leq \|\phi\|.$$

Thus $\phi(a) \geq 0$. □

4.1.8. Corollary. *Let A be a nonunital C*-algebra. Then any positive linear functional $\phi : A \to \mathbb{C}$ admits a unique extension $\tilde{\phi} : \tilde{A} \to \mathbb{C}$ with $\|\tilde{\phi}\| = \|\phi\|$.*

Proof. Exercise. □

4.1.9. Proposition. *Let A be a C*-algebra and let $B \subset A$ be a C*-subalgebra. Every positive linear functional $\phi : B \to \mathbb{C}$ admits an extension $\tilde{\phi} : A \to \mathbb{C}$. If $B \subset A$ is a hereditary C*-subalgebra then the extension is unique and, in that case, if $(u_\lambda)_\Lambda$ is an approximate unit for B then*

$$\tilde{\phi}(a) = \lim_\lambda \phi(u_\lambda a u_\lambda),$$

for any $a \in A$.

Proof. By the previous corollary, we may suppose that A is unital and $1_A \in B$. Then, applying the Hahn–Banach Theorem, there is an extension $\tilde{\phi} \in A^*$ such that $\|\tilde{\phi}\| = \|\phi\|$. Since $\|\phi\| = \phi(1_A) = \tilde{\phi}(1_A)$, it follows from Proposition 4.1.7 that $\tilde{\phi}$ is also positive.

If B is hereditary, then $\lim_\lambda \phi(u_\lambda) = \|\phi\| = \|\tilde{\phi}\| = \tilde{\phi}(1_A)$. Thus $\lim_\lambda \tilde{\phi}(u_\lambda - 1_A) = 0$. For any $a \in A$,

$$|\tilde{\phi}(a) - \phi(u_\lambda a u_\lambda)| \leq |\tilde{\phi}(a - a u_\lambda)| + |\tilde{\phi}(a u_\lambda - u_\lambda a u_\lambda)|$$
$$\leq \phi(a^* a)^{1/2} \phi((1_A - u_\lambda)^2)^{1/2} + \phi(u_\lambda a^* a u_\lambda)^{1/2} \phi((1_A - u_\lambda)^2)^{1/2}$$
$$\leq \phi((1_A - u_\lambda)^2)^{1/2}(\phi(a^* a)^{1/2} + \phi(u_\lambda a^* a u_\lambda)^{1/2})$$

which goes to zero in the limit. So $\tilde{\phi}(a) = \lim_\lambda \phi(u_\lambda a u_\lambda)$, as required. □

4.1.10. Proposition. *Let A be a nonzero C*-algebra and let $a \in A$ be a normal element. Then there is a state $\phi \in S(A)$ such that $\phi(a) = \|a\|$.*

Proof. Let $B \subset \tilde{A}$ be the C*-subalgebra generated by a and 1_A. Since B is commutative, we have $\hat{a} \in C(\Omega(B))$ and $\phi \in \Omega(B)$ such that $\phi(a) = \hat{a}(\phi) = \|a\|$. Since $\phi(1_A) = 1$, there is a positive extension to $\tilde{A}$. Then the restriction ϕ to A satisfies the requirements, since $\|\phi\| = \phi(1_A) = 1$. □

4.2   The Gelfand–Naimark–Segal construction

What we now call the Gelfand–Naimark–Segal (GNS) construction of a representation from a positive linear functional, Definition 4.2.3 below, was arrived at independently by Gelfand and Naimark in [49] and Segal in [111].

A *-representation,* or simply *representation*, of a C*-algebra A is a pair (H, π) consisting of a Hilbert space H and a *-homomorphism $\pi : A \to \mathcal{B}(H)$. If π is injective then we say (H, π) is *faithful*. If A has a faithful representation, then it is *-isomorphic to a closed self-adjoint subalgebra of $\mathcal{B}(H)$, that is, a concrete C*-algebra.

4.2.1. We begin by establishing that every C*-algebra has many representations. Let A be a C*-algebra. Given a positive linear functional ϕ, let

$$N_\phi := \{a \in A \mid \phi(a^*a) = 0\},$$

which is a closed left ideal in A. Define a map

$$\langle \cdot, \cdot \rangle_\phi : A/N_\phi \times A/N_\phi \to \mathbb{C}, \quad \langle a + N_\phi, b + N_\phi \rangle \to \phi(b^*a).$$

We leave it as an exercise to show that $\langle \cdot, \cdot \rangle_\phi$ defines an inner product on A/N_ϕ.

4.2.2. Let H_ϕ denote the Hilbert space obtained by completing A/N_ϕ with respect to the inner product described above. For any $a \in A$, we can define a linear operator $\pi_\phi(a) : A/N_\phi \to A/N_\phi$ by $\pi_\phi(a)(b + N_\phi) = ab + N_\phi$. Let $b + N_\phi \in A/N_\phi$ be an element with norm at most one so that $\langle b + N_\phi, b + N_\phi \rangle = \phi(b^*b) \leq 1$. Then

$$\|\pi_\phi(a)(b)\|^2 = \langle ab + N_\phi, ab + N_\phi \rangle = \phi(b^*a^*ab).$$

By Proposition 3.1.11, we have $b^*a^*ab \leq b^*\|a\|^2b$, so positivity of ϕ implies

$$\phi(b^*a^*ab) \leq \|a\|^2\phi(b^*b) \leq \|a\|^2.$$

Hence $\pi_\phi(a)$ is bounded and extends to an operator on H_ϕ.

Now define

$$\pi_\phi : A \to \mathcal{B}(H_\phi), \quad a \mapsto \pi_\phi(a).$$

Observe that π_ϕ is a *-homomorphism. In particular, (H_ϕ, π_ϕ) is a representation of A.

4.2.3. Definition. Let ϕ be a positive linear functional on a C*-algebra A. The representation (H_ϕ, π_ϕ) is called the Gelfand–Naimark–Segal representation associated to ϕ, or more commonly, the *GNS representation* associated to ϕ.

4.2.4. Let $(H_\lambda, \pi_\lambda)_{\lambda \in \Lambda}$ be a family of representations of a C*-algebra A. Define

$$\oplus_{\lambda \in \Lambda} \pi_\lambda : A \to \bigoplus_{\lambda \in \Lambda} H_\lambda$$

to be the map taking $a \in A$ to the element with $\pi_\lambda(a)$ in the λth coordinate. Then $(\bigoplus_{\lambda \in \Lambda} H_\lambda, \oplus_{\lambda \in \Lambda} \pi_\lambda)$ is a representation of A. It is faithful as long as, for each $a \in A \setminus \{0\}$, there is some λ such that $\pi_\lambda(a) \neq 0$.

4.2.5. Definition. Let A be a C*-algebra. The representation

$$\left(\bigoplus_{\phi \in S(A)} H_\phi, \ \oplus_{\phi \in S(A)} \pi_\phi \right)$$

is called the *universal representation* of A.

4.2.6. From the GNS construction above, we arrive at the following, usually called the Gelfand–Naimark Theorem.

Theorem (Gelfand–Naimark). *Let A be a C*-algebra. Then its universal representation is faithful.*

Proof. Exercise. □

4.3 More about representations

The previous section showed that every C*-algebra not only has at least as many representations as it has states, but that we can always find a *faithful* representation. This means that every abstract C*-algebra is *-isomorphic to a concrete C*-algebra. It is often useful to think of a particular C*-algebra as acting on some Hilbert space. In this section we will look more closely at representations to establish results we will require in the sequel.

4.3.1. Let (H, π) be a representation of a C*-algebra A. A vector $\xi \in H$ is called *cyclic* if the linear span of $\{\pi(a)\xi \in H \mid a \in A\}$, which we will denote by $\pi(A)\xi$ (that is to say, the orbit of ξ under $\pi(A)$), is dense in H. If such a ξ exists, then (H, π) is called a *cyclic representation*.

Theorem. *Let $\phi : A \to \mathbb{C}$ be a positive linear functional. The GNS representation associated to ϕ is cyclic with cyclic vector ξ_ϕ, where $\xi_\phi \in H_\phi$ is the unique vector satisfying*

$$\phi(a) = \langle \pi_\phi(a)\xi_\phi, \xi_\phi \rangle_\phi$$

for every $a \in A$. If ϕ is a state, then $\|\xi_\phi\| = 1$.

Proof. For $a \in A$, we will denote the vector $a + N_\phi \in H_\phi$ by ξ_a. Let $(u_\lambda)_\Lambda$ be an approximate unit for A. Then for $\lambda < \mu$ we have

$$\|\xi_{u_\mu} - \xi_{u_\lambda}\|^2 = \phi((u_\mu - u_\nu)^2) \leq \phi(u_\mu - u_\nu).$$

By Proposition 4.1.7, the net $(\phi(u_\lambda))_\Lambda$ converges to $\|\phi\|$, thus $(\xi_{u_\lambda})_\Lambda$ is Cauchy and hence converges to some $\xi_\phi \in H_\phi$. If $a \in A$ then, by continuity of the map $A \to H_\phi, x \mapsto \xi_x$, we have

$$\pi_\phi(a)\xi_\phi = \lim_\lambda \pi_\phi(a)\xi_{u_\lambda} = \lim_\lambda \xi_{au_\lambda} = \xi_a.$$

Thus ξ_ϕ is a cyclic vector. Furthermore,

$$\langle \pi_\phi(a^*a)\xi_\phi, \xi_\phi \rangle_\phi = \langle \pi_\phi(a)\xi_\phi, \pi_\phi(a)\xi_\phi \rangle_\phi = \langle \xi_a, \xi_a \rangle_\phi = \phi(a^*a),$$

for any $a \in A$.

For the final statement, by Corollary 4.1.8 we may assume that A is unital. In that case, $\phi(1_A) = 1$ and $\pi_\phi(1_A)(b) = b + N_\phi$ for every $b \in A$, so by continuity $\pi_\phi(1_A) = \text{id}$. Thus $\|\xi_\phi\|^2 = \langle \xi_\phi, \xi_\phi \rangle = \langle \pi_\phi(1_A)\xi_\phi, \xi_\phi \rangle = \phi(1_A) = 1$, and the result follows. $\square$

4.3.2. By a *unitary* between two Hilbert spaces H_1 and H_2 we mean a bounded surjective linear map $u : H_1 \rightarrow H_2$ such that $\langle u(\xi_1), u(\xi_2) \rangle_{H_2} = \langle \xi_1, \xi_2 \rangle_{H_1}$ for every $\xi_1, \xi_2 \in H_1$. If $H_1 = H_2$ then u is a unitary in $\mathcal{B}(H)$ in the sense of 2.1.1 (exercise).

Two representations (H_1, π_1) and (H_2, π_2) of a C^*-algebra A are *unitarily equivalent* if there is a unitary operator $u : H_1 \rightarrow H_2$ such that $\pi_2(a) = u\pi_1(a)u^*$ for every $a \in A$. As the name suggests, this is an equivalence relation on the representations of A.

4.3.3. Proposition. *Let (H_1, π_1) be a cyclic representation of A with cyclic vector ξ and let (H_2, π_2) be a cyclic representation of A with cyclic vector μ. Then the following are equivalent:*

(i) *(H_1, π_1) and (H_2, π_2) are unitarily equivalent with unitary $u : H_1 \rightarrow H_2$ satisfying $u\xi = \mu$,*

(ii) *$\langle \pi_1(a)\xi, \xi \rangle = \langle \pi_2(a)\mu, \mu \rangle$ for every $a \in A$.*

Proof. That (i) implies (ii) is clear. Conversely, let us suppose that $\langle \pi_1(a)\xi, \xi \rangle = \langle \pi_2(a)\mu, \mu \rangle$ for every $a \in A$. Define a linear map

$$u : \pi_1(A)\xi \rightarrow \pi_2(A)\mu, \qquad \pi_1(x)\xi \mapsto \pi_2(x)\mu.$$

Then

$$\|u\pi_1(x)\xi\|^2 = \|\pi_2(x)\mu\|^2 = \langle \pi_2(x^*x)\mu, \mu \rangle = \langle \pi_1(x^*x)\xi, \xi \rangle = \|\pi_2(x)\mu\|^2,$$

so, since ξ and μ are cyclic, we get that u extends to an isometry $H_1 \rightarrow H_2$. Moreover,

$$u\pi_1(x)\pi_1(y)\xi = \pi_2(xy)\mu = \pi_2(x)u\pi_1(y)\xi,$$

for every $x, y \in A$. Since ξ and μ are cyclic, we have $u\pi_1(x) = \pi_2(x)u$ on the dense subset $\{\pi_1(y)\xi \mid y \in A\}$ and hence on all of H_1. Thus H_1 and H_2 are unitarily equivalent and $u\xi = \mu$, as required. $\square$

4.3.4. A representation (H, π) of A is called *nondegenerate* if the linear span of $\{\pi(a)\xi \mid a \in A, \xi \in H\}$, denoted by $\pi(A)H$, is dense in H, or, equivalently, for each $\xi \in H \setminus \{0\}$ there is $a \in A$ such that $\pi(a)\xi \neq 0$. Clearly a cyclic representation is nondegenerate, but a nondegenerate representation need not be cyclic. However, it is always the direct sum of cyclic representations.

4.3.5. Theorem. *Let A be a C*-algebra. Any nondegenerate representation (H, π) is the direct sum of cyclic representations.*

Proof. Let $\xi \in H$ be nonzero and let $H_\xi = \overline{\pi(A)\xi}$. By the Kuratowski–Zorn Lemma, there is a maximal set $S \subset H \setminus \{0\}$ such that H_ξ, $\xi \in S$ are pairwise orthogonal. Let $\pi_\xi := \pi|_{H_\xi}$. Since each H_ξ is π-invariant, we see that each (H_ξ, π_ξ) is a cyclic representation. Moreover, since the H_ξ are pairwise orthogonal, their direct sum $\bigoplus_{\xi \in S} H_\xi$ is contained in H, and thus the direct sum of the representations (H_ξ, π_ξ), $\xi \in S$, is a representation on H. We will show that $\bigoplus_{\xi \in S} H_\xi = H$. Let $\eta \in (\bigoplus_{\xi \in S} H_\xi)^\perp$. Then it is easy to check that $H_\eta := \overline{\pi(A)\eta}$ is orthogonal to each H_ξ. We have $\pi(a)\eta \in H_\eta$ for every $a \in A$ and hence by maximality of S, we must have that $\pi(a)\eta = 0$ for every $a \in A$. Since (H, π) is nondegenerate this means $\eta = 0$. Hence $\bigoplus_{\xi \in S} H_\xi = H$. $\qquad\square$

4.4 Exercises

4.4.1. Show that the maps in Examples 4.1.2–4.1.4 give states. Show that the maps in Examples 4.1.2 and 4.1.3 are tracial, but that of Example 4.1.4 is not unless $H = \mathbb{C}$.

4.4.2. Let A and B be C*-algebras and $a \in A$.

  (i) If $\phi : A \to \mathbb{C}$ is a positive linear functional, show that $\phi(a^*) = \overline{\phi(a)}$.

  (ii) More generally, if $\psi : A \to B$ is any positive linear map, show that $\psi(a^*) = \psi(a)^*$.

4.4.3. Let A be a unital C*-algebra and $\phi \in A^*$ a positive linear functional. Without using approximate units or Proposition 4.1.7, show that $\phi(1_A) = \|\phi\|$.

4.4.4. Show that any positive linear functional $\phi : A \to \mathbb{C}$ admits a unique extension $\tilde{\phi} : \tilde{A} \to \mathbb{C}$ such that $\|\tilde{\phi}\| = \|\phi\|$.

4.4.5. Let A be a C*-algebra and $\phi : A \to \mathbb{C}$ a positive linear functional. Let $N_\phi := \{a \in A \mid \phi(a^*a) = 0\}$.

  (i) Show that N_ϕ is a closed left ideal in A.

  (ii) Show that A/N_ϕ is a vector space.

  (iii) Show that $\langle \cdot, \cdot \rangle_\phi : A/N_\phi \times A/N_\phi \to \mathbb{C}$ defined by $\langle a + N_\phi, b + N_\phi \rangle_\phi = \phi(b^*a)$ is an inner product on A/N_ϕ.

4.4.6. Recall that a unitary operator between two Hilbert spaces H_1 and H_2 we mean a bounded surjective linear map $u : H_1 \to H_2$ such that $\langle u(\xi_1), u(\xi_2) \rangle_{H_2} = \langle \xi_1, \xi_2 \rangle_{H_1}$ for every $\xi_1, \xi_2 \in H_1$. Show that if $H_1 = H_2$ then u is a unitary in $\mathcal{B}(H)$ in the sense of 2.1.1.

4.4.7. Let $A = M_n$ and $H = \mathbb{C}^m$ for $n, m \in \mathbb{N}$. For what values of m does A admit a faithful representation on H? Show that if $\pi_1, \pi_2 : M_n \to M_m$ are both faithful representations on $H = \mathbb{C}^m$, then they are unitarily equivalent.

4.4.8. Let A be a C*-algebra, $\phi : A \to \mathbb{C}$ a positive linear functional and let (H_ϕ, π_ϕ) be the GNS representation associated to ϕ. Suppose that $I \subset A$ is an ideal. Show that $\ker(\phi) \subset I$ if and only if $I \subset \ker(\pi_\phi)$.

4.4.9. Let A be a C*-algebra, $\phi : A \to \mathbb{C}$ a positive linear functional. Suppose that ϕ is faithful. Show that (H_ϕ, π_ϕ) is a faithful representation.

4.4.10. Prove that the universal representation of a C*-algebra A is always faithful. Thus abstract C*-algebras and concrete C*-algebras coincide. (Hint: use the characterisation of faithfulness given in 4.2.4.)

4.4.11. Let A be a finite-dimensional C*-algebra. Show that there exists a faithful representation of A on a finite-dimensional Hilbert space.

4.4.12. Let A be a C*-algebra and $\alpha : A \to A$ a *-automorphism. Suppose that $\phi : A \to \mathbb{C}$ is a positive linear functional that is α-invariant, that is, $\phi(\alpha(a)) = \phi(a)$ for every $a \in A$.

(i) Show that the map $u(a + N_\phi) = \alpha(a) + N_\phi$ extends to a unitary operator $u : H_\phi \to H_\phi$.

(ii) Show that $\pi_\phi(\alpha(a)) = u\pi_\phi(a)u^*$ for every $a \in A$.

4.4.13. Let A be a C*-algebra and $n \in \mathbb{N} \setminus \{0\}$. Show that for any C*-algebra A, $M_n(A)$ is also a C*-algebra.

4.4.14. Let A be a C*-algebra. Let $a, b \in A$. Then

$$x := \begin{pmatrix} a & 0 \\ 0 & b \end{pmatrix} \in M_2(A).$$

Determine $\mathrm{sp}(x)$ in terms of $\mathrm{sp}(a)$ and $\mathrm{sp}(b)$.

4.4.15. Show that $M_n \otimes A \cong M_n(A)$, as vector spaces where $\otimes$ denotes the tensor product of vector spaces.

(i) Show that $(a \otimes b)(c \otimes d) = (ac) \otimes (bd)$ and $(a \otimes b)^* = a^* \otimes b^*$, for $a, c \in M_n$ and $b, d \in A$ extends to multiplication and involution on $M_n \otimes A$, making it into a *-algebra.

(ii) By showing $M_n \otimes A \cong M_n(A)$ preserves the multiplication and involution, show that $M_n \otimes A$ can be made into a C*-algebra.

4.4.16. Let A be a simple C*-algebra. Show that A has a nonzero finite-dimensional representation if and only if $A \cong M_n$ for some $n \in \mathbb{N} \setminus \{0\}$.

5 Von Neumann algebras and irreducible representations

For any Hilbert space H, the bounded operators $\mathcal{B}(H)$ admit topologies distinct the norm topology. We will be interested in the weak operator and the strong operator topologies. A C*-algebra is closed in the norm topology, but if we instead close in either the weak operator or strong operator topology, we end up with a von Neumann algebra. Since a von Neumann algebra is still closed in the norm topology, it is an example of a C*-algebra. However, the study of von Neumann algebras and the study of C*-algebras tend to use different techniques, and thus are often studied separately. Nevertheless, the von Neumann algebra techniques in this chapter will prove useful in the sequel.

In this chapter we start by introducing the weak and strong operator topologies. In Section 5.2 we prove some of the basic facts about von Neumann algebras, such as von Neumann's Double Commutant Theorem, that von Neumann algebras contain lots of projections, and prove the result from Chapter 3 that an ideal I in a C*-algebra A has an approximate unit quasicentral for A. In Section 5.3 we return to the topic of representations and states. We show that a state is *pure* if and only if its GNS representation is irreducible.

5.1 Topologies on $\mathcal{B}(H)$

We begin by defining two topologies on $\mathcal{B}(H)$, the weak and strong operator topologies, and looking at some of their properties.

5.1.1. Definition. The *weak operator topology* on $\mathcal{B}(H)$ is the weakest topology in which the sets $W(a, \xi, \mu) = \{b \in \mathcal{B}(H) \mid |\langle (a - b)\xi, \mu \rangle| < 1\}$ are open.

The *strong operator topology* on $\mathcal{B}(H)$ is the weakest topology in which the sets $S(a, \xi) = \{b \in \mathcal{B}(H) \mid \|(a - b)\xi\| < 1\}$ are open.

The sets $W(a_i, \xi_i, \mu_i \mid 1 \leq i \leq n) = \bigcap_{i=1}^{n} W(a_i, \xi_i, \mu_i)$ form a base for the weak operator topology and similarly, sets of the form $W(a_i, \xi_i \mid 1 \leq i \leq n) = \bigcap_{i=1}^{n} S(a_i, \xi_i)$ are a base for the strong operator topology.

5.1.2. Proposition. *Let $(a_\lambda)_{\lambda \in \Lambda}$ be a net in $\mathcal{B}(H)$ and let $a \in \mathcal{B}(H)$. Then*

(i) *$(a_\lambda)_\Lambda$ converges to a in the weak operator topology, written $a_\lambda \xrightarrow{\text{WOT}} a$, if and only if $\lim_\lambda \langle a_\lambda \xi, \eta \rangle = \langle a\xi, \eta \rangle$ for all $\xi, \eta \in H$;*

(ii) *$(a_\lambda)_\Lambda$ converges to a in the strong operator topology, written $a_\lambda \xrightarrow{\text{SOT}} a$, if and only if $\lim_\lambda a_\lambda \xi = a\xi$ for all $\xi \in H$.*

Proof. Exercise. □

5.1.3. Convergence in the norm operator topology implies convergence in the strong topology implies convergence in the weak topology. The reverse implications do not hold unless H is finite-dimensional. (Exercise.)

© The Author(s), under exclusive license to Springer Nature Switzerland AG 2021
K. R. Strung, *An Introduction to C*-Algebras and the Classification Program*, Advanced Courses in Mathematics - CRM Barcelona, https://doi.org/10.1007/978-3-030-47465-2_5

5.1.4. Proposition. *Let $a \in \mathcal{B}(H)$. Left and right multiplication by a is both* SOT-*and* WOT-*continuous.*

Proof. Suppose that $(b_\lambda)_{\lambda \in \Lambda}$ is a net in $\mathcal{B}(H)$ that WOT-converges to b. Then, for $\xi, \mu \in H$, we have

$$\lim_\lambda \langle ab_\lambda \xi, \mu \rangle = \lim_\lambda \langle b_\lambda \xi, a^* \mu \rangle$$
$$= \langle b\xi, a^* \mu \rangle$$
$$= \langle ab\xi, \mu \rangle,$$

showing $ab_\lambda \xrightarrow{\text{WOT}} ab$. Thus multiplication on the left is WOT-continuous. The other calculations are similar and are left as exercises. $\qquad\square$

5.1.5. If $A \subset \mathcal{B}(H)$ is a subset of operators, then its *commutant* is defined as

$$A' := \{b \in \mathcal{B}(H) \mid ab = ba \text{ for every } a \in A\}.$$

Proposition. *Let S be a subset of $\mathcal{B}(H)$. Then the commutant*

$$S' = \{b \in \mathcal{B}(H) \mid bx = ab \text{ for all } a \in S\}$$

is closed in the weak operator topology.

Proof. Suppose that $(a_\lambda)_{\lambda \in \Lambda} \subset S'$ and that $a_\lambda \to a$ in the weak operator topology. Then, using the fact that left and right multiplication is WOT-continuous, for every $x \in S$ we have

$$xa = \text{WOT-}\lim_\lambda xa_\lambda = \text{WOT-}\lim_\lambda a_\lambda x = ax.$$

So $a \in S'$. $\qquad\square$

5.2 Brief interlude on von Neumann algebras

Even though we will not say too much about von Neumann algebras in this book, we will require some von Neumann algebra techniques. Every C*-algebra sits inside a von Neumann algebra, and it is often useful to take advantage of the extra space in a von Neumann algebra to deduce results about C*-algebras.

5.2.1. Definition. Let H be a Hilbert space and $A \subset \mathcal{B}(H)$ a C*-algebra. Then A is called a *von Neumann algebra* if A is equal to its *double commutant*, $A = A''$.

Observe that $1_{\mathcal{B}(H)} \in A''$ so a von Neumann algebra, unlike an arbitrary C*-algebra, is always unital. Since $(A'')'' = A''$ for any C*-algebra $A \subset \mathcal{B}(H)$, A'' is itself a von Neumann algebra. The double commutant of A is often called the *enveloping von Neumann algebra* of A in $\mathcal{B}(H)$. For an abstract C*-algebra A, its enveloping von Neumann algebra is defined to be the enveloping von Neumann algebra of $\pi_u(A) \subset \mathcal{B}(H_u)$, where (H_u, π_u) is the universal representation of Definition 4.2.5.

5.2.2. Since every C*-algebra A is contained in a von Neumann algebra A'', we may think of von Neumann algebras as "bigger" objects than C*-algebras in a certain sense. With more space, it can be easier to work in the enveloping von Neumann algebra, however often at the loss of finer C*-algebraic information. The analogue in the commutative setting is enclosing a topological space in a measure space. It is straightforward to see algebraically that $A \subset A''$, but, perhaps surprisingly, it can also be shown topologically. This is the essence of Theorem 5.2.7, but to prove it we require some preliminaries.

5.2.3. Recall that a *sesquilinear form* on a vector space V is a map

$$(\cdot, \cdot) : V \times V \to \mathbb{C}$$

which is linear in the first variable and conjugate-linear in the second. A sesquilinear form satisfies the *polarisation identity*

$$(\xi, \eta) = \frac{1}{4} \sum_{n=0}^{3} i^n (\xi + i^n \eta, \xi + i^n \eta).$$

We say that a sesquilinear form is bounded if there exists $C > 0$ such that $|(\xi, \eta)| \leq C \|\xi\| \|\eta\|$.

5.2.4. An inner product is of course an example of a sesquilinear form. If a sesquilinear form is on a Hilbert space, then we can relate it directly to the Hilbert space inner product, as the next lemma shows.

Lemma. *Let H be a Hilbert space and $(\cdot, \cdot) : H \times H \to \mathbb{C}$ a bounded sesquilinear form. There exists a unique operator $a \in \mathcal{B}(H)$ such that*

$$(\xi, \eta) = \langle a\, \xi, \eta \rangle_H, \qquad \text{for every } \xi, \eta \in H.$$

Proof. First of all, observe that if such an operator exists, it must be unique. To show existence, let $\eta \in H$. The map $\phi : H \to \mathbb{C}$ given by $\phi(\xi) = \langle \xi, \eta \rangle$ is a bounded, hence continuous, linear functional. By the Riesz Representation Theorem, there is a unique element $\zeta_\eta \in H$ such that

$$\phi(\xi) = \langle \xi, \zeta_\eta \rangle,$$

for every $\xi \in H$. Define $a : H \to H$ to be the adjoint of the linear operator

$$H \to H, \quad \eta \mapsto \zeta_\eta.$$

Then a is a linear operator satisfying the requirements of the lemma; the remaining details are left as an exercise for the reader. $\square$

5.2.5. An advantage of the strong operator topology is that it is closed under the ordering of self-adjoint elements in the following sense.

Proposition. *Let $(a_\lambda)_{\lambda \in \Lambda} \subset \mathcal{B}(H)$ be an increasing net of self-adjoint elements that is bounded above. Then $(a_\lambda)_{\lambda \in \Lambda}$ converges in the strong operator topology.*

Proof. We can find some $\lambda_0 \in \Lambda$ such that, for every $\lambda \geq \lambda_0$, we have $a_\lambda \geq b$ for some self-adjoint operator b. Let $b_\lambda = a_\lambda - b$ for $\lambda \geq \lambda_0$. Then $b \geq 0$ and $(b_\lambda)_{\lambda \geq \lambda_0}$ is a net of positive elements which is bounded above. Observe that if $(b_\lambda)_{\lambda \geq \lambda_0}$ converges in the strong operator topology, then so does the original net $(a_\lambda)_{\lambda \in \Lambda}$. Thus we may assume we have an increasing net of positive elements which is bounded. In particular, there is some $C > 0$ such that $\|b_\lambda\| \leq C$ for every λ, and so $\langle b_\lambda \xi, \xi \rangle \leq C\|\xi\|^2$ for every $\xi \in H$. In particular,

$$\langle b_\lambda \xi, \eta \rangle = 1/4 \sum_{n=0}^{3} i^n \langle b_\lambda(\xi + i^n \eta), \xi + i^n \eta \rangle$$

is bounded and hence converges, for every $\xi, \eta \in H$. Let

$$(\xi, \eta) := \lim_\lambda \langle b_\lambda \xi, \eta \rangle.$$

A straightforward calculation shows that $(\cdot, \cdot) : H \times H \to \mathbb{C}$ is a sesquilinear form which is moreover bounded. Then, by Lemma 5.2.4, there is an operator $c \in \mathcal{B}(H)$ such that

$$(\xi, \eta) = \langle c\xi, \eta \rangle \quad \text{for every } \xi, \eta \in H.$$

Now, since $\langle c\xi, \xi \rangle = \lim_\lambda \langle b_\lambda \xi, \xi \rangle \geq 0$ for any $\eta \in H$, we see that c is a positive operator, and therefore $b_\lambda - c$ is self-adjoint. Then

$$\|(b_\lambda - c)\xi\|^2 = \langle (b_\lambda - c)\xi, (b_\lambda - c)\xi \rangle = \langle (b_\lambda - c)^2 \xi, \xi \rangle \leq \|b_\lambda - c\| \langle (b_\lambda - c)\xi, \xi \rangle,$$

and since $\langle (b_\lambda - c)\xi, \xi \rangle = \langle b_\lambda \xi, \xi \rangle - \langle c\xi, \xi \rangle \to 0$, we have $\|(b_\lambda - c)\xi\| \to 0$, which is to say that b_λ SOT-converges to c. $\qquad \square$

5.2.6. Definition. If $A \subset \mathcal{B}(H)$ then we say that A *acts nondegenerately* on H if its null space

$$N_A := \{\xi \in H \mid a\xi = 0 \text{ for every } a \in A\}$$

is trivial.

5.2.7. Now we can prove von Neumann's Double Commutant Theorem. Often one of the equivalent conditions below is taken as the definition of a von Neumann algebra.

Theorem. *Let A be a C*-algebra acting nondegenerately on a Hilbert space H. The following are equivalent.*

(i) *A is a von Neumann algebra;*

(ii) $\overline{A}^{\text{WOT}} = A$;

(iii) $\overline{A}^{\text{SOT}} = A$.

Proof. We have $\overline{A}^{\text{SOT}} \subset \overline{A}^{\text{WOT}}$ since the strong operator topology is stronger than the weak topology. Since $A \subset A''$ and A'' is WOT-closed, we furthermore have $\overline{A}^{\text{WOT}} \subset A''$. It only remains to show that $A'' \subset \overline{A}^{\text{SOT}}$. Let $a \in A''$. It is enough to show that for any $n \in \mathbb{N} \setminus \{0\}$ and $\xi_1, \ldots, \xi_n \in H$ there is some $b \in A$ such that b is contained in $S(a, \xi_i \mid 1 \leq i \leq n)$. Notice that this is equivalent to finding $b \in A$ with $\sum_{i=1}^{n} \|(b-a)\xi_i\|^2 < 1$.

Consider $n = 1$. Let p be the orthogonal projection onto $\overline{A\xi_1}$. Then if $c \in A$ we have $pcp\xi_1 = cp\xi_1$ and if $\mu \neq \xi_1$ then $pcp\mu = cp\mu = 0$. Thus $pcp = cp$ for every $c \in A$, and we calculate

$$pc = (c^*p)^* = (pc^*p)^* = pcp = cp,$$

hence $p \in A'$. This means $p^\perp \in A'$, too.

If $\mu = p^\perp \xi_1$ then $A\mu = Ap^\perp \xi_1 = p^\perp A\xi_1 = 0$. Since A acts nondegenerately, we must therefore have $\mu = 0$. It follows that $\xi_1 \in \overline{A\xi_1}$. Since $a \in A''$ we have $pa = ap$ and $a\xi_1 \in \overline{A\xi_1}$. Thus we can find $b \in A$ satisfying $\|(a-b)\xi_1\| < 1$.

Now suppose $n \geq 2$. Let $H^{(n)} := H \oplus \cdots \oplus H$ be the direct sum of n copies of H. An arbitrary operator in $\mathcal{B}(H^{(n)})$ then looks like an $n \times n$ matrix $(x_{ij})_{ij}$ with each entry $x_{ij} \in \mathcal{B}(H)$. For $c \in A$, let $c^{(n)} \in \mathcal{B}(H^{(n)})$ be defined as

$$c^{(n)}(\xi_1, \ldots, \xi_n) := (c\xi_1, \ldots, c\xi_n),$$

and then set $A^{(n)} := \{c^{(n)} \mid c \in A\}$.

We claim $(A^{(n)})'' = (A'')^{(n)}$ where $(A'')^{(n)}$ is defined analogously to the above. It is easy to see that $c = (c_{ij})_{ij} \in (A^{(n)})'$ if and only if each $c_{ij} \in A'$. Thus $A^{(n)}$ contains all the matrix units e_{ij} where e_{ij} is the matrix with 1 in the (i,j)th entry and zero elsewhere. It follows that any $c \in (A^{(n)})''$ must commute with every e_{ij}. The only way this is possible is if all the diagonal entries of c are the same and the off-diagonal entries are zero, that is $c = c_{11}^{(n)}$ where $c_{11} \in A''$. It is clear that c commutes with each $b^{(n)}$ where $b \in A'$. Thus $c_{11} \in A''$ and $c \in (A'')^{(n)}$, proving the claim.

Now we apply the case for $n = 1$ to $a^{(n)} \in (A^n)''$ and $\xi = (\xi_1, \ldots, \xi_n)$ to find $b \in A$ with

$$1 > \|(a^{(n)} - b^{(n)})\xi\|^2 = \sum_{i=1}^{n} \|(a-b)\xi_i\|^2.$$

Thus b is in $S(a, \xi_i \mid 1 \leq i \leq n)$ and so $A'' \subset \overline{A}^{\text{SOT}}$, which proves the theorem. $\square$

5.2.8. Proposition. *The weak operator continuous linear functionals and the strong operator continuous linear functionals $\phi : \mathcal{B}(H) \to \mathbb{C}$ coincide and are always of the form*

$$\phi(a) = \sum_{i=1}^{n} \langle a\xi_i, \eta_i \rangle$$

for some $n \in \mathbb{N}$ and $\xi_i, \eta_i \in H$, $1 \leq i \leq n$.

Proof. It is easy to see that anything of the form $\phi(a) = \sum_{i=1}^{n} \langle a\xi_i, \mu_i \rangle$ will be a WOT-, and hence SOT-, continuous linear functional. Suppose now that ϕ is a SOT-continuous positive linear functional. Then the set $U := \{a \in \mathcal{B}(H) \mid |\phi(a)| < 1\}$ is open in the strong operator topology and contains zero. Thus we can find a basic set as given in Definition 5.1.1 containing zero that is completely contained in U. That is to say, there are $\xi_1, \ldots, \xi_n \in H$ such that

$$V := \{a \in \mathcal{B}(H) \mid \|a\xi_i\| < 1, 1 \leq i \leq n\} \subset \{a \in \mathcal{B}(H) \mid |\phi(a)| < 1\}.$$

If $a \in \mathcal{B}(H)$ satisfies $\sum_{i=1}^{N} \|a\xi_i\| < 1$ then clearly $a \in V$. As in the proof of Theorem 5.2.7, let $H^{(n)} := H \oplus \cdots \oplus H$ be the Hilbert space given by the direct sum of n copies of H. Set $\xi := (\xi_1, \ldots, \xi_n)$ and let $\psi_\xi : \mathcal{B}(H) \to H^{(n)}$ be the map given by

$$\psi_\xi(a) = (a\xi_1, \ldots, a\xi_n).$$

Now define

$$F : \psi_\xi(\mathcal{B}(H)) \to \mathbb{C}, \quad \psi_\xi(a) \mapsto \phi(a).$$

Note that if $\eta \in \psi_\xi(\mathcal{B}(H))$, then $\eta = (a\xi_1, \ldots, a\xi_n)$ for some $a \in A$ and thus if $\|\eta\| \leq 1$ we have $\sum_{i=1}^{n} \|a\xi_i\| \leq 1$. In that case $\phi(a) \leq 1$. Thus $\|F\| \leq 1$ and, by applying the Hahn–Banach Theorem, there is an extension to a continuous linear functional $\tilde{F} : H^{(n)} \to \mathbb{C}$. By the Riesz Representation Theorem, there is $\eta = (\eta_1, \ldots, \eta_n) \in H^{(n)}$ such that

$$\tilde{F}(\nu) = \langle \nu, \eta \rangle_{H^{(n)}} = \sum_{i=1}^{n} \langle \nu_i, \eta_i \rangle_H,$$

and so

$$\phi(a) = \tilde{F}(\psi_\xi(a)) = \sum_{i=1}^{n} \langle a\xi_i, \eta_i \rangle,$$

and the result follows since this is also WOT-continuous by Proposition 5.1.2. □

5.2.9. Unlike C*-algebras, which may have no nontrivial projections, von Neumann algebras are strong operator closed which implies they always contain many projections. In particular, if M is a von Neumann algebra and $a \in M$, then M contains the right and left support projections of a. By smallest projection satisfying a particular property we mean a projection p which satisfies that property and such that, given any other projection q also satisfying that property, $p \leq q$.

5.2.10. Definition. Let H be a Hilbert space and let $a \in \mathcal{B}(H)$ be an operator. The *left support projection* of a is the smallest projection $p \in \mathcal{B}(H)$ such that $pa = a$. The *right support projection* of a is the smallest projection $q \in M$ such that $aq = a$. Equivalently, the left support projection is the projection $p \in H$ onto $\overline{a(H)}$ and is sometimes called the *range projection* while the right support projection is the projection $q \in H$ onto $\overline{a^*(H)}$ and is sometimes called the *support projection* of a.

5.2.11. Proposition. *Let $M \subset \mathcal{B}(H)$ be a von Neumann algebra, and let $a \in M$. Let p denote the left support projection of a, and q the right support projection of a. Then $p, q \in M$.*

Proof. If a is not positive, let $a = u|a| = u(a^*a)^{1/2}$ be the polar decomposition of $a \in \mathcal{B}(\mathcal{H})$. By the functional calculus $(a^*a)^{1/2} \in M$. Furthermore, we have $\ker(a^*) = \ker((aa^*)^{1/2})$, so $\overline{a(H)}^{\perp} = \overline{(a^*a)^{1/2}(H)}^{\perp}$. Thus it is enough to show that support projections exist for positive elements.

Let a be positive and, without loss of generality, assume $\|a\| \leq 1$. By the functional calculus, $a^{1/2^n} \in M$ for every $n \in \mathbb{N}$. The sequence $(a_n)_{n \in \mathbb{N}}$, with $a_n = a^{1/2^n}$, is increasing and bounded and hence converges in the strong operator topology to some positive element p in the von Neumann algebra generated by a in M. Then, for $\xi \in H$ we have

$$\|(p^2 - a_n^2)\xi\| \leq \|(p^2 - a_n p)\xi\| + \|(a_n p - a_n^2)\xi\|$$
$$\leq \|(p - a_n)p\xi\| + \|a_n\|\|(p - a_n)\xi\|$$

and $\|(p - a_n)p\xi\| + \|a_n\|\|(p - a_n)\xi\| \to 0$ as $n \to \infty$. Thus $a_n^2 \to p$ strongly. But $a_n^2 = a_{n-1}$ for every $n \geq 0$. Thus $p = p^2$, so p is a projection.

Let us now check that p is the support projection of a. Define functions $f_n \in C_0(\mathrm{sp}(a))$ by $f_n(t) = t^{1+1/2^n}$, $n \in \mathbb{N}$. The sequence $(f_n)_{n \in \mathbb{N}}$ is increasing and converges pointwise to the identity function on $\mathrm{sp}(a)$. Thus by Dini's Theorem, the sequence also converges uniformly. Applying the functional calculus we see that

$$a = \lim_{n \to \infty} f_n(a) = \lim_{n \to \infty} a_n a = pa,$$

which shows that $\overline{a(H)} \subset p(H)$. Finally, since $p \in C^*(a)$ we have $p(H) \subset \overline{a(H)}$. $\square$

5.2.12. Proposition. *Let $A \subset \mathcal{B}(H)$ be a C*-algebra. Suppose that $K \subset H$ is an A-invariant subspace, that is, $AK = \{a\xi \mid a \in A, \xi \in K\} \subset K$. Let $p \in \mathcal{B}(H)$ be the orthogonal projection onto K. Then $p \in A'$.*

Proof. Since K is invariant $ap\xi \in K$ for every $a \in A$ and every $\xi \in H$. Thus $pap\xi = ap\xi$ for every $a \in A$ and $\xi \in H$. Hence

$$pa = (a^*p)^* = (pa^*p)^* = pap = ap,$$

so we have $p \in A'$. $\square$

5.2.13. Lemma. *Let $A \subset \mathcal{B}(H)$ be a C*-algebra and $I \subset A$ an ideal. Then the projection $p \in \mathcal{B}(H)$ onto the closed subspace IH is in A'.*

Proof. Since $A(IH) = (AI)H = AH$, the subspace IH is invariant. The rest follows from the previous proposition. $\square$

The following is a reformulation of the Hahn–Banach Separation Theorem which is required in the proof of Theorem 3.2.4.

5.2.14. Lemma. *Let V be a locally compact topological vector space and let $S \subset V$ be a closed convex subset. Then if $x \notin S$ the sets $\{x\}$ and S are strictly separated, that is, there is a continuous linear functional $\phi \in V^*$, $r \in \mathbb{R}$ and $\epsilon > 0$ such that*

$$\Re(\phi(x)) < r < r + \epsilon < \Re(\phi(s)) \text{ for every } s \in S,$$

where $\Re(\lambda)$ denotes the real part of the complex number λ.

5.2.15. At last we are able to prove Theorem 3.2.4: that every ideal I in a C*-algebra A has an approximate unit that is quasicentral for A.

Proof of Theorem 3.2.4. Let A be a C*-algebra, $I \subset A$ an ideal and let $\pi : A \to \mathcal{B}(H)$ be the universal representation of A. Since (H, π) is faithful, it is enough to show that the theorem holds for $\pi(A)$ and $\pi(I)$.

Let $(u_\lambda)_\Lambda$ be an approximate unit for $\pi(I) \subset \pi(A)$. Let $\mathcal{E}$ denote the convex hull of $\{u_\lambda \mid \lambda \in \Lambda\}$, that is,

$$\mathcal{E} := \left\{ \sum_{i=1}^n \mu_i x_i \mid \sum_{i=1}^n \mu_i = 1 \text{ and } x_i = u_\lambda \text{ for some } \lambda \in \Lambda \right\}.$$

If $\sum_{i=1}^n \mu_i x_i \in \mathcal{E}$ then there is some u_λ such that $x_i \leq u_\lambda$ for every $1 \leq i \leq n$. Thus $\sum_{i=1}^n \mu_i x_i \leq \sum_{i=1}^n \mu_i u_\lambda = u_\lambda$, which shows that $\mathcal{E}$ is upwards directed. For $e \in \mathcal{E}$, let $x_e := e$. Then we also have $\lim_\mathcal{E} x_e a = a = \lim a x_e$. Thus $(x_e)_{e \in \mathcal{E}}$ is an approximate unit for I.

We will show that for every $\lambda_0 \in \Lambda$ and finite set of elements $a_1, \ldots, a_n \in \pi(A)$ there is $e \in \mathcal{E}$, $e \geq u_{\lambda_0}$ with

$$\|a_i e - e a_i\| < 1/n \text{ for every } 1 \leq i \leq n.$$

Let $a_1, \ldots, a_n$ and u_{λ_0} be given. Let $H^{(n)}$ denote the Hilbert space given by the direct sum of n copies of H and for an element $a \in \pi(A)$, let $a^{(n)} \in M_n(\pi(A))$ denote the element with a copied n times down the diagonal. Notice that $(u_\lambda^{(n)})_\Lambda$ is an approximate unit for $M_n(\pi(I))$.

Let $\mathcal{E}_{\lambda \geq \lambda_0}$ denote the convex hull of $\{u_\lambda \mid \lambda \geq \lambda_0\}$, and set $c := a_1 \oplus \cdots \oplus a_n$. Put

$$\mathcal{S} := \{ce^{(n)} - e^{(n)}c \mid e \in \mathcal{E}_{\lambda \geq \lambda_0}\}.$$

We claim that $0 \in \overline{\mathcal{S}}$. To show this, suppose it is not. Since $\mathcal{E}_{\lambda \geq \lambda_0}$ is convex, it is easy to see that $\mathcal{S}$ is convex. Then by Lemma 5.2.14 there is a continuous linear functional $\phi \in (\mathcal{B}(H))^*$, $r \in \mathbb{R}$ and $\epsilon > 0$ such that

$$0 = \Re(\phi(0)) < r < r + \epsilon < \Re(\phi(s))$$

for every $s \in \mathcal{S}$. Rescaling if necessary, we may assume that $1 \leq \Re(\phi(s))$ for every $s \in \mathcal{S}$. Thus, by Proposition 5.2.8, there is $\xi, \eta \in H^{(n)}$ such that $\phi(a) = \langle a\xi, \eta \rangle$.

We have $u_\lambda \xrightarrow{\text{SOT}} p$ where p is the projection onto $\pi(I)H$. Since $p \in \pi(A)'$,

$$\phi(cu_\lambda^{(n)} - u_\lambda^{(n)}c) = \langle (cu_\lambda^{(n)} - u_\lambda^{(n)}c)\xi, \eta \rangle$$
$$\to \langle (cp^{(n)} - p^{(n)}c)\xi, \eta \rangle$$
$$= 0,$$

contradicting the fact that $\Re(\phi(s)) \geq 1$ for every $s \in \mathcal{S}$. Thus $0 \in \mathcal{S}$, which proves the claim.

It follows that for any $n \in \mathbb{N} \setminus \{0\}$, any finite subset $\mathcal{F} = \{a_1, \dots, a_n\}$ and any $\lambda_0 \in \Lambda$ there is $f_{\mathcal{F},\lambda_0} \in \mathcal{E}_{\lambda \geq \lambda_0}$ such that

$$\|(a_1 \oplus \cdots \oplus a_n)f_{\mathcal{F},\lambda_0}^{(n)} - f_{\mathcal{F},\lambda_0}^{(n)}(a_1 \oplus \cdots \oplus a_n)\| < 1/n.$$

Thus

$$\|a_i f_{\mathcal{F},\lambda_0} - f_{\mathcal{F},\lambda_0} a_i\| \leq \max_{1 \leq j \leq n} \|a_i f_{\mathcal{F},\lambda_0} - f_{\mathcal{F},\lambda_0} a_i\|$$
$$= \|(a_1 \oplus \cdots \oplus a_n)f_{\mathcal{F},\lambda_0}^{(n)} - f_{\mathcal{F},\lambda_0}^{(n)}(a_1 \oplus \cdots \oplus a_n)\|$$
$$< 1/n.$$

It follows that $(f_{\mathcal{F},\lambda})_{\mathcal{F},\lambda}$, where $\mathcal{F}$ runs over all finite subsets of $\pi(A)$ and $\lambda \in \Lambda$, is a quasicentral approximate unit. $\qquad\square$

5.3   Pure states and irreducible representations

With some von Neumann algebra theory in hand, we now return briefly to representations of C*-algebras. Recall for a C*-algebra A and positive linear functional $\phi : A \to \mathbb{C}$, we have the associated GNS representation of Definition 4.2.3. When one looks at representations associated to *pure states*, defined below, we see that the associated representations have no nontrivial closed vector subspaces, that is, they are *irreducible* (Definition 5.3.5).

5.3.1. Definition. A state $\phi : A \to \mathbb{C}$ on a C*-algebra is called a *pure state* if, for any positive linear function $\psi : A \to \mathbb{C}$ with $\psi \leq \phi$ (which is to say that the functional $\phi - \psi$ is positive), there exists $t \in [0, 1]$ such that $\psi = t\phi$.

5.3.2. Let X be a convex compact set in a vector space V. A point $x \in X$ is called an *extreme point* if, whenever $x = ty + (1-t)z$ for some $t \in (0,1)$, we have $x = y = z$. A subset $F \subset X$ is called a *face* of X if, whenever $x \in F$ and $x = ty + (1-t)z$ for some $t \in (0,1)$, we have $y, z \in F$. If x is an extreme point of F then x is also an extreme point of X.

Let A be a C*-algebra and let S denote the set of norm-decreasing positive linear functionals. Then S is a weak-* closed subset of the unit ball of A^*. Thus by the Banach–Alaoglu Theorem, S is weak-* compact. It is easy to see that S is a convex set. We will see that the pure states are extreme points in S.

The space S is often called the *quasi-state space* of A.

5.3.3. Proposition. *Let A be a C^*-algebra and let S denote the weak-* compact convex set of norm-decreasing linear functionals on A. Then the extreme points of S are the zero functional and the pure states of A.*

Proof. Suppose that $\phi, \psi \in S$ and there is $t \in (0, 1)$ such that $0 = t\phi + (1 - t)\psi$. Then $0 \le (1 - t)\psi(a^*a) = -t\phi(a^*a) \le 0$ for every $a \in A$. So $\psi(a^*a) = \phi(a^*a) = 0$ for every $a \in A$, from which it follows that ϕ and ψ are both the zero functional. Thus the zero functional is an extreme point.

Now suppose that ϕ is a pure state and $\phi = t\psi + (1 - t)\rho$ for $\psi, \rho \in S$ and $t \in (0, 1)$. Then $t\psi \le \phi$ so there is $t' \in [0, 1]$ such that $t\psi = t'\phi$. Since $1 = \|\phi\| = t\|\psi\| + (1 - t)\|\rho\|$ we must have $\|\psi\| = \|\rho\| = 1$. Thus $t = t'$, so $\psi = \phi$ whence also $\rho = \phi$. So ϕ is an extreme point of S.

Finally, suppose that ϕ is a nonzero extreme point of S. Since $\phi/\|\phi\|, 0 \in S$ and $\phi = \|\phi\|(\phi/\|\phi\|) + (1 - \|\phi\|)0$, we must have $\|\phi\| = 1$. Suppose that $\psi \in S$ and $\psi \le \phi$. Let $t := \|\psi\| \in [0, 1]$. Then $\phi = t(\psi/\|\psi\|) + (1 - t)(\phi - \psi)/\|\phi - \psi\|$ and since ϕ is an extreme point, we must have $\phi = \psi/\|\psi\|$. Thus $\psi = t\phi$, so ϕ is a pure state. $\qquad\square$

5.3.4. Let A be a C^*-algebra. Recall that $S(A)$ denotes the states on A. It is easy to see that $S(A)$ is a face in the set S of all norm-decreasing linear functionals on A. The state space $S(A)$ is nonempty, convex, and, when A is unital, it is moreover weak-* compact. It follows from the Krein–Milman Theorem (see [71] or any functional analysis textbook) that for a unital C^*-algebra A, the state space $S(A)$ is the weak-* closed convex hull of its extreme points, which, since $S(A)$ is a face, are exactly the pure states.

5.3.5. Definition. A representation $\pi : A \to \mathcal{B}(H)$ is *irreducible* if $K \subset H$ is a closed vector subspace with $\pi(a)K \subset K$, then $K \in \{0, H\}$.

In fact what we have defined above should rightfully be called a *topologically irreducible* representation. A representation (H, π) of A is *algebraically irreducible* if H has no nontrivial $\pi(A)$-invariant subspaces, closed or not. It is clear that any algebraically irreducible representation is topologically irreducible. A deep, and far from obvious result, says that any topologically irreducible representation is also algebraically irreducible, and in light of this, we are justified in using simply the term *irreducible*. The main ingredient in the proof of the fact that topologically irreducible implies algebraically irreducible is called *Kadison's Transitivity Theorem* [63], which we will not prove here. The interested reader can find a good exposition of Kadison's Transitivity Theorem in [83, Section 5.2].

5.3.6. This theorem is often called *Schur's Lemma*.

Theorem. *Let (H, π) be a nonzero representation of a C^*-algebra A. Then π is irreducible if and only if $\pi(A)' = \mathbb{C} \cdot \mathrm{id}_H$.*

Proof. Let π be a irreducible representation of A. Since π is nonzero, there is a nonzero positive $a \in \pi(A)'$. Since $\pi(A)' = (\pi(A)')''$, $\pi(A)'$ is a von Neumann algebra and therefore the support projection p of a is contained in $\pi(A)'$. But then $\pi(A)p\xi = p\pi(A)\xi$ for every $\xi \in H$, so $p(H)$ is a nonzero $\pi(A)$-invariant vector subspace. Hence $p(H) = H$, that is, $p = \mathrm{id}_H$. Suppose that $\mathrm{sp}(a)$ contains more than one element. Then we can define a continuous function $f \in C_0(\mathrm{sp}(a))$ such that $f|_E = 0$ for some proper subset $E \subset \mathrm{sp}(a)$. But then the support projection of the positive nonzero element $f(a)$ is strictly smaller than $p = \mathrm{id}$, which is impossible since a was arbitrary. So the spectrum of every nonzero positive element consists of a single nonzero positive number, meaning that every positive element in $\pi(A)'$ is invertible. Thus by Theorem 1.2.9, $\pi(A)' = \mathbb{C} \cdot \mathrm{id}_H$.

Conversely, suppose that $\pi(A)' = \mathbb{C}\cdot\mathrm{id}_H$ and that $K \subset H$ is a nonzero closed $\pi(A)$-invariant vector subspace. Let p be the orthogonal projection onto K. Then the $p \in \pi(A)' = \mathbb{C} \cdot \mathrm{id}_H$, so we must have $p = \mathrm{id}_H$ and $K = H$. $\qquad\square$

5.3.7. Given a positive linear function ϕ on a C*-algebra A, let (H_ϕ, π_ϕ) denote the associated GNS representation of A (Definition 4.2.3). We will denote by ξ_ϕ the cyclic vector given by Theorem 4.3.1.

Proposition. *Let A be a C*-algebra and $\phi : A \to \mathbb{C}$ a state. Suppose that $\psi : A \to \mathbb{C}$ is a positive linear functional with $\psi \leq \phi$. Then there exists $x \in \pi_\phi(A)'$ such that $0 \leq x \leq 1$ and*

$$\psi(a) = \langle \pi_\phi(a)x\xi_\phi, \xi_\phi \rangle,$$

for every $a \in A$.

Proof. As in 4.2.1, let $N_\phi := \{a \in A \mid \phi(a^*a)\}$. Since $\psi(a^*a) \leq \phi(a^*a)$ for every $a \in A$, we can define a sesquilinear form $(\cdot, \cdot) : A/N_\phi \times A/N_\phi \to \mathbb{C}$ by

$$(a + N_\phi, b + N_\phi) = \psi(b^*a).$$

Using the Cauchy–Schwarz inequality of Proposition 4.1.6 we have

$$|(a + N_\phi, b + N_\phi)|^2 = |\psi(b^*a)|^2 \leq \psi(b^*b)\psi(a^*a)$$
$$\leq \phi(b^*b)\phi(a^*a) = \|b + N_\phi\|^2\|a + N_\phi\|^2$$

for every $a, b \in A$. Thus $\|(\cdot, \cdot)\| \leq 1$ on A/N_ϕ and so we can extend it to a sesquilinear form $(\cdot, \cdot) : H_\phi \to \mathbb{C}$ which also has norm at most one. By Lemma 5.2.4, there exists a unique $x \in \mathcal{B}(H_\phi)$ such that

$$(\xi, \eta) = \langle x\xi, \eta \rangle, \quad \xi, \eta \in H_\phi.$$

Thus

$$\psi(b^*a) = (a + N_\phi, b + N_\phi) = \langle x(\pi_\phi(a)\xi_\phi), \pi_\phi(b)\xi_\phi \rangle,$$

so in particular $\langle x(\pi_\phi(a)\xi_\phi), \pi_\phi(a)\xi_\phi \rangle \geq 0$ for every $a \in A$, showing that x is positive. The norm bound on $(\cdot, \cdot)$ then gives us $0 \leq x \leq 1$.

Now let us show that $x \in \pi_\phi(A)'$. For $a, b, c \in A$ we have

$$\langle \pi_\phi(a)x(b + N_\phi), c + N_\phi \rangle = \langle x(b + N_\phi), a^*c + N_\phi \rangle = \psi(c^*ab)$$
$$= \langle x(ab + N_\phi), c + N_\phi \rangle = \langle x\pi(a)(b + N_\phi), c + N_\phi \rangle,$$

so $x\pi(a) = \pi(a)x$ for every $a \in A$, as required.

For any $a, b \in A$ we have

$$\psi(b^*a) = \langle x(\pi_\phi(a)\xi_\phi), \pi_\phi(b)\xi_\phi \rangle = \langle x\pi_\phi(b^*a)\xi_\phi, \xi_\phi \rangle,$$

so for $(u_\lambda)_\Lambda$ an approximate unit for A, we have $\psi(u_\lambda a) = \langle x\pi_\phi(u_\lambda a)\xi_\phi, \xi_\phi \rangle$ from which we obtain

$$\psi(a) = \langle \pi_\phi(a)x\xi_\phi, \xi_\phi \rangle,$$

for every $a \in A$.

It remains to show that x is unique. Suppose $y \in \pi_\phi(A)'$ and $\psi(a) = \langle \pi_\phi(a)y\xi_\phi, \xi_\phi \rangle$. Then, for every $a, b \in A$,

$$\langle y(b + N_\phi), a + N_\phi \rangle = \langle y\pi_\phi(b^*a)\xi_\phi, \xi_\phi \rangle = \langle \pi_\phi(b^*a)y\xi_\phi, \xi_\phi \rangle = \psi(b^*a),$$

and

$$\psi(b^*a) = \langle \pi_\phi(b^*a)x\xi_\phi, \xi_\phi \rangle = \langle x\pi_\phi(b^*a)\xi_\phi, \xi_\phi \rangle = \langle x(b + N_\phi), a + N_\phi \rangle,$$

so $x = y$ since A/N_ϕ is dense in H_ϕ. $\qquad \square$

5.3.8. Theorem. *Let ϕ be a state on a C*-algebra A. Then ϕ is pure if and only if the GNS representation (H_ϕ, π_ϕ) is irreducible.*

Proof. Suppose that ϕ is a pure state on A and let $x \in \phi(A)'$ be an element satisfying $0 \leq x \leq 1$. Define

$$\psi : A \to \mathbb{C}, \qquad a \mapsto \langle \pi_\phi(a)x\xi_\phi, \xi_\phi \rangle.$$

Then ψ is a positive linear functional on A and $\psi \leq \phi$ so there is a $t \in [0, 1]$ satisfying $\psi = t\phi$. Then $\langle \pi_\phi(a)x\xi_\phi, \xi_\phi \rangle = t\phi(a) = \langle t\pi_\phi(a)\xi_\phi, \xi_\phi \rangle$ for every $a \in A$. It follows that

$$\langle x(a + N_\phi), b + N_\phi \rangle = \langle x\pi_\phi(a)\xi_\phi, \pi_\phi(b)\xi_\phi \rangle = \langle x\pi(b^*a)\xi_\phi, \xi_\phi \rangle$$
$$= \langle t\pi_\phi(b^*a)\xi_\phi, \xi_\phi \rangle = \langle t(a + N_\phi), b + N_\phi \rangle,$$

for every a, b in A. Hence $x = t \cdot 1$. Since x was an arbitrary element of $\pi_\phi(A)'$, we have $\pi_\phi(A)' = \mathbb{C} \cdot \mathrm{id}_H$. Thus (H_ϕ, π_ϕ) is irreducible by Theorem 5.3.6.

Conversely, suppose that (H_ϕ, π_ϕ) is irreducible. Let ψ be a positive linear functional with $\psi \leq \phi$. By Proposition 5.3.7 there exists $x \in \pi_\psi(A)'$, $0 \leq x \leq 1$ such that $\psi(a) = \langle \pi_\phi(a)x\xi_\phi, \xi_\phi \rangle$ for every $a \in A$. Since (H_ϕ, π_ϕ) is irreducible, we have that $\pi_\phi(A)' = \mathbb{C} \cdot \mathrm{id}_H$ so $x = t$ for some $t \in [0, 1]$. Then $\psi(a) = \langle t\pi_\phi(a)\xi_\phi, \xi_\phi \rangle = t\phi(a)$ for every $a \in A$. So ϕ is a pure state. $\qquad \square$

5.3.9. A proof of the next lemma should be available in any functional analysis book. One can also be found in the Appendix of [83].

Lemma. *Let X be a nonempty convex set in a vector space V. Suppose that $\phi : X \to \mathbb{C}$ is a continuous linear functional. Let $M := \sup\{\Re(\phi(x)) \mid x \in X\}$. Then $F := \{x \in X \mid \Re(\phi(x)) = M\}$ is a compact face in X.*

5.3.10. The next theorem tells us that for every nonzero C*-algebra, there is a plentiful supply of pure states.

Theorem. *Let $A \neq 0$ be a C*-algebra and $a \in A_+$. Then there exists a pure state ϕ such that $\phi(a) = \|a\|$.*

Proof. Obviously if $a = 0$ then $\phi(a) = 0 = \|a\|$ for any pure state ϕ. So assume that $a \neq 0$. Let $\hat{a} : A^* \to \mathbb{C}$ be the function $\hat{a}(\psi) = \psi(a)$. Then $\hat{a}$ is both weak-* continuous and linear and thus it follows from Proposition 4.1.10 that $\|a\| = \sup\{\psi(a) \mid \psi \in S\}$. By the previous lemma, $F = \{\psi \in S \mid \psi(a) = \|a\|\}$ is a weak-* compact face in S. By the Krein–Milman Theorem, F has an extreme point ϕ. Since F is a face of S, ϕ is an extreme point of S as well. Since $\phi(a) = \|a\| \neq 0$, ϕ is not the zero functional. Thus ϕ is a pure state. $\qquad\square$

5.4 Exercises

5.4.1. Let $(a_\lambda)_{\lambda \in \Lambda}$ be a net in $\mathcal{B}(H)$ and let $a \in \mathcal{B}(H)$. Show that

(i) $(a_\lambda)_\Lambda$ converges to a in the weak operator topology if $\lim_\lambda \langle a_\lambda \xi, \eta \rangle = \langle a\xi, \eta \rangle$ for all $\xi, \eta \in H$;

(ii) $(a_\lambda)_\Lambda$ converges to a in the strong operator topology if $\lim_\lambda a_\lambda \xi = a\xi$ for all $\xi \in H$.

5.4.2. Let H be a Hilbert space. In $\mathcal{B}(H)$, show that convergence in the operator norm topology implies strong operator convergence which in turn implies weak operator convergence. Show that the reverse implications do not necessarily hold.

5.4.3. Fill in the details to the proof of Lemma 5.2.4.

5.4.4. Show that the commutant S' of a subset $S \subset \mathcal{B}(H)$ is closed in the weak operator topology. Show that if $S = S^*$ then S' is a unital C*-algebra.

5.4.5. Let A be a C*-algebra and let (H, π) be the universal representation of A. Show that $\overline{\pi(A)}^{\text{SOT}}$ is a von Neumann algebra.

5.4.6. Let $A \subset \mathcal{B}(H)$ be C*-algebra with approximate unit $(u_\lambda)_\Lambda$. Does $(u_\lambda)_\Lambda$ converge in the strong operator or weak operator topology?

5.4.7. Show that if $a \in \mathcal{B}(H)$ is positive then the sequence $a^{1/2}(a + n^{-1} 1_{\mathcal{B}(H)})^{-1/2}$ converges strongly to the support projection of a. (Why do we write *the* support projection without specifying right or left?)

5.4.8. Show that for any Hilbert space H there is an increasing net of finite rank projections converging to the identity in the strong operator topology.

5.4.9. Let $(a_\lambda)_\Lambda \subset \mathcal{B}(H)$ be a net that is WOT-convergent. Show that $(a_\lambda)_\Lambda$ must be norm bounded.

5.4.10. Let $A \subset \mathcal{B}(H)$ be a C*-algebra. Suppose that B is a strongly closed hereditary subalgebra of A. Show that there is a unique projection $p \in B$ such that $B = pAp$.

5.4.11. Recall (3.3.1) that an element in an arbitrary C*-algebra need not admit a polar decomposition. This is not the case in a von Neumann algebra. Let M be a von Neumann algebra on a Hilbert space H and $a \in M$ and let $a = v|a|$ be the polar decomposition of a as an element of $\mathcal{B}(H)$ as in Proposition 3.3.2.

(i) Show that if u is a unitary in M' then u commutes with v. (Hint: use the uniqueness of v.)

(ii) Show that v commutes with every element of M', hence $v \in M'' = M$.

5.4.12. Let A be a C*-algebra and let $a \in A$. By the previous exercise, there exists $u \in A''$ such that $a = u|a|$ and $u^*u = 1_{A''}$. Show that if a is invertible then u is a unitary in A.

5.4.13. Let A be a unital C*-algebra. Recall that $T(A)$ denotes the set of tracial states on A (4.1.1). Show that $T(A)$ is a closed, convex subset of $S(A)$. Hence, by the Krein–Milman Theorem, the tracial state space $T(A)$ is the closed convex hull of extreme tracial states.

5.4.14. Let $A := C(X)$ for X a compact metric space. The Riesz Representation Theorem tells us that if τ is a state – which is automatically tracial since A is commutative – there exists a regular Borel probability measure μ on X such that

$$\tau(f) = \int f d\mu, \quad f \in A.$$

Let μ be a Borel probability measure on X, and let τ be the associated tracial state given by integration with respect μ as above.

(i) Suppose that there exists a subset $A \subset X$ such that $0 < \mu(A) < 1$. Show that there exists $\epsilon, \delta > 0$ such that $\mu_1 = (1 + \epsilon)\mu|_A + (1 - \delta)\mu|_{X \setminus A}$ and $\mu_2 = (1 - \epsilon)\mu|_A + (1 + \delta)\mu|_{X \setminus A}$ are probability measures and $\mu = \frac{1}{2}\mu_1 + \frac{1}{2}\mu_2$. Deduce that τ is not an extreme point of $T(A)$.

(ii) Suppose that every set $A \subset X$ satisfies $\mu(A) \in \{0, 1\}$. Show that there exists $x \in X$ such that $\mu(A) = 1$ if and only if $x \in A$. Show that in this case, $\tau(f) = f(x)$.

(iii) Show that $\tau \in T(A)$ is an extreme point if and only if there exists $x \in X$ such that $\tau(f) = f(x)$ for every $f \in A$.

6 Tensor products for C*-algebras

In linear algebra, the tensor product construction allows us to take two vector spaces and "multiply" them together to get a new vector space with a universal property regarding bilinear maps. To get from a vector space tensor product to a C*-algebraic tensor product, we need to be able to define an algebra structure and complete with respect to a C*-norm. The tensor product of C*-algebras can be seen as the noncommutative analogue of the Cartesian product of topological spaces. Exercise 6.5.6 makes this statement more precise. We will see that defining a *-algebra structure on the vector space tensor product of two C*-algebras is straightforward, but defining a C*-norm is a tricky matter since in general there can be more than one. The two most common tensor products one encounters are the *minimal* C*-norm and the *maximal* C*-norm. Both have advantages and disadvantages. For example, the minimal tensor is "small enough" that the product of simple C*-algebras will again be simple, while this is no longer true for the maximal tensor product. On the other hand, the minimal tensor product has no universal property while the maximal does.

Fortunately, there are many cases when the minimal and maximal tensor products coincide. Nuclearity, a key concept in Part III, can be expressed in a few equivalent ways, one of which is given in terms of the tensor product: if A is nuclear, then for any B, there is only one C*-norm on $A \otimes B$. In the final chapters, when we will be almost exclusively interested in nuclear C*-algebras, we will be able to take advantage of properties of any of these norms.

In this chapter, we assume that the reader has a basic understanding of tensor products of vector spaces. A more comprehensive treatment, which includes vector space tensor products, can be found, for example, in [18, Chapter 3]. The first section constructs the minimal tensor product by representing an algebraic tensor product of C*-algebras on the tensor product of two Hilbert spaces. In the second section, we collect some statements that can be made about C*-norms in general and in the third section we show that the maximal tensor product satisfies a useful universal property. In the final section we look at nuclear C*-algebras and show that an extension of nuclear C*-algebra by a nuclear C*-algebra is again nuclear.

6.1 The minimal tensor product

The minimal tensor product, also called the *spatial tensor product*, of two C*-algebras is our starting point. It is defined via a tensor product of Hilbert spaces. The construction of the minimal tensor product also shows that there is indeed at least one tensor product, which will allow us to make sense of the definition of the maximal tensor product.

6.1.1. We begin with the tensor product of two Hilbert spaces. Let H and K be Hilbert spaces and form the vector space tensor product $H \odot K$.

© The Author(s), under exclusive license to Springer Nature Switzerland AG 2021
K. R. Strung, *An Introduction to C*-Algebras and the Classification Program*, Advanced Courses in Mathematics - CRM Barcelona, https://doi.org/10.1007/978-3-030-47465-2_6

Proposition. *Let H and K be Hilbert spaces. Then there is a unique inner product on $H \odot K$ such that*

$$\langle \xi \otimes \eta, \xi' \otimes \eta' \rangle = \langle \xi, \xi' \rangle_H \langle \eta, \eta' \rangle_K,$$

for every $\xi, \xi' \in H$ and $\eta, \eta' \in K$.

Proof. Let $\phi : H \to \mathbb{C}$ and $\psi : K \to \mathbb{C}$ be conjugate-linear functionals, and let $\bar{\phi}$ and $\bar{\psi}$ denote the complex conjugate of each map. Then

$$\bar{\phi} \otimes \bar{\psi} : H \times K \to \mathbb{C}, \qquad (\xi, \eta) \mapsto \bar{\phi}(\xi)\bar{\psi}(\eta),$$

extends from simple tensors to a bilinear map. Thus, we get a well-defined linear map

$$\bar{\phi} \otimes \bar{\psi} : H \odot K \to \mathbb{C},$$

whence a conjugate-linear map

$$\phi \otimes \psi : H \odot K \to \mathbb{C}.$$

Now, for any $\xi \in H$, define a conjugate-linear functional

$$\phi_\xi(\xi') := \langle \xi, \xi' \rangle_H, \quad \text{for } \xi' \in H,$$

and, similarly, for $\eta \in K$, define

$$\psi_\eta(\eta') := \langle \eta, \eta' \rangle_K, \quad \text{for } \eta' \in K.$$

From this, for every $\xi \in H$ and every $\eta \in K$, we obtain a conjugate-linear map $\phi_{\eta \otimes \eta'}(\xi \otimes \xi') : H \odot K \to \mathbb{C}$. The map from $H \times K$ into the space of conjugate-linear functionals on $H \odot K$, call it V, given by mapping $\eta \times \xi \mapsto \phi_\eta \otimes \psi_\xi$ is bilinear, giving a linear map $\Phi : H \odot K \to V$. Set

$$\langle \zeta, \zeta' \rangle := \Phi(\zeta)(\zeta'), \quad \zeta, \zeta' \in H \odot K.$$

Observe that this is a sesquilinear form. On simple tensors, we have

$$\langle \eta \otimes \xi, \eta' \otimes \xi' \rangle = \phi_{\eta \otimes \eta'}(\xi \otimes \xi') = \langle \xi, \xi' \rangle_H \langle \eta, \eta' \rangle_K.$$

Uniqueness follows immediately. The fact that the form is hermitian follows from uniqueness together with the fact that $\langle \cdot, \cdot \rangle_H$ and $\langle \cdot, \cdot \rangle_K$ are inner products. Thus it remains to prove that the form is positive definite.

Let $\nu \in H \odot K$. Then we can write $\nu = \sum_{i=1}^n \xi_n \otimes \eta_i$ for some $\xi_i \in H$, $\eta_i \in K$, $1 \le i \le n$. Let $\zeta_1, \ldots, \zeta_m$ be an orthonormal basis in K for the span of the η_i, $1 \le i \le n$. Then we can rewrite ν in terms of this basis, say $\nu = \sum_{i=1}^m \xi'_i \otimes \zeta_i$, for some $\xi'_i \in H$, $1 \le i \le m$. Now we compute

$$\langle \nu, \nu \rangle = \sum_{i,j=1}^m \langle \xi'_i \otimes \zeta_i, \xi'_j \otimes \zeta_j \rangle = \sum_{i,j=1}^m \langle \xi'_i, \xi'_j \rangle_H \langle \zeta_i, \zeta_j \rangle_K$$

$$= \sum_{i,j=1}^m \langle \xi'_i, \xi'_j \rangle_H \delta_{ij} = \sum_{i=1}^m \|\xi'_i\| \ge 0,$$

which shows positivity. Moreover, if $\langle \nu, \nu \rangle = 0$ then we must have that each $\|\xi_i'\| = 0$, which means $\xi_i' = 0$ and hence $\nu = 0$. So $\langle \cdot, \cdot \rangle$ is positive definite, and we have now shown it is an inner product. $\qquad\qquad\qquad\qquad\qquad\qquad\qquad\qquad\square$

6.1.2. The inner product above makes $H \odot K$ into a pre-Hilbert space, and so we complete it to a Hilbert space, denoted $H \hat{\otimes} K$. Observe that the norm on $H \hat{\otimes} K$ satisfies

$$\|\xi \otimes \eta\| = \|\xi\|\|\eta\|, \quad \text{for } \xi \in H, \ \eta \in K.$$

6.1.3. Lemma. *Let H and K be Hilbert spaces. Suppose we have two operators $a \in \mathcal{B}(H)$ and $b \in \mathcal{B}(K)$. Then there exists a unique operator in $\mathcal{B}(H \hat{\otimes} K)$, denoted $a \hat{\otimes} b$ such that*

$$(a \hat{\otimes} b)(\xi \otimes \eta) = a(\xi) \otimes b(\eta), \quad \text{for } \xi \in H \text{ and } \eta \in K.$$

Furthermore, the norm of $a \hat{\otimes} b$ is given by $\|a \hat{\otimes} b\| = \|a\|\|b\|$.

Proof. Let $a \otimes b$ denote the tensor product of the two operators defined on the algebraic tensor product $H \odot K$. We need to check that $a \otimes b$ is bounded so that it can be extended to $H \hat{\otimes} K$. Without loss of generality, since $\mathcal{B}(H)$ and $\mathcal{B}(K)$ are unital, we may assume that a and b are unitary operators (Proposition 3.1.6). Suppose $\zeta = \sum_{i=1}^n \xi_i \otimes \eta_i \in H \odot K$. As in the previous proof, we may assume that the η_i are orthogonal. Then, since a and b are unitary,

$$\|(a \otimes b)(\zeta)\|^2 = \left\| \sum_{i=1}^n a(\xi_i) \otimes b(\eta_i) \right\|^2 = \sum_{i=1}^n \|a(\xi_i) \otimes b(\eta_i)\|^2$$

$$= \sum_{i=1}^n \|a(\xi_i)\|^2 \|b(\eta_i)\|^2 = \|\zeta\|^2,$$

so we see that $\|a \otimes b\| = 1$. In particular, it is bounded and so we can extend it to an operator $a \hat{\otimes} b$ on $H \hat{\otimes} K$.

Using the density of $H \odot K$ in $H \hat{\otimes} K$, it is clear that the map $\mathcal{B}(H) \to \mathcal{B}(H \hat{\otimes} K)$ which sends $a \mapsto a \hat{\otimes} \operatorname{id}_K$ is an injective *-homomorphism, hence is necessarily isometric, as is the analogously defined map $\mathcal{B}(K) \to \mathcal{B}(H \hat{\otimes} K)$. Thus

$$\|a \hat{\otimes} b\| = \|(a \hat{\otimes} \operatorname{id}_K)(\operatorname{id}_H \hat{\otimes} b)\|$$
$$\leq \|a \hat{\otimes} \operatorname{id}_K\| \| \operatorname{id}_H \hat{\otimes} b\|$$
$$= \|a\|\|b\|.$$

Now, assume that $a, b \neq 0$. Then, for sufficiently small $\epsilon > 0$, we can find unit vectors $\xi \in H$ and $\eta \in K$ such that

$$\|a(\xi)\| > \|a\| - \epsilon > 0, \quad \text{and} \quad \|b(\eta)\| > \|b\| - \epsilon > 0.$$

Then

$$\|(a\hat{\otimes}b)(\xi \otimes y)\| = \|a(\xi)\|\|b(\eta)\|$$
$$> (\|a\| - \epsilon)(\|b\| - \epsilon).$$

Since ϵ was chosen arbitrarily, we have that $\|a\hat{\otimes}b\| \geq \|a\|\|b\|$. Combining this with the previous estimate, we see that $\|a\hat{\otimes}b\| = \|a\|\|b\|$. $\square$

6.1.4. It is straightforward to check that given operators $a, a' \in \mathcal{B}(H)$ and $b, b' \in \mathcal{B}(K)$, the operations $(a\hat{\otimes}b)(a'\hat{\otimes}b') = aa'\hat{\otimes}bb'$ and $(a\hat{\otimes}b)^* = a^* \otimes b^*$ make sense. In fact, using facts about vector space tensor products, it is straightforward to show that if A and B are *-algebras and $a, a' \in A$ and $b, b' \in B$ then

$$(a \otimes b)(a' \otimes b') = aa' \otimes bb'$$

defines a unique multiplication on the (vector space) tensor product $A \odot B$ and

$$(a \otimes b)^* = a^* \otimes b^*$$

defines a unique involution. When $A \odot B$ is equipped with these operations, we will call it the *-*algebra tensor product* of A and B.

It is also an easy exercise to show that if A, B, C and D are *-algebras, and $\varphi : A \to C$ and $\psi : B \to C$ are *-homomorphisms with commuting images, then there exists a unique *-homomorphism $\pi : A \odot B \to C$ defined by

$$\pi(a \otimes b) = \varphi(a)\psi(b).$$

In particular, if $\varphi : A \to C$ and $\psi : B \to D$ are *-homomorphisms, then the tensor product $\varphi \otimes \psi : A \odot B \to C \odot D$ is also a *-homomorphism.

6.1.5. Proposition. *Let A and B be C*-algebras with *-representations (H, φ) and (K, ψ) respectively. Then there exists a unique *-homomorphism*

$$\pi : A \odot B \to \mathcal{B}(H\hat{\otimes}K), \quad \pi(a \otimes b) = \varphi(a)\hat{\otimes}\psi(b), \quad \text{for } a \in A, b \in B.$$

If both φ and ψ are injective, then so is π.

Proof. Let $\varphi' : A \to \mathcal{B}(H\hat{\otimes}K)$ be the *-homomorphism defined by $\varphi'(a) = \varphi(a)\hat{\otimes}\text{id}_K$ and define $\psi' : B \to \mathcal{B}(H\hat{\otimes}K)$ by $\psi'(b) = \text{id}_H \hat{\otimes}\psi(b)$. Then φ' and ψ' have commuting images in $\mathcal{B}(H\hat{\otimes}K)$ and induce a unique map

$$\pi : A \odot B \to \mathcal{B}(H\hat{\otimes}K)$$

satisfying $\pi(a \otimes b) = \varphi'(a)\psi'(b) = \varphi(a)\hat{\otimes}\psi(b)$.

Suppose that φ and ψ are injective and suppose $c \in \ker(\pi)$. Let $c = \sum_{i=1}^m a_i \otimes b_i$ where the b_i are linearly independent. Then since ψ is injective, the $\psi(b_i)$, $1 \leq i \leq m$, are also linearly independent. We have $0 = \pi(c) = \sum_{i=1}^m \varphi(a_i) \otimes \psi(b_i)$, so $\varphi(a_i) = 0$ for every $1 \leq i \leq m$. Thus, by injectivity of φ, we also have $c = 0$. $\square$

6.1.6. Let A be a C*-algebra with universal representation (H, π_A) and let B be a C*-algebra with universal representation (K, π_B). Then, by the above, there is a a unique injective $*$-homomorphism

$$\pi : A \odot B \to \mathcal{B}(H \hat{\otimes} K)$$

such that

$$\pi(a \otimes b) = \pi_A(a) \otimes \pi_B(b).$$

Thus we may define a C*-norm (Definition 2.1.3) on $A \odot B$ by

$$\|c\|_{\min} = \|\pi(c)\|, \qquad c \in A \odot B.$$

Note that $\|a \otimes b\|_{\min} = \|a\| \|b\|$ for every $a \in A$ and $b \in B$.

Definition. The *minimal*, or *spatial* tensor product of A and B is given by

$$A \otimes_{\min} B := \overline{A \odot B}^{\|\cdot\|_{\min}}.$$

Since the norm $\|\cdot\|$ satisfies the C*-equality on the dense $*$-subalgebra $A \odot B$, by continuity the C*-equality holds for every $a \in A \otimes_{\min} B$. In particular, $A \otimes_{\min} B$ is indeed a C*-algebra.

6.2 General C*-norms on tensor products

In general, there may be more than one norm on $A \odot B$ which is a C*-norm, allowing us to complete $A \odot B$ into more than one C*-algebra. (Why doesn't this contradict Exercise 2.5.17?) These C*-tensor norms work well with respect to $*$-homomorphisms, as we will see. For $\gamma : A \odot B \to \mathbb{R}_+$ a norm satisfying the C*-equality, we denote the completion of $A \odot B$ with respect to γ as $A \otimes_\gamma B$.

6.2.1. Lemma. *Let A and B be C*-algebras and suppose that $\gamma : A \odot B \to \mathbb{R}_+$ is a C*-norm on $A \odot B$. Fix $a' \in A$ and $b' \in B$. Then the maps*

$$\varphi : A \to A \otimes_\gamma B, \quad a \mapsto a \otimes b', \quad \psi : B \to A \otimes_\gamma B, \quad b \mapsto a' \otimes b,$$

are both continuous.

Proof. We prove the result only for φ, as the corresponding result for ψ is completely analogous. Since φ is a linear map and A and $A \otimes_\gamma B$ are C*-algebras, thus in particular Banach spaces, we may employ the Closed Graph Theorem: if $(a_n)_{n \in \mathbb{N}}$ is a sequence in A converging to zero and $(\varphi(a_n))_{n \in \mathbb{N}}$ converges to $c \in A \otimes_\gamma B$, then φ is continuous if and only if $c = 0$.

Without loss of generality, we may assume that $b' \in B$ is positive and also that each a_n, $n \in \mathbb{N}$ is positive, so $c = \varphi(\lim_{n \to \infty} a_n) \geq 0$. Suppose $c \neq 0$. Then, by Proposition 4.1.10, there is a positive linear functional $\tau : A \otimes_\gamma B \to \mathbb{C}$ such that $\tau(c) = \|c\| > 0$. Define

$$\rho : A \to \mathbb{C}, \quad a \mapsto \tau(a \otimes b').$$

Then ρ is a positive linear functional on A, and hence is continuous. Thus we have $\rho(a_n) \to 0$ as $n \to \infty$, and so $\tau(c) = 0$, which is a contradiction. Thus $c = 0$, showing φ is continuous. $\square$

6.2.2. Proposition. *Suppose that A and B are nonzero C*-algebras and that $\pi : A \otimes_\gamma B \to \mathcal{B}(H)$ is a nondegenerate representation of $A \otimes_\gamma B$, where γ is a C*-norm. Then there are unique nondegenerate *-representations (H, π_A) and (H, π_B) of A and B respectively, satisfying*

$$\pi(a \otimes b) = \pi_A(a)\pi_B(b),$$

for every $a \in A$ and $b \in b$.

Proof. We will prove the theorem for A and B unital. The general case is Exercise 6.5.1. Let $K := \pi(A \odot B)H$ and note that K is dense in H since the representation is nondegenerate. Let $\eta \in K$. Then, since K is defined as the image of the algebraic tensor product of A and B, there exists $n \in \mathbb{N}$, $a_i \in A$, $b_i \in B$ and $\xi_i \in H$, $1 \le i \le n$, such that

$$\eta = \sum_{i=1}^{n} \pi(a_i \otimes b_i)\xi_i.$$

Suppose that we can also find $m \in \mathbb{N}$, $a_i' \in A$ $b_i' \in B$ and $\xi_i' \in H$, $1 \le i \le m$ such that

$$\eta = \sum_{i=1}^{m} \pi(a_i' \otimes b_i')\xi_i'.$$

Then

$$\pi(a \otimes 1_B)(\eta) = \sum_{i=1}^{n} \pi(aa_i \otimes b_i)\xi_i = \sum_{j=1}^{m} \pi(aa_i' \otimes b_i')\xi_i',$$

so for every $a \in A$, the map

$$\pi_a : K \to K, \quad \eta \mapsto \pi(a \otimes 1_B)(\eta)$$

where η is written in the form above, is a well-defined linear map. Also, by Lemma 6.2.1, $\|\pi(a \otimes b)\| \le M\|b\|$ for some positive constant M (which depends on a). It follows that $\|\pi_a(\eta)\| \le M\|\eta\|$ and so we can extend π_a uniquely to all of H. By abuse of notation, let us also denote the extension by π_a.

In a similar manner, for $b \in B$ we define a bounded linear operator $\pi_b : H \to H$, which, for $\eta = \sum_{i=1}^n \pi(a_i \otimes b_i)\xi_i \in K$ is exactly

$$\pi_b(\eta) = \sum_{i=1}^n \pi(a_i \otimes bb_i)\xi_i.$$

It is easy to check that the maps

$$\pi_A : A \to \mathcal{B}(H), \quad a \mapsto \pi_a,$$

and

$$\pi_B : B \to \mathcal{B}(H), \quad b \mapsto \pi_b,$$

are *-homomorphisms which satisfy $\pi(a \otimes b) = \pi_A(a)\pi_B(b) = \pi_B(b)\pi_A(a)$.

It remains to check uniqueness and nondegeneracy.

We will show that if $\pi'_A : A \to \mathcal{B}(H)$ and $\pi'_B : B \to \mathcal{B}(H)$ are *-homomorphisms satisfying $\pi(a \otimes b) = \pi'_A(a)\pi'_B(b) = \pi'_B(b)\pi'_A(a)$ for every $a \in A$ and $b \in B$, then π'_A and π'_B must be nondegenerate. This will clearly imply the nondegeneracy of π_A and π_B. Let $\eta \in H$ be a vector satisfying $\pi'_A(\eta) = 0$. Then $\pi(a \otimes 1_B)(\eta) = 0$ for every $a \in A$. But this implies that $\pi(a \otimes b)\eta = \pi((a \otimes 1_B)(1_A \otimes b))\eta = 0$, for every $a \in A$ and $b \in B$. Then $\pi(c)(\eta) = 0$ for every $c \in A \odot B$. The *-algebraic tensor product $A \odot B$ is dense in $A \otimes_\gamma B$, which implies that $\pi(c)\eta = 0$ for every $c \in A \otimes_\gamma B$. Since π is assumed to be nondegenerate, we have $\eta = 0$. Thus π'_A is nondegenerate. Similarly, π'_B is nondegenerate.

Finally, suppose that $\pi'_A : A \to \mathcal{B}(H)$ and $\pi'_B : B \to \mathcal{B}(H)$ are *-homomorphisms satisfying $\pi(a \otimes b) = \pi'_A(a)\pi'_B(b) = \pi'_B(b)\pi_A(a')$ for every $a \in A$ and $b \in B$. Then $\pi'_A(a) = \pi(a \otimes 1_B) = \pi_A(a)$ for every $a \in A$, and similarly $\pi'_B(b) = \pi(b)$ for every $b \in B$. $\qquad\square$

6.2.3. Recall that a seminorm on a space is defined just as a norm, except it need only be positive, rather than positive definite.

Definition. A C*-*seminorm* on a *-algebra A is a seminorm $\gamma : A \to \mathbb{R}_+$ such that, for all a and b in A, we have $\gamma(ab) \leq \gamma(a)\gamma(b)$, $\gamma(a^*) = \gamma(a)$ and $\gamma(a^*a) = \gamma(a)^2$.

6.2.4. Proposition. *Let A and B be C*-algebras and γ a C*-seminorm on $A \odot B$. Then*

$$\gamma(a \otimes b) \leq \|a\|\|b\|,$$

for every $a \in A$ and $b \in B$.

Proof. Observe that if γ is a seminorm, then $\max\{\gamma, \|\cdot\|_{\min}\}$ defines a norm. Thus, without loss of generality, we may assume that γ is in fact a norm. Let (H, π) be the universal representation of $A \otimes_\gamma B$. Since π is nondegenerate, it induces *-homomorphisms $\pi_A : A \to \mathcal{B}(H)$ and $\pi_B : B \to \mathcal{B}(H)$ satisfying $\pi(a \otimes b) =$

$\pi_A(a)\pi_B(b)$. Since π is also faithful, we have $\gamma(a \otimes b) = \|\pi(a \otimes b)\|$ for every $a \in A$ and $b \in B$. Thus

$$\gamma(a \otimes b) = \|\pi(a \otimes b)\| = \|\pi_A(a)\|\|\pi_B(b)\| \le \|a\|\|b\|,$$

for every $a \in A$ and $b \in B$. $\square$

6.2.5. The reader will note that we have not justified the term *minimal* for the norm defined in the previous section. We do, however, have the following theorem, due to Takesaki [116].

Theorem. *Let A and B be nonzero C^*-algebras and γ a C^*-norm on $A \odot B$. Then, for every $c \in A \odot B$ we have*
$$\|c\|_{\min} \le \gamma(c).$$

We will not prove this theorem, as it would require quite a bit more set-up. Readers can convince themselves of this fact via the mathematically dodgy "proof by nomenclature". Alternatively, for those interested, [83, §6.4] has an accessible proof using material we have already covered here. We do, however, have the following consequence of the theorem.

6.2.6. Proposition. *Let A and B be C^*-algebras and γ any C^*-norm for $A \odot B$. Then*
$$\gamma(a \otimes b) = \|a\|\|b\|,$$

for any $a \in A$ and any $b \in B$.

Proof. Let $a \in A$ and $b \in B$. Then, by 6.1.6 we have

$$\|a\|\|b\| = \|a \otimes b\|_{\min} \le \gamma(a \otimes b).$$

By Proposition 6.2.4, $\gamma(a \otimes b) \le \|a\|\|b\|$. Thus in fact $\gamma(a \otimes b) = \|a\|\|b\|$. $\square$

6.2.7. Another consequence of the minimality of the minimal tensor product norm is that the minimal tensor product of two simple C^*-algebras (as defined in 2.1.6) is again simple. For the sake of brevity, we do not include a proof since we have not developed all the prerequisites. The proof can be found as the corollary to Theorem 2 in Takesaki's original paper [116], or in Pedersen's book [89].

Theorem. *Let A and B be simple C^*-algebras. Then $A \otimes_{\min} B$ is simple.*

6.2.8. We defined the minimal norm with respect to universal representations, but the next lemma says that we get the minimal norm as long as the representations are faithful.

Lemma. *Let A and B be C^*-algebras. Suppose that (H, ρ_A) and (K, ρ_B) are faithful representations of A and B respectively. Then $\|\rho_A \hat{\otimes} \rho_B(c)\| = \|c\|_{\min}$.*

Proof. For $c \in A \odot B$, define $\gamma(c) = \|\rho_A \hat{\otimes} \rho_B(c)\|$. Since ρ_A and ρ_B are both injective, so is $\rho_A \otimes \rho_B$. Thus this defines a C*-norm on $A \odot B$. Then, for $a \in A$ and $b \in B$, we have $\|a\|\|b\| = \gamma(a \otimes b)$ as well as $\gamma(a \otimes b) \geq \|a \otimes b\|_{\min} = \|a\|\|b\|$. Thus, $\gamma(a \otimes b) = \|a \otimes b\|_{\min}$ and so, for every $c \in A \odot B$, we have $\|\rho_A \hat{\otimes} \rho_B(c)\| = \|c\|_{\min}$, as required. $\square$

6.2.9. Theorem. *Let $\varphi : A \to C$ and $\psi : B \to D$ be *-homomorphisms. Then there is a unique *-homomorphism*

$$\varphi \otimes \psi : A \otimes_{\min} B \to C \otimes_{\min} D,$$

satisfying

$$(\varphi \otimes \psi)(a \otimes b) = \varphi(a) \otimes \psi(b),$$

for every $a \in A$ and $b \in B$. Moreover, if φ and ψ are both injective, so is $\varphi \otimes \psi$.

Proof. There exists a unique *-homomorphism $\rho : A \odot B \to C \odot D \subset C \otimes_{\min} D$ such that $\rho(a \otimes b) = \varphi(a) \otimes \psi(b)$, for every $a \in A$ and $b \in B$. Then

$$
\begin{aligned}
\|\rho(a \otimes b)\|_{\min} &= \|\varphi(a) \otimes \psi(b)\|_{\min} \\
&= \|\varphi(a)\|\|\psi(b)\| \\
&\leq \|a\|\|b\| \\
&= \|a \otimes b\|_{\min}.
\end{aligned}
$$

Thus ρ is bounded with respect to the minimal norm on $A \odot B$ and so extends to a *-homomorphism, which we denote $\varphi \otimes \psi : A \otimes_{\min} B \to C \otimes_{\min} D$, satisfying $(\varphi \otimes \psi)(a \otimes b) = \varphi(a) \otimes \psi(b)$ for every $a \in A$ and $b \in B$.

Suppose now that both φ and ψ are injective. Let (H_C, π_C) and (H_D, π_D) denote the universal representations of C and D, respectively. Then $(H_C, \pi_C \circ \varphi)$ is a faithful representation of A and $(H_D, \pi_D \circ \psi)$ is a faithful representation of B. By the previous lemma,

$$\|c\|_{\min} = \|(\pi_C \circ \varphi) \hat{\otimes} (\pi_D \circ \psi)(c)\| = \|\pi_C \hat{\otimes} \pi_D(\varphi \otimes \psi)(c)\| = \|\varphi \otimes \psi(c)\|_{\min}.$$

So $\varphi \otimes \psi$ is isometric on $A \odot B$, hence on $A \otimes_{\min} B$, and thus is injective. $\square$

6.3 The maximal tensor product

We have shown the existence of at least one C*-tensor product on $A \odot B$, namely the minimal tensor product. It follows from Proposition 6.2.4 that every C*-norm is bounded. Thus, we are also able to define the maximal tensor product.

6.3.1. Definition. Let A and B be C*-algebras and let $\mathcal{T}$ denote the set of all C*-norms on $A \odot B$. For $\zeta \in A \odot B$, define

$$\|\zeta\|_{\max} = \sup_{\gamma \in \mathcal{T}} \gamma(\zeta).$$

The *maximal tensor product* of A and B is defined to be

$$A \otimes_{\max} B := \overline{A \odot B}^{\|\cdot\|_{\max}}.$$

Of course, in the preceding definition, one must verify that $\|\cdot\|_{\max}$ does indeed define a C*-norm on $A \otimes B$, but this is easy (exercise). As a result of the definition, it is clear that the maximal tensor norm bounds all other C*-norms on $A \odot B$, thus justifying the name *maximal*.

The maximal tensor product has a particularly useful universal property.

6.3.2. Theorem (Universal property of the maximal tensor product). *Suppose that A, B and C are C*-algebras. If $\varphi : A \to C$ and $\psi : B \to C$ are *-homomorphisms whose images commute in C, then there exists a unique *-homomorphism*

$$\pi : A \otimes_{\max} B \to C$$

which, on simple tensors, satisfies

$$\pi(a \otimes b) = \varphi(a)\psi(b), \quad a \in A, b \in B.$$

Proof. Let $\varphi \otimes \psi : A \odot B \to C$ be the unique *-algebra homomorphism which was defined in 6.1.4. Define $\gamma : A \odot B \to \mathbb{R}_+$ by $c \mapsto \|\varphi \otimes \psi(c)\|$. This is a C*-seminorm on $A \odot B$, and so $\max\{\gamma, \|\cdot\|_{\min}\}$ defines a norm on $A \odot B$ which is necessarily bounded by the maximal norm. Then

$$\|\varphi \otimes \psi(a \otimes b)\| \leq \max\{\gamma(a \otimes b), \|a \otimes b\|_{\min}\} \leq \|a \otimes b\|_{\max},$$

which shows that $\varphi \otimes \psi$ is bounded on $A \odot B$. Let π be its extension to all of $A \otimes_{\max} B$. It is straightforward to check that π satisfies the requirement. Moreover, since π is defined explicitly on the simple tensors, it is clear that π must be the unique such *-homomorphism. $\qquad\square$

6.4 Nuclear C*-algebras

The focus of Part III is nuclear C*-algebras. Nuclear C*-algebras enjoy many good structural properties. Here we introduce them via their original tensor product definition and show that an extension of a nuclear C*-algebra by a nuclear C*-algebra is again nuclear. In the next chapter we will see that the equivalent definition via the completely positive approximation property is very useful.

6.4.1. Definition. A C*-algebra is *nuclear* if for every C*-algebra B there is only one C*-norm on $A \odot B$. Equivalently, A is nuclear if for every C*-algebra B the minimal and maximal tensor products of A and B coincide.

6.4.2. Theorem. *Let A, B and C be C*-algebras and suppose $\pi : A \twoheadrightarrow B$ is a surjective *-homomorphism. Let $J := \ker(\pi)$ and let $j : J \to A$ denote the inclusion map. Suppose B or C is nuclear. Then,*

$$0 \longrightarrow J \otimes_{\min} C \xrightarrow{j \otimes \mathrm{id}} A \otimes_{\min} C \xrightarrow{\pi \otimes \mathrm{id}} B \otimes_{\min} C \longrightarrow 0$$

is a short exact sequence.

Proof. Since j is injective, Theorem 6.2.9 tells us that for any $x \in J \odot C$ we have $\|x\|_{\min} = \|(j \otimes \mathrm{id})(x)\|_{\min}$. We also have $(\pi \otimes \mathrm{id})(A \otimes_\gamma C) = \pi(A) \otimes_\gamma C$ for any C*-norm $\gamma \in \mathcal{T}$, so since π is surjective, so is $\pi \otimes \mathrm{id}$. Since the image of j is the kernel of π, $j(J)$ is an ideal in A. Thus $I := (j \otimes \mathrm{id})(J \otimes C)$ is an ideal of $A \otimes_{\min} C$. Let

$$D := (A \otimes_{\min} C)/I,$$

and let $q : A \otimes_{\min} C \to D$ be the quotient map. Since $(\pi \otimes \mathrm{id})(I) = 0$, there exists a unique *-homomorphism $\tilde{\pi} : D \to B \otimes_{\min} C$ satisfying $\tilde{\pi} \circ q = \pi \otimes \mathrm{id}$. We claim that $\tilde{\pi}$ is a *-isomorphism. It is clear by the definition of $\tilde{\pi}$ that it is surjective, so we need only show injectivity. Consider the map

$$B \times C \to D, \quad (b, c) = (\pi(a), c) \mapsto a \otimes c + I.$$

It is easy to see that it is bilinear and hence there is a unique linear map

$$\rho : B \odot C \to D,$$

satisfying $\rho(b \otimes c) = a \otimes c + I$. One checks that ρ is both *-preserving and multiplicative. Moreover,

$$\gamma : B \odot C \to \mathbb{R}_+, \qquad x \mapsto \max\{\|\rho(x)\|, \|x\|_{\min}\},$$

defines a C*-norm. By assumption, either B or C is nuclear so $B \odot C$ has a unique C*-norm. Thus $\gamma(x) = \|x\|_{\min}$ for every $x \in B \otimes C$. In particular, $\|\rho(x)\| \le \|x\|_{\min}$ for every $x \in B \odot C$, so ρ extends to a *-homomorphism

$$\rho : B \otimes_{\min} C \to D.$$

For every $a \in A$ and $c \in C$ we have

$$\rho \circ \tilde{\pi}(a \otimes c + I) = \rho(\pi(a) \otimes c) = a \otimes c + I,$$

whence $\rho \circ \tilde{\pi} = \mathrm{id}_D$. Thus $\tilde{\pi}$ is injective, proving the claim.

Now, since $\tilde{\pi}$ is a *-isomorphism we have $\ker(q) = \ker(\pi \otimes \mathrm{id}) = I = \mathrm{im}(j \otimes \mathrm{id})$. Thus the sequence is short exact. $\qquad\square$

6.4.3. Theorem. *Suppose that B and J are nuclear C*-algebras and*

$$0 \longrightarrow J \xrightarrow{j} A \xrightarrow{\pi} B \longrightarrow 0$$

is a short exact sequence. Then A is nuclear.

Proof. Let C be an arbitrary C*-algebra. Since B is nuclear, it follows from the previous theorem that

$$0 \longrightarrow J \otimes_{\min} C \xrightarrow{j \otimes \mathrm{id}} A \otimes_{\min} C \xrightarrow{\pi \otimes \mathrm{id}} B \otimes_{\min} C \longrightarrow 0$$

is a short exact sequence. Observe that the identity map $A \odot C \to A \otimes_{\min} C$ extends to a map $\varphi : A \otimes_{\max} C \to A \otimes_{\min} C$ since $\|\operatorname{id}(a)\|_{\min} \leq \|a\|_{\max}$. Thus to show that A is nuclear we need to show that φ is injective, since in that case we have an isomorphism $A \otimes_{\min} C \cong A \otimes_{\max} C$.

There is a unique *-homomorphism $\tilde{j} : J \odot C \to A \otimes_{\max} C$ that satisfies

$$\tilde{j}(a \otimes c) = j(a) \otimes c$$

for every $a \in J$ and $c \in C$. As in the proof of the previous lemma,

$$\gamma : J \odot C \to \mathbb{R}_+, \quad x \mapsto \max\{\|\tilde{j}(x)\|_{\max}, \|x\|_{\min}\}$$

defines a C*-norm on $J \odot C$, which, since J is nuclear, coincides with the minimal norm. Thus $\|\tilde{j}(x)\| \leq \|x\|_{\min}$, which implies $\tilde{j}$ extends to a *-homomorphism

$$\tilde{j} : J \otimes_{\min} C \to A \otimes_{\max} C,$$

satisfying

$$\varphi \circ \tilde{j} = j \otimes \mathrm{id}.$$

There is also a unique *-homomorphism $\tilde{\pi} : A \odot C \to B \otimes_{\min} C$ satisfying

$$\tilde{\pi}(a \otimes c) = ((\pi \otimes \mathrm{id}) \circ \varphi)(a \otimes c)$$

for every $a \in A$ and $c \in C$. Define

$$\gamma' : A \odot C \to \mathbb{R}_+, \quad x \mapsto \max\{\|\tilde{\pi}(x)\|_{\min}, \|x\|_{\max}\}.$$

Then γ' is a C*-norm on $A \odot C$. Since $\gamma'(x) \leq \|x\|_{\max}$ for every $x \in A \odot C$, we must have $\|\tilde{\pi}(x)\|_{\min} \leq \|x\|_{\max}$. Thus we can extend $\tilde{\pi}$ by continuity to get a map

$$\tilde{\pi} : A \otimes_{\max} C \to B \otimes_{\min} C.$$

Let $I := \operatorname{im}(\tilde{j}) \subset A \otimes_{\max} C$ and let $q : A \otimes_{\max} C \to (A \otimes_{\max} C)/I$ denote the corresponding quotient map. Put $D := (A \otimes_{\max} C)/I$.

As in the previous proof, since B is nuclear, there is a *-homomorphism

$$\rho : B \otimes_{\min} C \to D$$

that satisfies $\rho(\pi(a) \otimes x) = a \otimes x + I$ for every $a \in A$ and $c \in C$. In particular, we have the commutative diagram

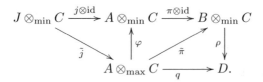

Suppose that $x \in \ker(\varphi)$. Then $0 = (\pi \otimes \mathrm{id}) \circ \varphi(x) = \tilde{\pi}(x)$, so $0 = \rho \circ \tilde{\pi}(x) = q(x)$, which means that $x \in I$. Let $y \in J \otimes_{\min} C$ satisfy $x = \tilde{j}(y)$. Then we have $0 = \varphi \circ \tilde{j}(y) = (j \otimes \mathrm{id})(y)$. Since $j \otimes \mathrm{id}$ is injective, $y = 0$. Thus also $x = 0$, showing that φ is injective and hence $A \otimes_{\max} C \cong A \otimes_{\min} C$. $\square$

6.5   Exercises

6.5.1. Prove Theorem 6.2.2 for A and B not necessarily unital. Hint: Use approximate units and Proposition 6.2.1. To show uniqueness, use the fact that any other maps π'_A, π'_B satisfying the required properties are necessarily nondegenerate and so if $(u_\lambda)_\Lambda$ is an approximate unit, we have $\pi'_A(u_\lambda)$ and $\pi'_B(u_\lambda)$ converging strongly to id_H.

6.5.2. Let H_1, H_2, H_3 be Hilbert spaces.

(i) Show that there exists a unique unitary

$$u : (H_1 \hat{\otimes} H_2) \hat{\otimes} H_3 \to H_1 \hat{\otimes} (H_2 \hat{\otimes} H_3)$$

such that

$$u((\xi_1 \otimes \xi_2) \otimes \xi_3) = \xi_1 \otimes (\xi_2 \otimes \xi_3), \quad \xi_i \in H_i, i = 1, 2, 3.$$

(ii) Show that for $a_i \in \mathcal{B}(H_i)$, $i = 1, 2, 3$ we have

$$u((a_1 \hat{\otimes} a_2) \hat{\otimes} a_3) u^* = a_1 \hat{\otimes} (a_2 \hat{\otimes} a_3).$$

(iii) Show that the minimal tensor product is associative up to *-isomorphism.

6.5.3. Show that the tensor product of two nuclear C*-algebras is nuclear.

6.5.4. Let A and B be C*-algebras. Show that there is a unique *-isomorphism $\sigma : A \otimes_{\min} B \to B \otimes_{\min} A$ satisfying $\sigma(a \otimes b) = b \otimes a$. The map σ is called the *flip map*.

6.5.5. Let A be a C*-algebra and X a locally compact Hausdorff space. We say that a continuous function $f : X \to A$ vanishes at infinity if $x \mapsto \|f(x)\|_A$ vanishes at infinity in the usual sense. Let $C_0(X, A)$ denote the C*-algebra of continuous functions $X \to A$ vanishing at infinity (with supremum norm and pointwise operations).

(i) If X is compact, show that any $f \in C(X, A)$ can be written as the limit of functions of the form $\sum_{i=1}^{k} f_i a_i$ where $f_i \in C(X)$ and $a_i \in A$. (Hint: for $n \in \mathbb{N}$ use a partition of unity $\gamma_1, \ldots, \gamma_m \in C(X)$ subordinate to a finite cover $U_1, \ldots, U_m$ of X by sets of the form

$$U_i = \{x \in X \mid \|f(x) - a_i\| \le 1/n\}$$

for some $a_1, \ldots, a_m \in A$ and consider the function $f_n := \sum_{i=1}^{m} \gamma_i a_i$.)

(ii) By extending $f \in C_0(X, A)$ to a function on the one-point compactification $\tilde{f} \in C(X \cup \{\infty\}, A)$ with $\tilde{f}(\infty) = 0$, show that f can be written as the limit of functions of the form $\sum_{i=1}^{k} f_i a_i$ where $f_i \in C_0(X)$ and $a_i \in A$. (Hint: Replace each γ in the partition of unity of $X \cup \{\infty\}$ above with $\tilde{\gamma} = (\gamma - \gamma(\infty))|_X$.)

(iii) Show that the map

$$C_0(X) \times A \to C_0(X, A), \quad (f, a) \mapsto fa,$$

induces an injective *-homomorphism $\varphi : C_0(X) \odot A \to C_0(X, A)$.

(iv) Show that $C_0(X) \odot A \to \mathbb{R}_+$, $f \mapsto \|\varphi(f)\|$ defines a C*-norm on $C_0(X) \odot A$.

(v) Use the fact that $C_0(X)$ is nuclear (due to Takesaki [116]; also see Exercise 7.3.11) to show that the *-homomorphism φ extends uniquely to an isometric *-homomorphism $\tilde{\varphi} : C_0(X) \otimes_{\min} A \to C_0(X, A)$.

(vi) Deduce that $C_0(X) \otimes_{\min} A \cong C_0(X, A)$. (Of course, since $C_0(X)$ is nuclear, we need not include the subscript "min" on the tensor product.)

6.5.6. Let X and Y be compact Hausdorff spaces. Show that

$$C(X) \otimes_{\min} C(Y) \cong C(X \times Y).$$

As above, we can drop min since $C(X)$ is nuclear. (Hint: Look at the Gelfand spectrum of $C(X) \otimes C(Y)$.)

7 Completely positive maps

We have seen that C*-algebras boast a number of good structural properties that distinguish them from arbitrary Banach algebras. There are various types of maps one could consider between C*-algebras which aim at preserving particular C*-algebraic properties. Perhaps the most obvious demand is that we preserve the algebraic structure and the involution, which in turn guarantees continuity; these are of course exactly *-homomorphisms. However, it is often too much to ask for the existence of a *-homomorphism. For example, we saw that a commutative C*-algebra A admits many characters, which are just *-homomorphisms $A \to \mathbb{C}$. The existence of characters is of fundamental importance to the Gelfand Theorem. However, for noncommutative C*-algebras, we may not have any characters at our disposal. Nevertheless, maps into $\mathbb{C}$ play an important role, as we saw in our study of positive linear functionals and representation theory. In this case, we have relaxed the condition that multiplication is preserved and are interested only in the linear structure and the order structure. The maps of interest in this section, *completely positive maps*, similarly abandon any attempt to preserve multiplication and focus on the linear and order structure of a C*-algebra. In fact, completely positive maps might be thought of as "operator-valued linear functionals", that is, we now allow for the range to be C*-algebras more general than $\mathbb{C}$.

Completely positive maps with matrix-valued domains are particularly useful. In the next chapter, we will introduce the approximately finite C*-algebras, which are particularly tractable thanks to fact that they come equipped with many *-homomorphisms to and from matrix algebras. For an arbitrary C*-algebra, such *-homomorphisms are too much too expect, however, the large class of nuclear C*-algebras allow for similar approximations by finite-dimensional algebras via completely positive maps. They also have many application beyond the study of C*-algebras, for example to nonself-adjoint operator algebras, operator systems, and quantum information theory.

Completely positive maps were described for concrete C*-algebras in Exercise 3.4.5. Of course, now we know that every abstract C*-algebra is *-isomorphic to a concrete C*-algebra, so the definition makes sense for all C*-algebras. We begin the chapter by introducing completely positive maps for C*-algebras and certain subspaces called *operator systems* and examining some of their properties. Looking at linear maps, we will show necessary and sufficient conditions for complete positivity. In the second section, we focus on the *completely positive approximation property*, which is a condition equivalent to nuclearity for a given C*-algebra, but one which is often easier to verify and work with as it can be seen as establishing the existence of noncommutative partitions of unity.

7.1 Completely positive maps

In this first section, we introduce operator systems and completely positive maps. We will also characterise completely positive maps as cutdowns of *-homomor-

K. R. Strung, *An Introduction to C*-Algebras and the Classification Program*, Advanced Courses in Mathematics - CRM Barcelona, https://doi.org/10.1007/978-3-030-47465-2_7

phism (Stinespring's Theorem) and see a number of different ways of verifying that a given map is completely positive.

7.1.1. Definition. Let A be a C*-algebra. An *operator system* $E \subset A$ is a closed self-adjoint linear subspace containing the unit of A. Note in particular that any unital C*-subalgebra $B \subset A$ is an operator system.

7.1.2. Definition. Let A and B be C*-algebras. Suppose A is unital and $E \subset A$ is an operator system. A linear map $\varphi : E \to B$ is *positive* if $\varphi(E \cap A_+) \subset B_+$. For any map $\varphi : E \to B$ and $n \in \mathbb{N}$ we can define $\varphi^{(n)} : M_n(E) \to M_n(B)$ by applying φ entry-wise. If $\varphi^{(n)} : M_n(E) \to M_n(B)$ is positive for every $n \in \mathbb{N}$ then we say φ is *completely positive* (c.p.).

If φ is also contractive, then we say φ is *completely positive contractive* (c.p.c.), and if φ is unital, we say it is a *unital completely positive map* (u.c.p.).

It follows from Exercise 3.4.5 that every *-homomorphism $\varphi : A \to B$ is completely positive. It also follows from Exercise 3.4.5 that a cutdown of a *-homomorphism is completely positive, that is, if $v \in B$ and $\varphi : A \to B$ is a *-homomorphism, than the map

$$v^*\varphi v : A \to B, \qquad a \mapsto v^*\varphi(a)v,$$

is also completely positive. Observe that, unless v is a unitary, $v^*\varphi v$ will not be a *-homomorphism. So completely positive maps are more general. However, they are intimately related to *-homomorphisms. In fact, every completely positive map can be seen as the cutdown of a *-homomorphism, as we see in Stinespring's Theorem.

7.1.3. Theorem (Stinespring). *Let A be a unital C*-algebra and $\varphi : A \to \mathcal{B}(H)$ a c.p. map. There exist a Hilbert space K, a *-representation $\pi : A \to \mathcal{B}(K)$ and an operator $v : H \to K$ such that*

$$\varphi(a) = v^*\pi(a)v, \qquad \text{for every } a \in A.$$

Proof. This is proved using a variation of the GNS construction (Section 4.2). Let $A \odot H$ denote the vector space tensor product of A and H. For elements $x = \sum_{i=1}^n a_i \otimes \xi_i$ and $y = \sum_{j=1}^n b_j \otimes \eta_j$ of $A \odot H$, define

$$\langle x, y \rangle := \left\langle \varphi(b_j^* a_i)\xi_i, \eta_j \right\rangle_H.$$

It is straightforward to check that this defines a sesquilinear form (5.2.3) on $A \odot H$. Moreover, one can show that $\langle x, x \rangle \geq 0$ for every $x \in A \odot H$, that is, $\langle \cdot, \cdot \rangle$ is positive semidefinite. Let

$$N = \{x \in A \odot H \mid \langle x, x \rangle = 0\},$$

which is a closed subspace of $A \odot H$. Then $\langle \cdot, \cdot \rangle$ induces an inner product on $(A \odot H)/N$. Define

$$K := \overline{(A \odot H)/N}^{\langle \cdot, \cdot \rangle}.$$

If $x \in A \odot H$, let $[x]$ denote the image of x in K under the quotient map. Define

$$v : H \to K, \qquad \eta \mapsto [1_A \otimes \eta].$$

Now fix $a \in A$. For $x = \sum_{i=1}^n b_i \otimes \xi_i \in A \odot H$, define $\pi(a) : A \odot H \to K$ by

$$\pi(a)(x) = \pi(a) \left(\sum_{i=1}^n b_i \otimes \xi_i \right) = \left[\sum_{i=1}^n (ab_i) \otimes \xi_i \right],$$

and extend to $\pi(a) : K \to K$. Then $\pi : A \to \mathcal{B}(K)$ is a *-representation. For every $a \in A$ and $\xi, \eta \in H$ we have

$$\begin{aligned}
\langle v^* \pi(a) v(\xi), \eta \rangle_H &= \langle \pi(a)([1_A \otimes \xi]), [1_A \otimes \eta] \rangle_K \\
&= \langle [a \otimes \xi], [1_A \otimes \eta] \rangle_K \\
&= \langle \varphi(a)\xi, \eta \rangle_H.
\end{aligned}$$

Thus $v^* \pi(a) v = \varphi(a)$ for every $a \in A$.   □

The next proposition is more or less contained in the proof of Stinespring's Theorem, but it is also useful on its own.

7.1.4. Proposition. *Let A be a C*-algebra and $\varphi : A \to \mathcal{B}(H)$. Then φ is completely positive if and only if for every $n > 1$, every $a_1, \ldots, a_n \in A$, and every $\xi_1, \ldots, \xi_n \in H$, we have*

$$\sum_{i,j=1}^n \langle \varphi(a_j^* a_i) \xi_j, \xi_i \rangle \geq 0.$$

Proof. As in the proof of Stinespring's Theorem, for elements $x = \sum_{i=1}^n a_i \otimes \xi_i$ and $y = \sum_{j=1}^n b_j \otimes \eta_j$ of $A \odot H$, define

$$\langle x, y \rangle := \sum_{i,j=1}^n \langle \varphi(b_j^* a_i) \xi_j, \eta_i \rangle_H.$$

If φ is completely positive, then this sesquilinear form is positive, so

$$\sum_{i,j=1}^n \langle \varphi(a_j^* a_i) \xi_j, \xi_i \rangle \geq 0.$$

In the other direction, we have $\langle x, x \rangle = \sum_{i,j=1}^n \langle \varphi(a_j^* a_i) \xi_j, \xi_i \rangle \geq 0$, so $\langle \cdot, \cdot \rangle$ defines a positive semidefinite sesquilinear form. Proceeding as in the proof of Stinespring's Theorem, we can construct the Hilbert space K, a contraction $v : H \to K$ and *-homomorphism $\pi : K \to K$ such that $v^* \pi(a) v = \varphi(a)$. So φ is a cutdown of a *-homomorphism and therefore completely positive.   □

7.1.5. Proposition. *Let A be a C^*-algebra and let $\varphi : A \to \mathcal{B}(H)$ be a linear map that is $*$-preserving. Then φ is completely positive if and only if for every $a_1, \ldots, a_n, x_1, \ldots, x_n \in A$ and $\xi \in H$ we have*

$$\sum_{i,j=1}^{n} \langle \varphi(x_i^*)\varphi(a_i^* a_j)\varphi(x_j)\xi, \xi \rangle \geq 0.$$

Proof. If φ is completely positive then, applying the previous proposition, we have

$$\sum_{i,j=1}^{n} \langle \varphi(x_i^*)\varphi(a_i^* a_j)\varphi(x_j)\xi, \xi \rangle = \sum_{i,j=1}^{n} \langle \varphi(a_i^* a_j)\varphi(x_j)\xi, \varphi(x_i)\xi \rangle \geq 0.$$

Suppose that $\sum_{i,j=1}^{n} \langle \varphi(x_i^*)\varphi(a_i^* a_j)\varphi(x_j)\xi, \xi \rangle \geq 0$. Then since φ is $*$-preserving we have $\sum_{i,j=1}^{n} \langle \varphi(a_i^* a_j)\varphi(x_j)\xi, \varphi(x_i)\xi \rangle \geq 0$. By breaking up H into a direct sum of cyclic subspaces (Theorem 4.3.5), the result follows from the preceding proposition. $\qquad\square$

7.1.6. We can use Stinespring's Theorem to show that c.p.c. maps behave well with respect to C^*-tensor products, just as we saw for $*$-homomorphisms in Theorem 6.2.9.

Theorem. *Let $\varphi : A \to C$ and $\psi : B \to D$ be completely positive maps. Then, for any C^*-tensor product, there is a unique completely positive map*

$$\varphi \otimes \psi : A \otimes B \to C \otimes D,$$

satisfying

$$\varphi \otimes \psi(a \otimes b) = \varphi(a) \otimes \psi(b),$$

for every $a \in A$ and $b \in B$. Moreover $\|\varphi \otimes \psi\| = \|\varphi\|\|\psi\|$. In particular if φ and ψ are c.p.c., so is their tensor product.

Proof. Exercise. $\qquad\square$

7.1.7. When considering linear maps from M_n into C^*-algebras, there is a simple criterion to test if a given map is positive. One needs only check what happens on the matrices e_{ij}, where e_{ij} is the $n \times n$ matrix with 1 in the (i,j)th entry and 0 elsewhere.

Proposition. *Let $\varphi : M_n \to A$ be a linear map into a C^*-algebra A. Let $e_{ij} \in M_n$ be as above. Then φ is completely positive if and only if $(\varphi(e_{ij}))_{ij} \in M_n(A)$ is positive.*

Proof. We leave it as an exercise to show that if φ is completely positive, then $(\varphi(e_{ij}))_{ij} \in M_n(A)$ is positive. Suppose that $(\varphi(e_{ij}))_{ij} \in M_n(A)$ is positive. Then $(\varphi(e_{ij}))_{ij}$ has a positive square root $b = (b_{ij})_{ij} \in M_n(A)$ and

$$\varphi(e_{ij}) = \sum_{k=1}^{n} b_{ik}^* b_{kj}.$$

Let $H = \mathbb{C}^n$ with standard orthonormal basis $\xi_1, \ldots, \xi_n$. Let $\pi : A \to \mathcal{B}(K)$ be a faithful representation on a Hilbert space K. For $a \in M_n$, the map

$$a \mapsto a \otimes \mathrm{id}_H \otimes \mathrm{id}_K$$

is a *-homomorphism from $M_n \to \mathcal{B}(H \hat{\otimes} H \hat{\otimes} K)$. Define $v : K \to H \hat{\otimes} H \hat{\otimes} K$ by

$$v\eta = \sum_{i,j=1}^{n} \xi_j \otimes \xi_i \otimes b_{ij}\eta, \qquad \eta \in K.$$

Then, for any $a = (a_{ij})_{ij} \in M_n$ we have

$$\langle v^*(a \otimes \mathrm{id}_H \otimes \mathrm{id}_K)v\,\eta, \eta' \rangle = \langle (a \otimes \mathrm{id}_H \otimes \mathrm{id}_K)v\,\eta, v\,\eta' \rangle$$

$$= \left\langle (a \otimes \mathrm{id}_H \otimes \mathrm{id}_K)\left(\sum_{i,j=1}^{n} \xi_j \otimes \xi_i \otimes b_{ij}\eta\right), \sum_{k,l=1}^{n} \xi_l \otimes \xi_k \otimes b_{kl}\eta' \right\rangle$$

$$= \left\langle \sum_{i,j=1}^{n} a(\xi_j) \otimes \xi_i \otimes b_{ij}\eta, \sum_{k,l=1}^{n} \xi_l \otimes \xi_k \otimes b_{kl}\eta' \right\rangle$$

$$= \sum_{i,j,k,l} \langle a(\xi_j), \xi_l \rangle \langle \xi_i, \xi_k \rangle \langle b_{ij}\eta, b_{kl}\eta' \rangle = \sum_{j,k,l} \langle a(\xi_j), \xi_l \rangle \langle b_{kj}\eta, b_{kl}\eta' \rangle$$

$$= \sum_{j,l} \langle a(\xi_j), \xi_l \rangle \langle \varphi(e_{lj})\eta, \eta' \rangle = \sum_{j,l} a_{lj} \langle \varphi(e_{lj})\eta, \eta' \rangle$$

$$= \sum_{j,l} \langle \varphi(a_{lj}e_{lj})\eta, \eta' \rangle = \langle \varphi(a)\eta, \eta' \rangle.$$

The fourth step follows from Proposition 6.1.1. Thus $v^*(a \otimes \mathrm{id}_H \otimes \mathrm{id}_K)v = \varphi(a)$, which implies φ is completely positive, as we saw in Exercise 3.4.5 (c). $\qquad\square$

7.1.8. Let $\varphi : A \to M_n$ be a linear map. For $a \in A$, let $\varphi(a)(i,j)$ denote the (i,j)th entry of the matrix $\varphi(a)$. Define

$$\hat{\varphi} : M_n(A) \to \mathbb{C}$$

by

$$\hat{\varphi}((a_{ij})_{ij}) := \sum_{i,j=1}^{n} \varphi(a_{ij})(i,j), \quad (a_{ij})_{ij} \in M_n(A).$$

Then $\hat{\varphi}$ is a linear functional. As a consequence of the previous proposition, we have the following characterisation of c.p. maps in terms of the associated linear functional.

Proposition. *Let A be a unital C*-algebra and $\varphi : A \to M_n$ a linear map. Then φ is completely positive if and only if the linear functional $\hat{\varphi} : M_n(A) \to \mathbb{C}$ is positive.*

Proof. Suppose that φ is completely positive. As before, let $H = \mathbb{C}^n$ with orthonormal basis $\xi_1, \ldots, \xi_n$. Then, for $(a_{ij})_{ij} \in M_n(A)$ we have

$$\hat{\varphi}((a_{ij})_{ij}) = \sum_{i,j=1}^n \varphi(a_{ij})(i,j) = \langle (\varphi^{(n)}(a_{ij}))_{ij}\xi, \xi \rangle,$$

where $\xi := (\xi_1, \ldots, \xi_n)^T \in \mathbb{C}^{n^2}$. Since φ is completely positive, so too is the map $\varphi^{(n)} : M_n(A) \to M_{n^2}$, and hence $\hat{\varphi}$ is positive.

Now suppose that $\hat{\varphi}$ is a positive linear functional. Let $\pi_{\hat{\varphi}} : M_n(A) \to \mathcal{B}(H_{\hat{\varphi}})$ be the GNS representation associated to $\hat{\varphi}$ (Definition 4.2.3). Denote the cyclic vector of the representation by μ (Theorem 4.3.1). Let $\{e_{ij}\}_{1 \le i,j \le n}$ denote the standard matrix units generating $M_n \subset M_n(A)$. Let $\xi_1, \ldots, \xi_n$ be an orthonormal basis for $\mathbb{C}^n$ and define $v : \mathbb{C}^n \to H_{\hat{\varphi}}$ by

$$v\xi_i = \pi_{\hat{\varphi}}(e_{1,i})\mu,$$

Then, for $a \in A$ and $i, j \in \{1, \ldots, n\}$, we have

$$\left\langle v^* \pi_{\hat{\varphi}}\left(\begin{pmatrix} a & & \\ & \ddots & \\ & & a \end{pmatrix}\right) v \cdot \xi_i, \xi_j \right\rangle = \left\langle \pi_{\hat{\varphi}}\left(\begin{pmatrix} a & & \\ & \ddots & \\ & & a \end{pmatrix}e_{1,i}\right)\mu, \pi_{\hat{\varphi}}(e_{1,j})\mu \right\rangle$$

$$= \left\langle \pi_{\hat{\varphi}}\left(e_{j,1}\begin{pmatrix} a & & \\ & \ddots & \\ & & a \end{pmatrix}e_{1,i}\right)\mu, \mu \right\rangle$$

$$= \langle \pi_{\hat{\varphi}}(ae_{ji})\mu, \mu \rangle = \hat{\varphi}(ae_{ji})$$

$$= \sum_{k,l=1}^n \varphi(ae_{ji})(k,l) = \varphi(a)(j,i)$$

$$= \langle \varphi(a)\xi_i, \xi_j \rangle.$$

Thus $\varphi(a) = v^* \pi_{\hat{\varphi}}\left(\begin{pmatrix} a & & \\ & \ddots & \\ & & a \end{pmatrix}\right) v$, hence is completely positive. $\square$

7.1.9. Conversely, if $\psi : A \otimes M_n \cong M_n(A) \to \mathbb{C}$ is a linear functional, we can define a map $\varphi : A \to M_n$ by $(\varphi(a))_{ij} = (\psi(a \otimes e_{ij}))_{ij}$. We leave it as an exercise to check that $\hat{\varphi} = \psi$. Thus the previous proposition gives a one-to-one correspondence between completely positive maps $A \to M_n$ and positive linear functionals $M_n(A) \to \mathbb{C}$.

7.1.10. Lemma. *Suppose $E \subset A$ is an operator system in a unital C*-algebra A. If $\varphi : E \to \mathbb{C}$ is a positive linear functional, then $\|\varphi\| = \varphi(1_A)$.*

Proof. Let $\epsilon > 0$ and find $x \in E$ with $\|x\| \leq 1$ and $|\varphi(x)| \geq \|\varphi\| - \epsilon$. Without loss of generality, we may assume that x is self-adjoint. Then $\varphi(x) \in \mathbb{R}$. Since x is self-adjoint, we have that $x \leq \|x\| \cdot 1$, so also $\varphi(x) \leq \|x\|^{-1}\varphi(1)$. In particular, this holds when $x = 1_A$. Thus $\varphi(1_A) = 1$. $\square$

7.1.11. Lemma. *Let A be a unital C*-algebra and $E \subset A$ an operator system. Then, for any $n \in \mathbb{N} \setminus \{0\}$ and any completely positive map $\varphi : E \to M_n$, there exists a completely positive map $\tilde{\varphi} : A \to M_n$ extending φ.*

Proof. Given $\varphi : E \to M_n$ we define a positive linear functional $\hat{\varphi} : M_n(E) \to \mathbb{C}$ as in Proposition 7.1.8. By the previous lemma, we have $\hat{\varphi}(1_{M_n(A)}) = 1$. Use the Hahn–Banach Theorem to extend $\hat{\varphi}$ to a linear functional $\Phi : M_n(A) \to \mathbb{C}$ with $\|\Phi\| = \|\hat{\varphi}\| = \hat{\varphi}(1) = \Phi(1)$. By Proposition 4.1.7, Φ is positive and so by the correspondence given in 7.1.9 there is a completely positive map $\Psi : A \to M_n$ with $\hat{\Psi} = \Phi$. Then $\tilde{\varphi} := \Psi$ is the required map. $\square$

7.1.12. Arveson's Extension Theorem ([5, Theorem 1.2.3]) is an important generalisation of the last theorem. It can be thought of as a noncommutative version of the Hahn–Banach Theorem and is used frequently in the literature. The proof requires some tools we have not developed here, but since the reader can find a good exposition of the necessary background in [18], we will skip the details.

Theorem (Arveson's Extension Theorem). *Let A be a unital C*-algebra and $E \subset A$ an operator system. Suppose $\varphi : E \to \mathcal{B}(H)$ is a c.p.c. map. Then there is a c.p.c. map $\tilde{\varphi} : A \to \mathcal{B}(H)$ extending φ.*

7.2 Completely positive approximation property

The completely positive approximation property allows us to approximate a C*-algebra by finite-dimensional C*-algebras via c.p.c maps. Not every C*-algebra has the completely approximation property. It is a deep result that a C*-algebra has this property if and only if it is nuclear (Theorem 7.2.2, below).

7.2.1. Let A be a C*-algebra. Let $\mathcal{F} \subset A$ be a finite subset and let $\epsilon > 0$. An $(\mathcal{F}, \epsilon)$ *c.p. approximation* for A is a triple (F, ψ, ϕ) consisting of a finite-dimensional C*-algebra F and completely positive maps $\psi : A \to F$ and $\phi : F \to A$ satisfying

$$\|\phi \circ \psi(a) - a\| < \epsilon \text{ for every } a \in \mathcal{F}.$$

Definition. A C*-algebra A has the *completely positive approximation property* (c.p.a.p.) if, for every finite subset $\mathcal{F} \subset A$ and every $\epsilon > 0$, there is an $(\mathcal{F}, \epsilon)$ *c.p. approximation* for A with ψ and ϕ both contractive.

An important theorem, which we will not prove here, is that the c.p.a.p. is equivalent to nuclearity. The result is due to Choi and Effros [25] as well as Kirchberg [66].

7.2.2. Theorem. *Let A be a C*-algebra. Then A is nuclear if and only if A has the completely positive approximation property.*

Completely positive contractive maps from finite-dimensional C*-algebras have a useful property, called the *lifting property*.

7.2.3. Definition. Let A and B be C*-algebras and let $J \subset B$ be an ideal. Denote by $\pi : B \to B/J$ the quotient map. We say that a completely positive contractive map $\varphi : A \to B/J$ has the *lifting property* if there exists a c.p.c. map $\psi : A \to B$ such that $\pi \circ \psi = \varphi$.

7.2.4. Proposition. *Let A be a C*-algebra with ideal $J \subset A$. Then every c.p.c. map $\varphi : M_n \to A/J$ is liftable. Moreover, if φ is unital, the lift can be chosen to be unital.*

Proof. Let e_{ij}, $1, \leq i, j \leq n$ be matrix units for M_n. Since $\varphi : M_n \to A/J$ is c.p.c., we have that $a := (\varphi(e_{ij}))_{ij} \in M_n(A/J)$ is positive by Proposition 7.1.7. Let $\pi_n : M_n(A) \to M_n(A/J)$ be the quotient map, which is given by the quotient map $\pi : A \to A/J$ applied entry-wise. Since π is surjective, there exists some $b = (b_{ij})_{ij} \in M_n(A)$ such that $\pi_n(b) = a$. Since $\|a\| = \inf_{x \in M_n(J)} \|a + x\|$ and J is closed, we may choose b to have the same norm as a. Replacing b with $(b^*b)^{1/2}$ if necessary, we may assume that $b \geq 0$. Define $\psi : M_n \to A$ on matrix units by

$$\psi(e_{ij}) = b_{ij}.$$

By construction, this is a contractive lift of φ and by Proposition 7.1.7, ψ is completely positive. For the second statement, observe that if φ is unital then we can replace ψ with $\psi + 1_A - \psi(1_{M_n})$. $\qquad\qquad\square$

7.2.5. In fact, c.p.c. maps with more general domains may also have the lifting property. This is a first glimpse of the utility of the c.p.a.p. Since the c.p.a.p. means there are good approximations through matrices, we can extend this lifting property from finite-dimensional domains to the case that the domain is separable and nuclear. This is known as the Choi–Effros Lifting Theorem [22], which we will also prove below (Theorem 7.2.7).

By *point-norm topology* on a set of maps between C*-algebras A and B (or more generally, operator systems), we mean the topology of pointwise convergence, that is, a sequence $(\varphi_n : A \to B)_{n \in \mathbb{N}}$ converges to $\varphi : A \to B$ in the point-norm topology if and only if $\|\varphi_n(a) - a\| \to 0$ as $n \to \infty$, for every $a \in A$.

7.2.6. Lemma. *Let A and B be separable C*-algebras and $J \subset B$ an ideal. Then the set of liftable completely positive contractive maps $A \to B/J$ is closed with respect to the point-norm topology.*

Proof. Let $\pi : A \to B/J$ denote the quotient map. Let $\varphi : A \to B/J$ be a completely positive contractive map. Suppose that $(\psi_n : A \to B)_{n \in \mathbb{N}}$ is a sequence of c.p.c. maps such that $\lim_{n \to \infty} \pi \circ \psi_n(a) = \varphi(a)$ for every $a \in A$. Since A is separable, there is a sequence $(a_k)_{k \in \mathbb{N}} \subset A$ that is dense in A. Without loss of generality we may assume that

$$\|\pi \circ \psi_n(a_k) - \varphi(a_k)\| \le 1/2^n,$$

whenever $k \le n$.

We will show by induction that there are c.p.c. maps $\varphi_n : A \to B$ such that

$$\|\pi \circ \varphi_n(a_k) - \varphi(a_k)\| < 1/2^n, \text{ and } \|\varphi_{n+1}(a_k) - \varphi_n(a_k)\| < 1/2^n,$$

for $k \le n$.

The above holds for $\varphi_1 = \psi_1$. Now suppose that, for $n \ge 1$, there are c.p.c. maps $\varphi_1, \ldots, \varphi_n$ satisfying $\pi \circ \varphi_n(x_k) - \varphi(x_k) \le 1/2^n$. By Theorem 3.2.4 there exists an approximate unit $(u_\lambda)_\lambda$ of J that is quasicentral for B.

Choose λ_0 to be large enough so that, for every $k \le n$, we have

$$\|(1 - u_{\lambda_0})^{1/2} \psi_{n+1}(a_k)(1 - u_{\lambda_0})^{1/2}\| < 1/2^{n+2}.$$

By taking λ_0 to be even larger if necessary, we may also arrange

$$\|u_{\lambda_0}^{1/2} \varphi_n(a_k) u_{\lambda_0}^{1/2} - \varphi_n(a_k)\| < 1/2^{n+2}.$$

(Note that both these estimates are using quasicentrality of u_λ.)

Let $u := u_{\lambda_0}$ and set

$$\varphi_{n+1}(a) := (1 - u)^{1/2} \psi_{n+1}(a)(1 - u)^{1/2} + u^{1/2} \varphi_n(a) u^{1/2}, \quad a \in A.$$

We have $\pi(\varphi_{n+1})(a) = \psi_{n+1}(a)$. Also,

$$\begin{aligned}
\|\varphi_{n+1}&(a_k) - \varphi_n(a_k)\| \\
&= \|(1 - u)^{1/2} \psi_{n+1}(a_k)(1 - u)^{1/2} + u^{1/2} \varphi_n(a_k) u^{1/2} - \varphi_n(a_k)\| \\
&\le \|(1 - u)^{1/2} \psi_{n+1}(a_k)(1 - u)^{1/2}\| + \|u^{1/2} \varphi_n(a_k) u^{1/2} - \varphi_n(a_k)\| \\
&< 1/2^{n+1},
\end{aligned}$$

so φ_n is the desired sequence. Since the sequence $(a_k)_{k \in \mathbb{N}}$ is dense in A, the sequence of c.p.c. maps $(\varphi_n)_{n \in \mathbb{N}}$ converges to a c.p.c map $\tilde{\varphi} : A \to B$, which is easily seen to be a lift of φ. $\qquad\square$

7.2.7. Theorem (Choi–Effros Lifting Theorem). *Let A and B be C^*-algebras with A separable and nuclear, and let $J \subset B$ be an ideal. Then every c.p.c. map $\rho : A \to B/J$ is liftable.*

Proof. Let $(a_i)_{i \in \mathbb{N}}$ be a dense sequence in A. Since A is nuclear, for every integer $k > 0$ there are c.p.c. maps $\psi_k : A \to M_{n_k}$ and $\varphi_k : M_{n_k} \to A$ such that

$$\|\varphi_k \circ \psi_k(a_i) - a_i\| < 1/k,$$

for every $i \leq k$. For each k, $\rho \circ \phi_k : M_{n_k} \to B/J$ is a c.p.c. map, and so, by Lemma 7.2.4, it is liftable to a c.p.c. map $\nu_k : M_{n_k} \to B$. Then $\nu_k \circ \psi_k : A \to B$ is a c.p.c. lift of $\rho \circ \varphi_k \circ \psi_k$ converging pointwise to some map $\tilde{\rho} : A \to B$ as $k \to \infty$. It follows from the previous lemma that $\tilde{\rho}$ is a c.p.c. lift of ρ. $\square$

7.2.8. The next corollary is an easy exercise to prove.

Corollary. *Let A and B be C^*-algebras with B separable and nuclear. Suppose $\pi : A \to B$ is a surjective $*$-homomorphism. Then there is a completely positive contractive lift $\sigma : B \to A$, that is, σ is a c.p.c. map satisfying $\sigma \circ \pi = \mathrm{id}_B$.*

7.3   Exercises

7.3.1. Show that the map

$$\varphi : M_2 \to M_2, \quad \begin{pmatrix} a & b \\ c & d \end{pmatrix} \mapsto \begin{pmatrix} a & c \\ b & d \end{pmatrix}$$

is positive but is not completely positive.

7.3.2. Let A be a C^*-algebra and $\varphi : M_n \to A$ a linear map. For $1 \leq i, j \leq n$, let $e_{ij} \in M_n$ denote the $n \times n$ matrix with 1 in the (i, j)th entry and 0 in every other entry. Show that if φ is completely positive, then

$$(\varphi(e_{ij}))_{ij} = \begin{pmatrix} \varphi(e_{11}) & \varphi(e_{12}) & \cdots & \varphi(e_{1n}) \\ \varphi(e_{21}) & \varphi(e_{22}) & \cdots & \varphi(e_{2n}) \\ \vdots & \vdots & \ddots & \vdots \\ \varphi(e_{n1}) & \varphi(e_{n2}) & \cdots & \varphi(e_{nn}) \end{pmatrix}$$

is a positive element in $M_n(A)$.

7.3.3. Show that any state on a C^*-algebra A is completely positive.

7.3.4. Show that the composition of two c.p.(c.) maps is again c.p.(c.).

7.3.5. Let A and B be C^*-algebras and $\varphi : A \to B$ a completely positive contractive map. Use Stinespring's Theorem to show that, for any $a, b \in A$, the following inequality holds:

$$\|\varphi(ab) - \varphi(a)\varphi(b)\| \leq \|\varphi(aa^*) - \varphi(a)\varphi(a^*)\|^{1/2}\|b\|.$$

7.3.6. Let A, B and φ be as above and suppose that $C \subset A$ is a C^*-subalgebra such that $\varphi|_C$ is multiplicative. Let $c \in C$ and $a \in A$. Show that $\varphi(ca) = \varphi(c)\varphi(a)$ and $\varphi(ac) = \varphi(a)\varphi(c)$.

7.3.7. Let A be a separable unital C*-algebra and suppose that for every finite subset $\{a_1, \ldots, a_n\} \subset A$ and every $\epsilon > 0$ there exist a projection $p \in A$, $n \in \mathbb{N}\backslash\{0\}$, and a finite-dimensional C*-subalgebra $B \cong M_n$ with $1_B = p$ and $b_1, \ldots, b_n \in B$ such that $\|pa_ip - b_i\| < \epsilon$ for $i = 1, \ldots, n$.

(i) Show that for every finite subset $\mathcal{F} \subset A$ and every $\epsilon > 0$ there exists a natural number $n \in \mathbb{N} \setminus \{0\}$ and a u.c.p. map $\varphi : pAp \to M_n$ such that $\|\varphi(pap)\| > \|pap\| - \epsilon$.

(ii) Show that for every finite subset $\mathcal{F} \subset A$ and every $\epsilon > 0$ there exists $n \in \mathbb{N} \setminus \{0\}$ and a c.p.c. map $\psi : A \to M_n$ such that $\|\psi(a)\| > \|a\| - \epsilon$.

7.3.8. Let $\phi : A \to C$ and $\psi : B \to D$ be completely positive contractive maps. Show that for any C*-tensor product there is a unique completely positive contractive map

$$\phi \otimes \psi : A \otimes B \to C \otimes D,$$

satisfying

$$\phi \otimes \psi(a \otimes b) = \phi(a) \otimes \psi(b),$$

for every $a \in A$ and $b \in B$. Show that $\|\phi \otimes \psi\| = \|\phi\|\|\psi\|$.

7.3.9. Let $A := C(X)$ for some compact metric space X and let B be another C*-algebra.

(i) Let $x_0, x_1, \ldots, x_n \in X$ be a finite subset of points and suppose there are projections $p_0, p_1, \ldots, p_n \in B$ satisfying $\sum_{i=0}^{n} p_i = 1$. Show that the map

$$\varphi : A \to B, \quad f \mapsto \sum_{i=0}^{n} f(x_i)p_i,$$

is a completely positive map.

(ii) Suppose that $\varphi : A \to B$ is a positive, unital and linear map. Show that φ is the point-norm limit of maps of the form given in (i). Show that φ is moreover completely positive.

7.3.10. Let X be a compact Hausdorff space. Suppose that $\mathcal{U} = \{U_i\}_{i=1}^{d}$ is a finite open cover of X and let $g_1, \ldots, g_d : X \to [0, 1]$ be a partition of unity subordinate to $\mathcal{U}$ and let $x_i \in U_i$ satisfy $g_i(x_i) = 1$. Show that the map

$$\varphi : \mathbb{C}^d \to C(X), \quad (\lambda_1, \ldots, \lambda_d) \mapsto \sum_{i=1}^{d} \lambda \cdot g_i$$

is completely positive and contractive.

7.3.11. Let X be a compact Hausdorff space. Show that $C(X)$ has the completely positive approximation property, without appealing to Theorem 7.2.2.

7.3.12. Let A and B be C*-algebras with B separable and nuclear, and suppose that $\pi : A \to B$ is a surjective *-homomorphism. Show that there is a completely positive contractive lift $\sigma : B \to A$, that is, σ is a c.p.c. map satisfying $\sigma \circ \pi = \mathrm{id}_B$.

7.3.13. Let A and B be separable, unital C*-algebras with nonempty tracial state spaces $T(A)$ and $T(B)$ (4.1.1).

(i) Let $\tau_A \in T(A)$ and $\tau_B \in T(B)$. Define $\tau_A \otimes \tau_B : A \otimes_{\min} B \to \mathbb{C}$ on simple tensors by $\tau_A \otimes \tau_B(a \otimes b) = \tau_A(a)\tau_B(b)$. Show that $\tau_A \otimes \tau_B \in T(A \otimes_{\min} B)$.

(ii) Show that for any $\tau \in T(A \otimes_{\min} B)$, the maps

$$a \mapsto \tau(a \otimes 1_B), \quad b \mapsto \tau(1_A \otimes b),$$

define tracial states on A and B respectively.

(iii) Suppose that in addition to the assumptions above, A and B are both simple, and B has a unique tracial state τ_B. Show that, for any $a \in A_+$ and any $\tau \in T(A)$ the map

$$\tau_a : B \to \mathbb{C}, \quad b \mapsto \frac{\tau(a \otimes b)}{\tau(a \otimes 1_B)},$$

is a well-defined tracial state on B, and therefore $\tau_a := \tau_B$.

(iv) With A and B as in (iii), show that if $a \in A_+$ is positive, then for any $\tau \in T(A \otimes_{\min} B)$ there exists $\tau_A \in T(A)$ such that $\tau(a \otimes b) = \tau_A(a)\tau_B(b)$. Deduce that any tracial state $\tau \in T(A \otimes_{\min} B)$ is of the form $\tau_A \otimes \tau_B$ as defined in (i). Thus $T(A \otimes_{\min} B) \cong T(A)$.

(v) For X a compact metric space and $n \in \mathbb{N} \setminus \{0\}$, let $A := C(X, M_n)$. Denote by tr_n the unique normalised tracial state on M_n. Show that $\tau \in T(A)$ is an extreme point if and only if $\tau(f) = \mathrm{tr}_n(f(x))$ for some $x \in X$. (Hint: see Exercise 5.4.14.)

Part II

Simple Examples

8 Inductive limits and approximately finite (AF) C*-algebras

Chapter 6 gave us a way to produce new C*-algebras from old ones by taking their tensor product. In this chapter, we will also construct new C*-algebras from old ones, this time by starting with a sequence of C*-algebras and taking their limit. We will mainly focus on two classes of examples, the approximately finite (AF) C*-algebras and a subclass of these called the uniformly hyperfinite (UHF) C*-algebras, both of which are built from finite-dimensional C*-algebras. AF algebras are some of the easiest examples of infinite-dimensional C*-algebras to describe. Thanks to their inductive limit structure and good approximation properties, they are quite tractable. Despite being a relatively straightforward extension of the class of finite-dimensional C*-algebras, they form a large class of C*-algebras. We will see this already in this chapter when we determine the isomorphism classes of UHF algebras. Later on, in Chapter 13 we will see that AF algebras can also be classified by a particular isomorphism invariant now called the *Elliott invariant*. Their classification can be seen as the launching point of the classification program, the focus of Part III.

In this chapter, we first collect some facts about finite-dimensional C*-algebras. In particular, we obtain a characterisation of finite-dimensional C*-algebras as direct sums of matrix algebras. In the second section, we introduce inductive limits in a general setting, before looking specifically at UHF algebras and then AF algebras. We prove a classification result for UHF algebras. We then prove a number of technical perturbation results, which will allow us to show in the final section that AF algebras can also be described by a local approximation property. We then look at some further properties of AF algebras.

© The Author(s), under exclusive license to Springer Nature Switzerland AG 2021
K. R. Strung, *An Introduction to C*-Algebras and the Classification Program*, Advanced
Courses in Mathematics - CRM Barcelona, https://doi.org/10.1007/978-3-030-47465-2_8

8.1   Finite-dimensional C*-algebras

Finite-dimensional C*-algebras are the building blocks of both UHF and AF algebras. Of course, we have encountered many finite-dimensional C*-algebras, in particular matrix algebras, already. Here, we characterise all finite-dimensional C*-algebras.

8.1.1. Proposition. *Let A be a finite-dimensional C*-algebra. Then A is simple if and only if $A \cong M_n$ for some $n \in \mathbb{N}$.*

Proof. Exercise.   □

8.1.2. Theorem. *Let F be a finite-dimensional C*-algebra. Then there are k, n_1, ..., $n_k \in \mathbb{N}$ such that*

$$F \cong M_{n_1} \oplus M_{n_2} \oplus \cdots \oplus M_{n_k}.$$

Proof. The proof is by induction on the dimension of F. If the dimension of F is one, then we have $F \cong \mathbb{C}$, so the theorem holds. Suppose that the dimension of F is $m > 1$ and that the theorem holds for all finite-dimensional C*-algebras with dimension less than m.

   If F is simple, then by Proposition 8.1.1, $F \cong M_n$ with $n^2 = m$. Otherwise, F has a proper nonzero ideal. Let I be an ideal with minimum dimension. Then it follows from Exercise 3.4.12 that I itself contains no nontrivial ideals, and so from Proposition 8.1.1 we have $I \cong M_{n_1}$ for some $n_1 \in \mathbb{N}$. In particular, I has a unit p, which is a projection in F, and we have $I = pFp$.

   Let $a \in F$. Then $x = pa - ap \in I$ and so $x = pxp = p(pa - ap)p = 0$, that is, p commutes with every element of F. It follows that $F \cong pFp \oplus (1 - p)F(1 - p)$ where 1 is the unit of F if F unital or of $\tilde{F}$ if F is nonunital (in the latter case, note that indeed $(1-p)F(1-p) \subset F$). In either case, $(1-p)F(1-p)$ has dimension less than F, so by the induction hypothesis, there are $k, n_2, n_3, \ldots, n_k \in \mathbb{N}$ such that $(1 - p)F(1 - p) \cong M_{n_2} \oplus \cdots \oplus M_{n_k}$. Thus $F = M_{n_1} \oplus M_{n_2} \oplus \cdots \oplus M_{n_k}$, as required (and in fact F was unital all along).   □

8.2   Inductive limits

Now we turn to the inductive limit, sometimes called direct limit, construction. The starting point for our construction is the enveloping C*-algebra of a *-algebra that is equipped with a C*-seminorm. Recall that the definition for a C*-seminorm was given in Definition 6.2.3.

8.2.1. Let $p : A \to \mathbb{R}_+$ be a C*-seminorm on a *-algebra A. Then $N = \ker(p)$ is a self-adjoint ideal in A, and this induces a C*-norm on the quotient A/N given by

$$\|a + N\| = p(a).$$

Let $B = \overline{A/N}^{\|\cdot\|}$ be the completion with respect to this norm. Define the multiplication and involution in the obvious way. This makes B into a C*-algebra (exercise). We call B the *enveloping* C*-*algebra* of (A, p).

The map $i : A \to B : a \to a + N$ is called the *canonical map* and the image of A under i is a dense *-subalgebra of B.

8.2.2. An *inductive sequence* of C*-algebras $(A_n, \varphi_n)_{n \in \mathbb{N}}$ consists of a sequence of C*-algebras $(A_n)_{n \in \mathbb{N}}$ and a sequence of connecting *-homomorphisms

$$(\varphi_n : A_n \to A_{n+1})_{n \in \mathbb{N}}.$$

8.2.3. Proposition. *Let* $(A_n, \varphi_n)_{n \in \mathbb{N}}$ *be an inductive sequence of* C*-*algebras. Let*

$$\mathcal{A} := \left\{ a = (a_j)_{j \in \mathbb{N}} \subset \prod_{j \in \mathbb{N}} A_j \,\middle|\, \exists N \in \mathbb{N} \text{ such that } a_{j+1} = \varphi_j(a_j) \text{ for all } j \geq N \right\}.$$

Then $\mathcal{A}$ *is a* *-algebra under pointwise operations and*

$$p : \mathcal{A} \to \mathbb{R}_+, \quad a \mapsto \lim_{k \to \infty} \|a_k\|_{A_k},$$

is a C*-*seminorm on* $\mathcal{A}$.

Proof. Exercise.   □

8.2.4. Definition. Let $(A_n, \varphi_n)_{n \in \mathbb{N}}$ be an inductive sequence of C*-algebras. The *inductive limit* of $(A_n, \varphi_n)_{n \in \mathbb{N}}$, written $\varinjlim(A_n, \varphi_n)$, (or simply $\varinjlim A_n$ if it is clear what the maps should be) is the enveloping C*-algebra of $(\mathcal{A}, p)$, where $\mathcal{A}$ and p are the *-algebra and C*-seminorm, respectively, as defined in Proposition 8.2.3.

8.2.5. Let $(A_n, \varphi_n)_{n \in \mathbb{N}}$ be an inductive sequence of C*-algebras and let $A = \varinjlim A_n$ be the inductive limit. It will be useful to describe maps between nonadjacent C*-algebras in the inductive sequence, as well as from each A_n in the sequence to A. Thus we define, for $n < m$

$$\varphi_{n,m} : A_n \to A_m$$

to be given by the composition

$$\varphi_{n,m} := \varphi_{m-1} \circ \cdots \circ \varphi_{n+1} \circ \varphi_n.$$

If $a \in A_n$, then define $(a_j)_{j \in \mathbb{N}} \subset \prod_{j \in \mathbb{N}} A_j$ by

$$a_j = \begin{cases} 0 & j < n, \\ a & j = n, \\ \varphi_{n,j-1}(a) & j > n. \end{cases}$$

Clearly $(a_j)_{j \in \mathbb{N}} \in \mathcal{A}$. From this we define the map

$$\varphi^{(n)} : A_n \to A$$

by $\varphi^{(n)}(a) = \iota((a_j)_{j \in \mathbb{N}})$ where $\iota : \mathcal{A} \to \mathcal{A}/N_p$ is the canonical map from $\mathcal{A}$ into its enveloping C*-algebra A. From this we get, for every $n, m \in \mathbb{N}$ with $n < m$, the commutative diagram

Observe that for $A = \varinjlim (A_n, \varphi_n)$ as above, we have $A \cong \overline{\bigcup_{n \in \mathbb{N}} \varphi^{(n)}(A)}$. We also get the following universal property.

8.2.6. Theorem. *Let* $(A_n, \varphi_n)_{n \in \mathbb{N}}$ *be an inductive sequence of* C*-algebras with limit $A = \varinjlim A_n$. *Suppose there is a* C*-algebra B *and for every* $n \in \mathbb{N}$ *there are* *-homomorphisms* $\psi^{(n)} : A_n \to B$ *making the diagrams*

$$
\begin{array}{ccc}
A_n & \xrightarrow{\varphi_n} & A_{n+1} \\
& \searrow{\scriptstyle \psi^{(n)}} & \downarrow{\scriptstyle \psi^{(n+1)}} \\
& & B
\end{array}
$$

commute. Then there is a unique *-homomorphism* $\psi : A \to B$ *making the diagrams*

$$
\begin{array}{ccc}
A_n & \xrightarrow{\varphi^{(n)}} & A \\
& \searrow{\scriptstyle \psi^{(n)}} & \downarrow{\scriptstyle \psi} \\
& & B
\end{array}
$$

commute.

Proof. Let B and the *-homomorphisms $\psi^{(n)} : A_n \to B$ be given. If $(a_j)_{j \in \mathbb{N}} \in \mathcal{A}$, then there is $N \in \mathbb{N}$ such that for every $j \geq N$ we have $a_{j+1} = \varphi_j(a_j)$. By commutativity of the first diagram we have $\psi^{(N)}(a_N) = \psi^{(j)}(a_j)$ for every $j \geq N$.

Suppose that $a \in A_n$, $b \in A_m$ and $\varphi^{(n)}(a) = \varphi^{(m)}(b) \in \varinjlim A_n$. If $n \leq m$ then $\varphi^{(m)} \circ \varphi_{n,m}(a) = \varphi^{(m)}(b)$ by commutativitiy of the diagram in 8.2.5. It follows that $\lim_{k \to \infty, k \geq m} \| \varphi_{n,k}(a) - \varphi_{m,k}(b) \| = 0$. Thus

$$\lim_{k \to \infty} \| \psi^{(k)}(\varphi_{n,k}(a)) - \psi^{(k)}(\varphi_{m,k}(b)) \| = 0.$$

The above shows that $\psi : \iota(\mathcal{A}) \to B$, defined on each $(a_j)_{j \in \mathbb{N}}$ by $\iota \circ \psi^N(a_N)$ for sufficiently large N, is well defined and extends to a *-homomorphism $\psi : A \to B$, making the second diagram commute. $\qquad \square$

8.2.7. Proposition. *Let $A = \varinjlim(A_n, \varphi_n)$ be the inductive limit of a sequence of C*-algebras and let B be a C*-algebra. For every $n \in \mathbb{N}$, let $\psi^{(n)} : A_n \to B$ be a *-homomorphism satisfying $\psi^{(n+1)} \circ \varphi_n = \psi^{(n)}$ and let $\psi : A \to B$ be the induced *-homomorphism. Then*

(i) *ψ is injective if and only if $\ker(\psi^{(n)}) \subset \ker(\varphi^{(n)})$ for every $n \in \mathbb{N}$;*

(ii) *ψ is surjective if and only if $B = \overline{\cup_{j=1}^{\infty} \psi^{(n)}(A_n)}$.*

Proof. First we show (i). Since ψ is a *-homomorphism, it is injective if and only if it is isometric. The increasing union $\bigcup_{n \in \mathbb{N}} \varphi^{(n)}(A_n)$ is dense in A, so it is enough to show that ψ is isometric when restricted to $\bigcup_{n \in \mathbb{N}} \varphi^{(n)}(A_n)$. This in turn holds if and only if ψ is isometric when restricted to $\varphi^{(n)}(A_n)$ for every $n \in \mathbb{N}$. Let $n \in \mathbb{N}$. By Theorem 8.2.6, we have $\psi^{(n)} = \psi \circ \varphi^{(n)}$. Let $y = \varphi^{(n)}(x) \in \varphi^{(n)}(A_n)$. Then $\psi|_{\varphi^{(n)}(A_n)}(y) = 0$ if and only if $\psi^{(n)}(x) = 0$. So $x \in \ker(\psi^{(n)}) \subset \ker(\varphi^{(n)})$, which is to say $y = 0$ and $\psi|_{\varphi^{(n)}(A_n)}$ is injective. Conversely, if ψ is injective, then so is $\psi|_{\varphi^{(n)}(A_n)} = \psi \circ \varphi^{(n)}$. Thus if $y \in \ker(\psi^{(n)})$ we have $\psi|_{\varphi^{(n)}(A_n)}(y) = \psi \circ \varphi^{(n)}(y) = 0$ implies $\varphi^{(n)}(y) = 0$. Thus $\ker(\psi^{(n)}) \subset \ker(\varphi^{(n)})$.

For (ii), the image of A under ψ is

$$\psi(A) = \psi\left(\overline{\bigcup_{n \in \mathbb{N}} \varphi^{(n)}(A_n)}\right) = \overline{\bigcup_{n \in \mathbb{N}} \psi \circ \varphi^{(n)}(A_n)} = \overline{\bigcup_{n \in \mathbb{N}} \psi^{(n)}(A_n)}.$$

Thus ψ is surjective if and only if $B = \overline{\bigcup_{n \in \mathbb{N}} \psi^{(n)}(A_n)}$. $\qquad\square$

8.2.8. Theorem. *Let $(A_n, \varphi_n)_{n \in \mathbb{N}}$ be an inductive limit of simple C*-algebras. Then $\varinjlim A_n$ is simple.*

Proof. A C*-algebra A is simple if and only if, whenever B is another nonzero C*-algebra and $\psi : A \to B$ is a surjective *-homomorphism, then φ is injective (Exercise 2.5.24). Suppose then that $\psi : \varinjlim A_n \to B$ is a surjection onto a nonzero C*-algebra B. For any $n \in \mathbb{N}$, $\varphi^{(n)}(A_n) \subset A$ is the image of a simple C*-algebra and so is also simple. Thus $\psi|_{\varphi^{(n)}(A_n)} : \varphi^{(n)}(A_n) \to B$ is either zero or injective. Since $A \cong \overline{\bigcup_{n \in \mathbb{N}} \varphi^{(n)}(A)}$ and $\bigcup_{n \in \mathbb{N}} \varphi^{(n)}(A)$ is dense, there must be some $N \in \mathbb{N}$ and $a \in \varphi^{(N)}(A)$ such that $\psi(a) \neq 0$. In this case $\psi|_{\varphi^{(N)}(A_N)}$ must be injective.

Now, if $k < N$ then $\varphi^{(k)}(A_k) \subset \varphi^{(N)}(A_N)$. Thus $\psi|_{\varphi^{(k)}(A_k)}$ is nonzero and hence injective. If $k > N$ then $\varphi^N(A_N) \subset \varphi^k(A_k)$. Then $a \in \varphi^k(A_k)$, so again $\psi|_{\varphi^{(k)}(A_k)}$ is nonzero and hence injective. Thus $\psi : A \to B$ is injective on a dense subset, hence injective. $\qquad\square$

8.2.9. We can also have simple inductive limits of nonsimple algebras, as we will see later. For that, we will want to know something about the ideal structure of an inductive limit.

Proposition. *Let $(A_n, \varphi_n)_{n \in \mathbb{N}}$ be an inductive limit with injective connecting maps. Let $A = \varinjlim(A_n, \varphi_n)$ and suppose that $J \subset A$ is an ideal. Then*

$$J \cong \overline{\bigcup_{n \in \mathbb{N}} J \cap \varphi^{(n)}(A_n)}.$$

Proof. Exercise.   □

8.3 UHF algebras

In this section we will see our first examples of inductive limits, which we will build from the easiest building blocks we can find: the matrix C*-algebras, M_n for $n < \infty$. These also allow us a first glimpse at a classification result for a given class of simple C*-algebras.

8.3.1. Consider the C*-algebras M_n and M_m. Suppose that we have a unital *-homomorphism $\varphi : M_n \to M_m$. Let tr_n and tr_m denote the unique normalised tracial states on M_n and M_m respectively. Then, since $\mathrm{tr}_m \circ \varphi$ is a tracial state, we must have $\mathrm{tr}_m \circ \varphi = \mathrm{tr}_n$. Let f be a rank one projection in M_n. Then $\varphi(f)$ is a projection M_m so we can find m orthogonal rank one projections e_i in M_m, $1 \le i \le m$, such that, for some $1 \le k \le m$ we have $\varphi(f) = \sum_{i=1}^m e_i$. Then

$$1/n = \mathrm{tr}_m \circ \varphi(f) = \mathrm{tr}_m \left(\sum_{i=1}^k e_i \right) = k/m.$$

So we must have $kn = m$, that is, n divides m. Thus φ maps the $n \times n$-matrix a to an $m \times m$ block matrix by copying a k times down the diagonal:

$$\varphi : M_n \to M_m, \quad a \mapsto \left. \begin{pmatrix} a & 0 & \cdots & 0 \\ 0 & a & \cdots & 0 \\ \vdots & & \ddots & \vdots \\ 0 & 0 & \cdots & a \end{pmatrix} \right\} k \text{ times.}$$

8.3.2. Definition. Let $(n_i)_{i \in \mathbb{N}}$ be a sequence of positive integers such that, for each i, n_i divides n_{i+1}, and let $\varphi_i : M_{n_i} \to M_{n_{i+1}}$ be unital *-homomorphisms. The inductive limit of the inductive system (M_{n_i}, φ_i),

$$\mathcal{U} := \varinjlim(M_{n_i}, \varphi_i)$$

is called a *uniformly hyperfinite* (UHF) algebra.

8.3.3. The following proposition is immediate from results of the previous section.

Proposition. *A UHF algebra is simple.*

8.3.4. It is also true that not only are UHF algebras simple like the matrix algebras, but also, like matrix algebras, they have a unique tracial state.

Proposition. *If $\mathcal{U}$ is a UHF algebra, then $\mathcal{U}$ has a unique tracial state.*

Proof. Let $\mathcal{U} := \varinjlim (M_{n_i}, \varphi_i)$ be an inductive limit decomposition for $\mathcal{U}$. Let $a := (a_i)_{i \in \mathbb{N}}$ be a sequence where each $a_i \in M_{n_i}$ and such that, for some $N > 0$, we have $\varphi_i(a_i) = a_{i+1}$. Then, for every $i \geq N$, we have $\mathrm{tr}_{n_{i+1}}(\varphi_i(a_i)) = \mathrm{tr}_{n_{i+1}}(a_{i+1})$. Since $\mathrm{tr}_{n_{i+1}} \circ \varphi_i$ is a tracial state on M_{n_i}, by uniqueness we have $\mathrm{tr}_{n_{i+1}} \circ \varphi_i = \mathrm{tr}_{n_{i+1}}$, and so $\mathrm{tr}_{n_{i+1}}(a_{i+1}) = \mathrm{tr}_{n_i}(a_i)$. Thus we may define

$$\tau(a) := \lim_{n \to \infty} \mathrm{tr}_{n_i}(a_i),$$

that is evidently a continuous linear functional with $\tau(1_\mathcal{U}) = 1$ and that satisfies the tracial condition $\tau(ab) = \tau(ba)$. Since sequences of this type are dense, we may extend τ to all of $\mathcal{U}$, giving us a tracial state on $\mathcal{U}$.

We leave it as an exercise to show that τ must be unique. □

8.3.5. A *supernatural number* $\mathfrak{p}$ is given by the infinite product $\mathfrak{p} = \prod_{p \text{ prime}} p^{k_p}$ where $k_p \in \mathbb{N} \cup \{\infty\}$. Every natural number is thus a supernatural number. A supernatural number is of *infinite type* if, for every prime p, we have either $k_p = 0$ or $k_p = \infty$.

To every UHF algebra $\mathcal{U} := \varinjlim M_{n_i}$, we may associate a supernatural number $\mathfrak{p}$ as follows. Given a prime p, let

$$k_p = \sup\{k \in \mathbb{N} \mid \text{ there exists m such that } p^k \text{ divides } n_m n_{m-1} \cdots n_1\}.$$

Then $\mathfrak{p} := \prod_{p \text{ prime}} p^{k_p}$ is a supernatural number. We will see that this supernatural number uniquely determines $\mathcal{U}$, up to *-isomorphism, among all UHF algebras. First, we need two technical lemmas.

8.3.6. Lemma. *Let A be a unital C*-algebra and suppose that $p, q \in A$ are projections satisfying $\|p - q\| < 1$. Let $v := 1_A - p - q + 2qp$. Then,*

(i) *v^*v is invertible in A,*

(ii) *$u := v(v^*v)^{-1/2}$ is a unitary satisfying $\|1_A - u\| \leq \sqrt{2}\|p - q\|$,*

(iii) *$q = upu^*$.*

Proof. We have

$$v^*v = (1_A - p - q + 2pq)(1_A - p - q + 2qp) = 1_A - (p - q)^2,$$

and also

$$vv^* = 1_A - (q - p)^2 = v^*v,$$

so v itself is a normal element. Since $\|p - q\| < 1$, we have $\|(p - q)^2\| < 1$ and thus v^*v is invertible by Theorem 1.2.4, showing (i).

Now, as v is normal, we have

$$v(v^*(v^*v)^{-1}) = v^*v(v^*v)^{-1} = 1_A \quad \text{and} \quad ((v^*v)^{-1}v^*)v = 1_A,$$

showing that u is invertible. Thus $uu^* = v(v^*v)^{-1/2}(v^*v)^{-1/2}v^* = v(v^*v)^{-1}v^* = 1_A$ and $u^*u = (v^*v)^{-1/2}v^*v(v^*v)^{-1/2} = 1_A$, so u is a unitary.

Let $v = v_1 + iv_2$ and $u = u_1 + iu_2$ where v_i, u_i, $i = 1, 2$ are the self-adjoint elements as given by Exercise 3.4.2. Then

$$v_1 = (1/2)(v^* + v) = (1/2)(1_A - p - q + 2pq + 1_A - p - q + 2qp) = 1_A - (p-q)^2 = v^*v,$$

so

$$u_1 = v_1(v^*v)^{-1/2} = (v^*v)^{1/2}.$$

Thus, since $v^*v \leq 1$, we have $0 \leq 1 - (v^*v)^{-1/2} \leq 1_A - v^*v$ and

$$\begin{aligned}
\|1_A - u\|^2 &= \|(1_A - u)^*(1_A - u)\| = \|2 - u^* - u\| \\
&= \|2 - 2u_1\| = 2\|1_A - (v^*v)^{1/2}\| \\
&\leq 2\|1_A - v^*v\| = 2\|(p-q)^2\| = 2\|p-q\|^2.
\end{aligned}$$

It follows that $\|1_A - u\| \leq \sqrt{2}\|p - q\|$, showing (ii).

Finally, for (iii), note that

$$vp = (1_A - p - q + 2qp)p = qp = q(1_A - p - q + 2pq) = qv,$$

so

$$pv^*v = (vp)^*v = (qv)^*v = v^*qv = v^*vp,$$

which is to say that p commutes with v^*v. Thus p commutes with $(v^*v)^{-1/2}$ (Exercise 2.5.25), so

$$upu^* = v(v^*v)^{-1/2}pu^* = vp(v^*v)^{-1/2}u^* = vp(v^*v)^{-1}v^* = qv(v^*v)^{-1}v^* = q,$$

showing (iii). □

8.3.7. Lemma. *Let A be a C*-algebra and suppose that $a \in A$ with $\|a\| \geq 1/2$ is self-adjoint and satisfies $\|a^2 - a\| < 1/4$. Then there exists a projection p in the C*-subalgebra generated by a such that $\|a - p\| \leq 2\|a^2 - a\|$.*

Proof. Let $\delta := \|a - a^2\| < 1/4$. If $\lambda \in \mathbb{R}$ satisfies $\|\lambda - \lambda^2\| \leq \delta < 1/4$ then $\lambda \in [-2\delta, 2\delta] \cup [1 - 2\delta, 1 + 2\delta]$. Thus $\text{sp}(a) \subseteq [-2\delta, 2\delta] \cup [1 - 2\delta, 1 + 2\delta]$. The function f given by

$$f(t) = \begin{cases} 0 & t \leq 2\delta, \\ \text{linear} & 2\delta \leq t \leq 1 - 2\delta, \\ 1 & t \geq 1 - 2\delta; \end{cases}$$

is continuous on $\text{sp}(a)$. Let $p = f(a)$. Then $f = f^2 = \overline{f}$, so p is a projection. Moreover, $|f(t) - t| \leq 2\delta$ for all $t \in [-2\delta, 2\delta] \cup [1 - 2\delta, 1 + 2\delta]$, so

$$\|p - a\| \leq 2\delta = 2\|a - a^2\|,$$

which completes the proof. $\square$

8.3.8. Theorem. *Suppose $\mathcal{U}_1 = \varinjlim(M_{n_i}, \varphi_i)$ and $\mathcal{U}_2 = \varinjlim(M_{m_i}, \psi_1)$ are two UHF algebras and let $\mathfrak{p}$ and $\mathfrak{q}$ be the respective associated supernatural numbers as in 8.3.5. Then $\mathcal{U}_1 \cong \mathcal{U}_2$ implies $\mathfrak{p} = \mathfrak{q}$. Thus UHF algebras are classified by their associated supernatural numbers.*

Proof. Suppose that $\mathcal{U}_1 \cong \mathcal{U}_2$ and denote by $\rho : \mathcal{U}_1 \to \mathcal{U}_2$ some *-isomorphism. Let τ_i denote the unique tracial states on $\mathcal{U}_i$, $i \in \{1, 2\}$.

Put

$$k_p := \sup\{k \in \mathbb{N} \mid \text{ there exists } r \text{ such that } p^k \text{ divides } n_r n_{r-1} \cdots n_1\},$$

and

$$l_p := \sup\{k \in \mathbb{N} \mid \text{ there exists } r \text{ such that } p^k \text{ divides } m_r m_{r-1} \cdots n_1\},$$

so that $\mathfrak{p} = \prod_{p \text{ prime}} p^{k_p}$ and $\mathfrak{q} = \prod_{p \text{ prime}} p^{l_p}$. We will show that $k_p \leq l_p$ for every prime p. Then, by symmetry, we will have $\mathfrak{p} = \mathfrak{q}$. To do so, observe that it is in turn enough to show that for each $i \in \mathbb{N}$ there is $j \in \mathbb{N}$ such that $r_1(i) := n_i n_{i-1} \cdots n_1$ divides $r_2(j) := m_j m_{j-1} \cdots m_1$.

Let $i \in \mathbb{N}$ and let p be a rank one projection in M_{n_i}. Observe that $\tau_1 \circ \varphi^{(i)}$ is a tracial state on M_{n_i}, hence $\tau_1 \circ \varphi^{(i)} = \text{tr}_{n_i}$. Thus

$$\tau(\varphi^{(i)}(p)) = r_1(i)^{-1}.$$

Since $\varphi^{(i)}(p)$ is a projection, we have $\rho \circ \varphi^{(i)}(p) \in \mathcal{U}_2$ is also a projection. Now, $\bigcup_{i \in \mathbb{N}} \psi^{(i)}(M_{n_i})$ is dense in $\mathcal{U}_2$. Thus there is some $j \in \mathbb{N}$ and some element $a \in M_{m_j}$, which we may assume is self-adjoint, satisfying $\|\psi^{(j)}(a) - \rho \circ \varphi^{(i)}(p)\| < 1/16$. Then

$$\|\psi^{(j)}(a)^2 - \psi^{(j)}(a)\| \leq \|\psi^{(j)}(a)^2 - \rho \circ \varphi^{(i)}(p)\| + \|\rho \circ \varphi^{(i)}(p) - \psi^{(j)}(a)\| < 1/8.$$

By Lemma 8.3.7, there exists a projection $q \in M_{m_j}$ satisfying $\|a - q\| < 1/2$. It follows that

$$\begin{aligned}
\|\rho(\varphi^{(i)}(p)) - \psi^{(m)}(q)\| &\leq \|\rho(\varphi^{(i)}(p)) - \psi^{(m)}(a)\| + \|\psi^{(m)}(a) - \psi^{(m)}(q)\| \\
&< 1/8 + 1/2 \\
&< 1.
\end{aligned}$$

Now we may apply Lemma 8.3.6 to see that the projections $\rho(\varphi^{(i)}(p))$ and $\psi^{(m)}(q)$ are unitarily equivalent, which means that

$$\tau_2(\psi^{(m)}(q)) = \tau_2(\rho(\varphi^{(i)}(p))) = \tau_1(\varphi^{(i)}(p)) = r_1(i)^{-1}.$$

But M_{m_j} has a unique tracial state tr_{m_j}, so $\tau_2(\psi^{(m)}(q)) = \mathrm{tr}_{m_j}(p)$. As in 8.3.1, we see that $\tau_2(\psi^{(m)}(q)) = d/(r_2(j))$ for some d. Thus, $r_2(j) = dr_1(i)$, which is to say, $r_1(i)$ divides $r_2(j)$. □

8.3.9. Approximately finite-dimensional (AF) algebras generalise UHF algebras. Where UHF algebras are inductive limit of matrix algebras, AF algebras are inductive limits of arbitrary finite-dimensional C*-algebras, which, as we saw in Theorem 8.1.2, are finite direct sums of matrix algebras. This class of inductive limits turns out to be quite a lot larger than the subclass of UHF algebras. For example, an AF algebra need not be either simple or unital.

Definition. An approximately finite-dimensional (AF) algebra is the inductive limit of a sequence (F_n, φ_n) where F_n is a finite-dimensional C*-algebra for every $n \in \mathbb{N}$.

8.3.10. Note that a UHF algebra is an AF algebra, but the opposite need not be the case. For example, the C*-algebra $\mathcal{K}$ of compact operators on a separable Hilbert space is an AF algebra but is not even unital. Still, AF algebras are a very nice class of C*-algebras and are quite tractable: we can say a lot about their structure. We saw that a UHF algebra is uniquely determined by its associated supernatural number. For AF algebras, the situation is slightly more complicated. In the simple unital case, we use the pointed, ordered K_0-group which will be introduced in Chapter 12. Since K_0 respects inductive limits (Theorem 12.2.4), this is a computable invariant for AF algebras. We will have more to say on this later on, in the meantime, we have another nice structural property of AF algebras: they are *stably finite*.

8.3.11. Recall from 3.3.1 that a partial isometry v in a C*-algebra A is an element such that $v = vv^*v$.

Definition. Let A be a C*-algebra. If $p, q \in A$ are projections, then p is *Murray–von Neumann equivalent* to q if there is a partial isometry $v \in A$ such that $v^*v = p$ and $vv^* = q$. The projection p is *Murray–von Neumann subequivalent* to q if there is a partial isometry $v \in A$ such that $v^*v = p$ and $vv^* \leq q$.

   A projection $p \in A$ is called *finite* if p is not Murray–von Neumann equivalent to a proper subprojection of itself, that is, there is no $v \in A$ with $v^*v = p$ and $vv^* \leq p$ but $vv^* \neq p$. Otherwise p is said to be *infinite*.

8.3.12. Definition. A unital C*-algebra A is called *finite* if 1_A is finite. A unital C*-algebra A is called *stably finite* if $M_n(A)$ is finite for every $n \in \mathbb{N}$.

   In the case that $A = \mathcal{B}(H)$, Murray–von Neumann equivalence measures the dimension of the range of the projection in the sense that projections $p, q \in$

$\mathcal{B}(H)$ are Murray–von Neumann equivalent if and only if $\dim(pH) = \dim(qH)$ (exercise). Note that this includes the possibility that the dimensions are both infinite. Moreover, we have $p \in \mathcal{B}(H)$ is finite if and only if $\dim(pH) < \infty$. Thus we see that $\mathcal{B}(H)$ is finite if and only if H is finite. However there are many examples of C*-algebras which are finite – in fact stably finite – and infinite-dimensional. In particular, this is true of AF algebras, the proof of which is not so difficult once we have the next proposition in hand.

8.3.13. An *isometry* in a unital C*-algebra A is an element $s \in A$ with $s^*s = 1_A$. Clearly any unitary in A is an isometry, but an isometry need not be a unitary in general. (See exercises.) However, if this is the case, then A is finite.

Proposition. *Let A be a unital C*-algebra. Then A is finite if and only if every isometry in A is a unitary.*

Proof. Suppose that there is $p \in A$ with $p \le 1_A$ and $v \in A$ such that $v^*v = 1_A$ and $vv^* = p$. Since $v^*v = 1_A$, v is an isometry and hence unitary. Thus we have $p = vv^* = 1_A$ which shows that 1_A is finite.

For the converse, suppose A is finite and that v is an isometry, $v^*v = 1_A$. It is easy to check that $1_A - vv^*$ is itself a projection, so in particular $1_A - vv^* \ge 0$. Then $vv^* \le 1_A$ and since 1_A is finite, we must have $vv^* = 1_A$, which is to say, v is a unitary. □

8.3.14. Theorem. *If A is a unital AF algebra, then A is stably finite.*

Proof. Exercise. (See hints at the end of the chapter.) □

8.4 A collection of perturbation arguments

We will prove that AF algebras can be characterized not only by their inductive limit structure, but also by a local condition, as long as the AF algebra is separable. (In fact, here we have only defined separable AF algebras. How would you define an AF algebra which is not necessarily separable?) To get there, we will require a number of perturbation lemmas. These will be useful not only in this chapter but throughout the rest of the notes, so we collect them here in their own section. Many of the lemmas here will rely on similar tricks; readers may wish to attempt to prove a statement themselves before reading through the given proof. These fun but technical perturbation arguments are ubiquitous in the literature, so it is important to get a feel for them.

8.4.1. Lemma. *Let A be a C*-algebra. Suppose that $b \in A$ is self-adjoint and $p \in A$ is a projection such that $\|p - b\| < 1/2$. Then there is a projection q in the C*-subalgebra generated by b, $\mathrm{C}^*(b) \subset A$, with $\|q - b\| \le 2\|p - b\|$. Moreover p and q are Murray–von Neumann equivalent in A, and if A is unital there is a unitary $u \in A$ such that $upu^* = q$.*

Proof. Let $\delta := \|b - p\|$. Suppose that $\lambda \in \mathbb{R}$ and $\mathrm{dist}(\lambda, \mathrm{sp}(p)) > \delta$. Then, by Exercise 2.5.23, we have $\|(\lambda - p)^{-1}\| = 1/(\mathrm{dist}(\lambda, \mathrm{sp}(p)))$. Suppose $a \in A$ satisfies $\|a - (\lambda - p)\| < \mathrm{dist}(\lambda, \mathrm{sp}(p))$. Then $a(\lambda - p)^{-1}$ is invertible since $\|a(\lambda - p)^{-1} - 1\| < 1$, which in turn implies that a itself is invertible in $\tilde{A}$. In particular,

$$\|(\lambda - b) - (\lambda - p)\| = \|b - p\| = \delta < \mathrm{dist}(\lambda, \mathrm{sp}(p)).$$

Thus $\lambda \notin \mathrm{sp}(b)$ and it follows that

$$\mathrm{sp}(b) \subset [-\delta, \delta] \cup [1 - \delta, 1 + \delta].$$

Define

$$f(t) = \begin{cases} 1 & t \geq 1 - \delta \\ \text{linear} & t \in [\delta, 1 - \delta] \\ 0 & t \leq \delta. \end{cases}$$

Then f is continuous on $\mathrm{sp}(b)$ and $q := f(b) \in \mathrm{C}^*(b)$ is a projection. Moreover, since $\|t - f(t)\| \leq \delta$, we have $\|q - b\| \leq \delta$. Thus

$$\|p - q\| = \|p - b + q - b\| \leq 2\delta = 2\|p - b\|.$$

Finally, since $\|p - b\| < 1/2$, we have $\|p - q\| < 1$. Thus, by Lemma 8.3.6, there is a unitary $u \in M(A)$, the multiplier algebra of A (2.2.7), such that $upu^* = q$. Since $A \subset M(A)$ is an ideal, $v := up \in A$ and satisfies $vv^* = q$ and $v^*v = p$, showing that p and q are Murray–von Neumann equivalent. If A was unital to begin with, then $A = M(A)$ (Theorem 2.2.10) so $u \in A$, and the proof is complete. $\square$

8.4.2. Lemma. *For any $\epsilon > 0$ and any $n \in \mathbb{N}$ there is a $\delta := \delta(\epsilon, n) > 0$ such that the following holds: For any C^*-algebra A and any C^*-subalgebra $B \subset A$, if $p_1, \ldots, p_n \in A$ is a set of n mutually orthogonal projections and $b_1, \ldots, b_n \in B$ satisfy $\|p_i - b_i\| < \delta$ for every $1 \leq i \leq k$, then there are mutually orthogonal projections $q_1, \ldots, q_n \in B$ with $\|p_i - q_i\| < \epsilon$ for every $0 \leq i \leq k$.*

Moreover, if A is unital, $1_A \in B$ and $\sum_{i=1}^n p_i = 1_A$, then we can choose $q_1, \ldots, q_n$ so that $\sum_{i=1}^n q_i = 1_A$.

Proof. The proof is by induction. We may assume that $\epsilon < 1/2$. Let $n = 1$. In the case that A is unital $p_1 = 1_A$ and $1_A \in B$. So we put $q_1 = 1_A$ and we are done. Otherwise, let $\delta < \min\{\epsilon/2, 1/2\}$, and suppose $b_1 \in B$ and $p_1 \in A$ is a projection with $\|p_1 - b_1\| < \delta$. Then by Lemma 8.4.1, there is a projection $q_1 \subset \mathrm{C}^*(b_1) \subset B$ with $\|p_1 - q_1\| < \epsilon$.

Now let $n > 1$ and suppose the lemma holds for $n - 1$. Let

$$\delta(\epsilon, n) := \min\{\epsilon/16, \delta(\epsilon/4, n - 1), 1\},$$

and let $b_1, \ldots, b_n \in B$. Suppose there are projections $p_1, \ldots, p_n$ in A such that $\|p_i - b_i\| < \delta(\epsilon/(4 + 4\|b_n\|), n - 1)$ for $1 \leq i \leq n$. Then there are mutually orthogonal projections $q_1, \ldots, q_{n-1}$ in B which satisfy $\|p_i - q_i\| < \epsilon/4$. Define

$p := \sum_{i=1}^{n-1} p_i$ and $q := \sum_{i=1}^{n-1} q_i$. We will find a projection q_n in $(1_{\tilde{A}} - q)B(1_{\tilde{A}} - q)$ that is close to b_n. This way q_n will be orthogonal to each of $q_1, \ldots, q_{n-1}$. We have

$$
\begin{aligned}
\|p_n - (1_{\tilde{A}} - q)b_n(1_{\tilde{A}} - q)\| &= \|(1_{\tilde{A}} - p)p_n(1_{\tilde{A}} - p) - (1_{\tilde{A}} - q)b_n(1_{\tilde{A}} - q)\| \\
&\leq \|(1_{\tilde{A}} - p)p_n(1_{\tilde{A}} - p) - (1_{\tilde{A}} - q)p_n(1_{\tilde{A}} - p)\| \\
&\quad + \|(1_{\tilde{A}} - q)p_n(1_{\tilde{A}} - p) - (1_{\tilde{A}} - q)b_n(1_{\tilde{A}} - p)\| \\
&\quad + \|(1_{\tilde{A}} - q)b_n(1_{\tilde{A}} - p) - (1_{\tilde{A}} - q)b_n(1_{\tilde{A}} - q)\| \\
&\leq \|p - q\| + \|b_n\|\|p - q\| + \|p_n - b_n\|.
\end{aligned}
$$

Since $\|b_n\| \leq \|b_n - p_n\| + \|p_n\| < 3$, we have

$$
\begin{aligned}
\|p - q\| + \|b_n\|\|p - q\| + \|p_n - b_n\| &< 4\|p - q\| + \|p_n - q_n\| \\
&< \epsilon/4 + \epsilon/4 \\
&= \epsilon/2.
\end{aligned}
$$

Thus by Lemma 8.4.1 there is a projection $q_n \in (1_{\tilde{A}} - q)B(1_{\tilde{A}} - q)$ with $\|p_n - q_n\| < \epsilon$. If A is unital and $\sum_{i=1}^{n} p_n = 1_A$, then

$$
\left\| 1_A - \sum_{i=1}^{n} q_i \right\| \leq n\epsilon.
$$

So, again by Lemma 8.4.1, provided ϵ is small enough, $\sum_{i=1}^{n} q_i$ is unitarily equivalent to 1_A, which is to say, $\sum_{i=1}^{n} q_i = 1_A$. $\qquad\square$

8.4.3. Lemma. *Let A be a unital C*-algebra and suppose that $p_1, \ldots, p_n$ and $q_1, \ldots, q_n$ are two families of projections with $p_i p_j = 0$ and $q_i q_j = 0$ for $1 \leq i \neq j \leq n$. For every $0 < \epsilon < 1$ there exists $\delta(\epsilon, n) > 0$ such that the following holds:*

If $\|q_i - p_i\| \leq \delta(\epsilon, n)$, then there exists a partial isometry v in the C-subalgebra generated by $p_1, \ldots, p_n, q_1, \ldots, q_n, 1_A$ such that*

$$
v^* p_i v = q_i, \quad v q_i v^* = p_i, \quad \text{and} \quad \|p_i - p_i v q_i\| < \epsilon/n, \quad 1 \leq i \leq n.
$$

Moreover, if $\sum_{i=1}^{n} p_i = 1_A$, then v is a unitary satisfying $\|v - 1_A\| < \epsilon$.

Proof. Let $\delta := \delta(\epsilon, n) = \sqrt{2}\epsilon/(4n)$. Applying Lemma 8.3.6, there exist unitaries $u_i \in C^*(p_i, q_i, 1_A)$ such that

$$
\|1_A - u_i\| < \sqrt{\delta}, \quad u_i q_i u_i^* = p_i, \quad 1 \leq i \leq n.
$$

Let $v := \sum_{i=1}^{n} p_i u_i q_i$. Then

$$
v^* v = \left(\sum_{i=1}^{n} q_i u_i^* p_i \right) \left(\sum_{i=1}^{n} p_i u_i q_i \right) = \sum_{i=1}^{n} q_i u_i^* p_i u_i q_i = \sum_{i=1}^{n} q_i,
$$

and similarly,

$$vv^* = \left(\sum_{i=1}^{n} p_i u_i q_i \right) \left(\sum_{i=1}^{n} q_i u_i^* p_i \right) = \sum_{i=1}^{n} p_i.$$

Since the sum of mutually orthogonal projections is again a projection, this shows that v is a partial isometry. For each i, $1 \le i \le n$ we have

$$v^* p_i v = \left(\sum_{j=1}^{n} q_j u_j^* p_j \right) p_i \left(\sum_{j=1}^{n} p_j u_j q_j \right) = q_i u_i^* p_i u_i q_i = q_i,$$

and a similar calculation gives $v q_i v^* = p_i$.

For $1 \le i \le n$ we estimate

$$\begin{aligned}
\|p_i - p_i v q_i\| &= \|p_i - p_i u_i q_i\| \\
&= \|p_i - p_i u_i p_i + p_i u_i p_i - p_i u_i q_i\| \\
&\le \|p_i\| \|1_A - u_i\| + \|p_i\| \|u_i\| \|p_i - q_i\| \\
&< \sqrt{\delta} + \delta \\
&= 2\epsilon/(4n) + \sqrt{2}\epsilon/(4n) < \epsilon/n.
\end{aligned}$$

Now suppose that $\sum_{i=1}^{n} p_i = 1$. Since $\epsilon < 1$, we have

$$\left\| 1_A - \sum_{i=1}^{n} q_i \right\| = \left\| \sum_{i=1}^{n} p_i - \sum_{i=1}^{n} q_i \right\| < \sqrt{2}\epsilon/4 < 1,$$

and so there is a unitary such that $\sum_{i=1}^{n} q_i = u^* 1_A u = 1_A$. Thus $vv^* = 1_A = v^* v$, which is to say, v is a unitary. Finally,

$$\begin{aligned}
\|1_A - v\| &= \left\| \sum_{i=1}^{n} p_i - \sum_{i=1}^{n} p_i u_i q_i \right\| \\
&\le n \cdot \left(\max_{i=1,\dots,n} \|p_i - p_i u_i q_i\| \right) \\
&< \epsilon,
\end{aligned}$$

which completes the proof. $\square$

8.4.4. Lemma. *For any $\epsilon > 0$ and $n \in \mathbb{N} \setminus \{0\}$, there is $\delta := \delta(\epsilon, n) > 0$ satisfying the following: For any C*-algebra A and projections $p_1, \dots, p_n \in A$ such that $\|p_i p_j\| < \delta$, $1 \le i \ne j \le n$, there exist mutually orthogonal projections $q_1, \dots, q_n \in A$ such that $\|p_i - q_i\| < \epsilon$, $1 \le i \le n$.*

Proof. The proof is by induction on n. If $n = 1$ then we simply take $p_1 = q_1$ and we are done. Let $n \ge 1$ and assume that the lemma holds for every $k \le n$. Without loss of generality, we may assume that $\epsilon < 1/3$. Define

$$\delta(\epsilon, n+1) := \min\{1/3, \epsilon/(12n), \delta(\epsilon/(12n), n)\}.$$

Let $p_1, \ldots, p_{n+1}$ be projections in A satisfying $\|p_i p_j\| < \delta(\epsilon, n+1)$, $1 \leq i \neq j \leq n+1$. Since $\delta(\epsilon, n+1) \leq \delta(\epsilon/(12n), n)$ we can find pairwise orthogonal projections $q_1, \ldots, q_n$ with $\|q_i - p_i\| < \epsilon/(12n)$, $1 \leq i \leq n$. Let $q := \sum_{i=1}^n q_i$, which itself is a projection in A since the q_i, $1 \leq i \leq n$ are pairwise orthogonal. We have

$$
\begin{aligned}
\|p_{n+1} - (1_{\tilde{A}} - q)p_{n+1}(1_{\tilde{A}} - q)\| &= \|qp_{n+1} - qp_{n+1}q + p_{n+1}q\| \\
&\leq 3\|qp_{n+1}\| \\
&\leq 3\sum_{i=1}^n \|q_i p_{n+1}\| \\
&< 3\left(\sum_{i=1}^n \|p_i p_{n+1}\| + \epsilon/(12n)\right) \\
&< \epsilon/4 + \epsilon/(4n) \\
&< \epsilon/2.
\end{aligned}
$$

Since $\epsilon < 1/3$, by Lemma 8.4.1, there is a projection $q_{n+1} \in (1_{\tilde{A}} - q)A(1_{\tilde{A}} - q)$ such that $\|q_{n+1} - p_{n+1}\| < \epsilon$. Finally, since $q_{n+1} = (1_{\tilde{A}} - q)q_{n+1}(1_{\tilde{A}} - q)$, we have $q_{n+1}q_i = 0$ for every $1 \leq i \leq n$. $\qquad\square$

8.4.5. Lemma. *Let A be a C*-algebra and $B \subset A$ a C*-subalgebra. Let $0 < \epsilon < 1$ and $\delta = \min\{1/5, \epsilon/(8 - 5\epsilon)\}$. Suppose that $p_1, p_2 \in A$ are orthogonal projections such that $v^*v = p_1$ and $vv^* = p_2$ for some $v \in A$. If there are orthogonal projections $q_1, q_2 \in B$ such that $\|p_i - q_i\| < \delta$ for $i = 1, 2$ and $b \in B$ with $\|b\| \leq 1$ and $\|v - b\| < \delta$, then there exists $w \in B$ such that*

$$
w^*w = q_1, \quad ww^* = q_2, \quad \|w - v\| < \epsilon.
$$

Proof. Let $x := q_2 b q_1$. Then $\|x\| \leq 1$ so $(x^*x)^{1/2} \geq x^*x$. Also,

$$
\begin{aligned}
\|q_1 - (x^*x)^{1/2}\| &\leq \|q_1 - x^*x\| \\
&= \|q_1 - q_1 b^* q_2 b q_1\| \\
&\leq \|q_1 - q_1 v^* q_2 v q_1\| + \|q_1 v^* q_2 v q_1 - q_1 b^* q_2 v q_1\| \\
&\quad + \|q_1 b^* q_2 v q_1 - q_1 b^* q_2 b q_1\| \\
&< \|q_1 - q_1 v^* p_2 v q_1 + q_1 v^* p_2 v q_1 - q_1 v^* q_2 v q_1\| + 2\delta \\
&< \|q_1 - q_1 p_1 q_1\| + 3\delta \\
&< \|q_1 - q_1 q_1 q_1\| + 4\delta \\
&= 4\delta.
\end{aligned}
$$

In particular, $\|q_1 - (x^*x)^{1/2}\| < 1$ and so $(x^*x)^{1/2}$ is invertible in the unital C*-algebra $q_1 B q_1$. Let y denote its inverse. Then $q_1 - (x^*x)^{1/2} < 4\delta$ implies $q_1 - 4\delta < (x^*x)^{1/2}$ so $\|y\| < 1/(1 - 4\delta)$. Let $w := xy$. Then

$$
w^*w = yx^*xy = q_1, \quad ww^* \in q_2 B q_2.
$$

Since w^*w is a projection, so is ww^*. Similarly to the above, xx^* is invertible in $q_2 B q_2$. Let z denote its inverse. Now,

$$xx^* ww^* = xx^* xy^2 x^* = xq_1 x^* = xx^*,$$

so

$$q_2 = zxx^* = zxx^* ww^* = q_2 ww^* = ww^*.$$

Finally,

$$\|v - x\| = \|p_2 v p_1 - x\|$$
$$\leq \|q_2 v q_1 - x\| + 2\delta$$
$$\leq 3\delta,$$

and

$$\|q_1 - y\| = 4\delta/(1 - 4\delta),$$

from which it follows that

$$\|x - w\| < \|x\|\|q_1 - y\| < (1 + \delta)4\delta/(1 - 4\delta).$$

Thus $\|v - w\| < 3\delta + (1 + \delta)4\delta/(1 - 4\delta) < 8\delta(1 - 4\delta) \leq \epsilon.$ □

8.4.6. Let A be a C*-algebra. Let $m, n_1, \ldots, n_m \in \mathbb{N} \setminus \{0\}$. A set of elements $\{e_{ij}^{(k)}\}_{1 \leq i,j \leq n_k, 1 \leq k \leq m} \subset A$ is called a *system of matrix units* if they satisfy the following identities:

$$e_{il}^{(k)} e_{lj}^{(k)} = e_{ij}^{(k)} \quad \text{and} \quad (e_{il}^{(k)})^* = e_{li}^{(k)}.$$

The C*-subalgebra of A generated by a system of matrix units $\{e_{ij}^{(k)}\}_{1 \leq k \leq m, 1 \leq i,j \leq n}$ is a finite-dimensional C*-algebra isomorphic to $M_{n_1} \oplus M_{n_2} \oplus \cdots \oplus M_{n_m}$ (exercise).

8.4.7. Lemma. *For any $\epsilon > 0$ there is a $\delta := \delta(\epsilon, n) > 0$ with the following property: Suppose A is C*-algebra with C*-subalgebra $B \subset A$ and that $\{e_{ij}\}_{i,j=1,\ldots,n}$ is a system of matrix units in A. If $\{a_{ij}\}_{i,j=1,\ldots,n} \subset B$ satisfy*

$$\|e_{ij} - a_{ij}\| < \delta, \text{ for every } 1 \leq i,j \leq n,$$

then there is a system of matrix units $\{f_{ij}\}_{i,j=1,\ldots,n} \subset B$ such that

$$\|f_{ij} - e_{ij}\| < \epsilon, \text{ for every } 1 \leq i,j \leq n.$$

Proof. Let $\delta := \min\{1/5, \epsilon/(8-5\epsilon), \delta(\epsilon, n)\}$, where $\delta(\epsilon, n)$ is given by Lemma 8.4.2. Then there are pairwise orthogonal projections $f_{ii} \in B$, $1 \leq i \leq n$ with

$$\|f_{ii} - e_{ii}\| < \epsilon.$$

Applying Lemma 8.4.5, since $\|a_{ij} - e_{ij}\| < \delta$, $e_{ij}^* e_{ij} = e_{jj}$ and $e_{ij} e_{ij}^* = e_{ii}$, $1 \leq i \neq j \leq n$, there are partial isometries $f_{ij} \in B$, $1 \leq i \neq j \leq n$ with $f_{ij}^* f_{ij} = f_{jj}$, $f_{ij} f_{ij}^* = f_{ii}$ and $\|f_{ij} - e_{ij}\| < \epsilon$. Thus $\{f_{ij}\}_{i,j=1,\ldots,n}$ is a system of matrix units satisfying the requirements. □

8.4.8. Lemma. *For any $\epsilon > 0$ and $n \in \mathbb{N}$ there exists $\delta = \delta(\epsilon, n) > 0$ such that the following holds: Let B and F be C*-subalgebras in a unital C*-algebra A, where $\dim F \le n$. Suppose there exists a system of matrix units $\{e_{ij}^{(k)}\}_{1 \le i,j \le n_k, 1 \le k \le m}$ generating F such that $\mathrm{dist}(e_{ij}^{(k)}, B) < \delta$ for every $1 \le i, j \le n_k, 1 \le k \le m$. Then there is a unitary in the C*-subalgebra of A generated by B and F such that $\|u - 1\| < \epsilon$ and $uFu^* \subset B$.*

Proof. If B and F do not contain 1_A then we take their unitisations with respect to 1_A, and put $\tilde{F} := F \oplus \mathbb{C} \cdot (1_A - 1_F)$ and $\tilde{B} := B \oplus \mathbb{C} \cdot 1_A$. Expand the system of matrix units of $\tilde{F}$ to include $(1_A - 1_F)$. In that case, since $\mathrm{dist}(1_F, B) < n\delta$, we have $\mathrm{dist}(1_A - 1_F, \tilde{B}) < n\delta$. Thus, possibly by scaling δ by $1/n$, without loss of generality, we assume that $1_A \in F$ and $1_A \in B$.

Suppose first that $k = 1$ so $F = M_n$ is a matrix algebra. Let

$$0 < \epsilon_0 < \sqrt{2}\epsilon/(16n^2).$$

Let $\delta := \delta(\epsilon_0, n) > 0$ be that given by Lemma 8.4.7. Then, since $\mathrm{dist}(e_{ij}^{(k)}, B) < \delta$, there are matrix units $f_{ij} \in B$, $1 \le i, j \le n$ which satisfy

$$\|f_{ij} - e_{ij}\| < \sqrt{2}\epsilon/(16n^2).$$

By Lemma 8.4.3, there is a unitary $u \in \mathrm{C}^*(F, B)$ with $\|v - 1_A\| < \epsilon/(4n)$ such that $v^* e_{ii} v = f_{ii}$ for every $1 \le i \le n$. Define

$$u := \sum_{i=1}^{n} e_{i1} v f_{1i}.$$

Then

$$\|e_{ii} - e_{i1} v f_{1i}\| \le \|e_{ii} - e_{i1} v e_{1i}\| + \|e_{i1} v e_{1i} - e_{i1} v f_{1i}\|$$
$$< \|e_{ii} - e_{i1} 1_A e_{1i}\| + \|e_{i1}\| \|v - 1_A\| \|e_{1i}\| + \sqrt{2}\epsilon/(16n^2)$$
$$< \epsilon/(4n) + \epsilon/(4n)$$
$$< \epsilon/2n.$$

Thus

$$\|1_A - u\| < \epsilon/2.$$

Furthermore, for every $1 \le k, l \le n$, we have

$$u^* e_{kl} u = \left(\sum_{i=1}^{n} f_{i1} v^* e_{1i} \right) e_{kl} \left(\sum_{i=1}^{n} e_{i1} v f_{1i} \right)$$
$$= f_{k1} v^* e_{1k} e_{kl} e_{l1} v f_{1l}$$
$$= f_{k1} v^* e_{11} v f_{1l}$$
$$= f_{k1} f_{11} f_{1l}$$
$$= f_{kl},$$

so $u^* F u \subset B$, which proves the lemma when $F = M_n$.

For the general case, we have $F = M_{n_1} \oplus \cdots \oplus M_{n_m}$. For $1 \le k \le m$, let $p_k := 1_{M_{n_k}}$. Then $p_1, \ldots, p_m \in F$ are mutually orthogonal projections so provided δ is small enough, Lemma 8.4.2 tells us there are mutually orthogonal projections $q_1, \ldots, q_m$ in B with $\|p_k - q_k\| < \delta/2$. For each k, we have matrix units in $p_k F p_k \cong M_{n_k}$ which are within δ of $q_k B q_k$. Applying the above, we obtain $u_1, \ldots, u_m$ such that $u_k^* u_k = u_k u_k^* = p_k$. Then we simply take $u = (u_1, \ldots, u_m)$, which is clearly unitary and satisfies the requirements. $\qquad\square$

8.5   Locally finite-dimensional C*-algebras

As we saw in the previous section, there is a lot of wiggle room with projections and partial isometries. Thanks to this, we can characterise separable approximately finite-dimensional C*-algebras as ones in which we can always find arbitrarily large finite-dimensional C*-subalgebras, without worrying too much about their position. The theorem making this precise is due to Bratteli [13]. The fact that we ask that the C*-algebra is separable is crucial; that the theorem does not hold in the nonseparable case was shown by Farah and Katsura [44].

8.5.1. Definition. A C*-algebra A is called *locally finite-dimensional* if, for every finite subset $\{a_0, \ldots, a_k\} \subset A$ and every $\epsilon > 0$, there is a finite-dimensional C*-subalgebra $F \subset A$ and $b_0, \ldots, b_k \in F$ such that

$$\|a_i - b_i\| < \epsilon,$$

for every $1 \le 1 \le k$.

8.5.2. Theorem (Bratteli). *A separable C*-algebra is AF if and only if it is locally finite-dimensional.*

Proof. It is easy to see that an AF algebra is locally finite-dimensional, so we need only show one direction of the theorem. Let A be a separable locally finite-dimensional C*-algebra. Fix a countable dense subset $(a_i)_{i \in \mathbb{N}}$ of the unit ball of A with $a_0 = 0$. Let $(\epsilon_i)_{i \in \mathbb{N}}$ be a sequence of positive numbers that is monotone decreasing to zero. If A is nonunital, we may adjoin an identity, since A is AF if and only if $\tilde{A}$ is AF. We proceed by induction. Since A is locally finite-dimensional, there is a finite-dimensional C*-subalgebra A_0 and since $a_0 = 0$, clearly $a_0 \in A_0$.

Assume now that for $k \ge 0$ we have finite-dimensional subalgebras A_k with $A_{i-1} \subset A_i$, $1 \le i \le k$, and $\mathrm{dist}(a_i, A_k) < \epsilon_k$ for every $0 \le i \le k$. Fix a system of matrix units $\{e_{ij}^{(r)} \mid 1 \le i, j \le n_r, 1 \le r \le m\}$ for A_k. Let $\delta := \delta(\epsilon_{k+1}/3, \dim(A_k))$ be given as in Lemma 8.4.8. Since A is locally finite-dimensional, there is a finite-dimensional C*-subalgebra $F \subset A$ with $\mathrm{dist}(e_{ij}^{(r)}, F) < \delta$ for every $1 \le i, j \le n_r, 1 \le r \le m$ and $\mathrm{dist}(a_i, F) < \epsilon_{k+1}/3$ for every $0 \le i \le k+1$. Thus, there is a unitary $u \in A$ with $u^* A_k u \subset F$ and $\|u - 1_A\| < \epsilon_{k+1}/3$. Let

$$A_{k+1} := uFu^*,$$

which is evidently a finite-dimensional C*-subalgebra and satisfies $A_k \subset A_{k+1}$. Moreover,

$$\mathrm{dist}(a_i, A_{k+1}) = \mathrm{dist}(u^* a_i u, F) \le 2\|u - 1_A\| + \epsilon_{k+1}/3 < \epsilon_{k+1}.$$

By construction, we have that the closure of $\bigcup_{i\in\mathbb{N}} A_i$ contains the dense subsequence $(a_i)_{i\in\mathbb{N}}$, so $A = \overline{\bigcup_{i\in\mathbb{N}} A_i} = \varinjlim(A_n, \iota_n)$, where $\iota_n : A_i \to A_{n+1}$ is the inclusion map. Thus A is AF. $\square$

In the general case, when our approximations are not by finite-dimensional C*-algebras but rather more general classes, such a local condition will not necessarily imply an inductive limit structure, even in the separable case.

8.5.3. For $n \in \mathbb{N}$, a C*-algebra is called *n-homogeneous* if all its irreducible representations have dimension n. We call a C*-algebra *homogeneous* if it is n-homogeneous for some n. A typical example is a C*-algebra of the form

$$p(C(X) \otimes M_n)p$$

for some compact Hausdorff space X and projection $p \in C(X) \otimes M_n$. An *approximately homogeneous* (AH) algebra is an inductive limit of direct sums of homogenous C*-algebras of the type $p(C(X) \otimes M_n)p$, with p and X as above.

Dădărlat and Eilers constructed a separable locally homogeneous C*-algebra which is not AH [34]. In the simple case, however, they do turn out to be the same, provided we put some restrictions on how fast we allow the covering dimension (see Definition 8.6.2 below) of the spaces to grow in comparison to the dimension growth of the matrix algebras. In the case that all X have dim $X \le d$ for some $d < \infty$ ("no dimension growth"), this was done in [112]. For the case of "slow-dimension growth", this was shown in [78]. The proof of this is, however, completely different to the one for AF algebras. It relies on the heavy machinery of the classification program, which we will encounter in Part III.

8.6   AF algebras as noncommutative zero-dimensional spaces

As was mentioned, because of the fact that the Gelfand transform is a *-isomorphism, C*-algebras are often thought of as "noncommutative" locally compact Hausdorff spaces. In the topological setting, there are several notions of the topological dimension of a space. A point should be zero-dimensional, an interval one-dimensional, an n-cube n-dimensional, and so forth.

8.6.1. In what follows, we will denote the indicator function of an open set U by χ_U, which is to say

$$\chi_U(x) = \begin{cases} 1 & \text{if } x \in U, \\ 0 & \text{if } x \notin U. \end{cases}$$

8.6.2. Definition. Let X be a locally compact Hausdorff space. We say that X has *covering dimension d*, written $\dim(X) = d$ if d is the least integer such that the following holds: For every open cover $\mathcal{O}$ of X there is a finite refinement $\mathcal{O}'$ such that, for every $x \in X$, $\sum_{U \in \mathcal{O}'} \chi_U(x) \leq d + 1$. If no such d exists, we say that $\dim(X) = \infty$.

The other standard notions of topological dimension are the large and small inductive dimensions (we won't need their definitions). In the case that we restrict ourselves to locally compact metrisable spaces, these three definitions coincide with the covering dimension, so we will usually just refer to the covering dimension of a compact metrisable space X as the dimension of X.

8.6.3. There is an alternative definition of covering dimension for a compact metric space X which is often useful. A proof of the equivalence of these definitions can be found in [87].

Proposition. *Let X be a compact metric space. Then $\dim(X) = d$ if and only if d is the least integer such that every continuous function $f : X \to \mathbb{R}^{n+1}$ can be approximated arbitrarily well by another function $g : X \to \mathbb{R}^{n+1}$ for which $0 \notin g(X)$.*

Moving to the noncommutative setting, we would like to find an analogue of the dimension of a space. At the commutative level, X metrisable corresponds to the C*-algebra being separable, so we content ourselves with trying to establish noncommutative versions of covering dimension for separable C*-algebras. These should extend covering dimension in the sense that the noncommutative dimension of $C_0(X)$ should be the same as the covering dimension. There are a few such extensions. We describe two of them below; a third is found in Chapter 17.

8.6.4. Definition ([16]). Let A be a unital separable C*-algebra A. We say that A has *real rank* d, written $RR(A) = d$, if d is the least integer such that, whenever $0 \leq n \leq d + 1$ the following holds: For every n-tuple $(x_1, \ldots, x_n)$ of self-adjoint elements in A and every $\epsilon > 0$ there exists an n-tuple $(y_1, \ldots, y_n) \subset A_{sa}$ such that $\sum_{k=1}^{n} y_k^* y_k$ is invertible and $\| \sum_{k=1}^{n} (x_k - y_k)^*(x_k - y_k) \| < \epsilon$. If there is no such d, then we say the real rank of A is infinite.

8.6.5. Theorem. *Let X be a compact metric space. Then $RR(C(X)) = \dim X$.*

Proof. Suppose that $\dim X = d$. Let $\epsilon > 0$. Suppose $(f_0, \ldots, f_d)$ is an $d+1$-tuple of self-adjoint elements in $C(X)$. Then each f_i is real-valued, so we may define $f : X \to \mathbb{R}^{d+1}$ to be $f(x) = (f_0(x), \ldots, f_d(x))$. Since $\dim X = d$, there exists a function $g \in C(X, \mathbb{R}^{d+1})$ such that $0 \notin g(X)$ and $\|f - g\| < \epsilon/(d+1)$. Since $g \in C(X, \mathbb{R}^{d+1})$ there are continuous functions $g_0, \ldots, g_d : X \to \mathbb{R}$ such that $g(x) = (g_0(x), \ldots, g_d(x))$ for every $x \in X$. Let $h := \sum_{k=0}^{d} g_k^2$. Then $h \in C(X)_{sa}$ and $h^2 > 0$ so h is invertible. Furthermore,

$$\left\| \sum_{k=0}^{d} (f_k - g_k)^*(f_k - g_k) \right\| \leq (d+1)\|f_k - g_k\| < \epsilon.$$

Since ϵ was arbitrary, this shows that $RR(C(X)) \leq d = \dim X$.

We leave the reverse argument as an exercise. $\square$

8.6.6. The stable rank has a very similar definition, dropping the fact that the n-tuples need be self-adjoint, and we only look at n-tuples from $1 \leq n \leq d$ (rather than $d+1$).

Definition ([98]). Let A be a unital separable C*-algebra A. We say that A has *stable rank* d, written $SR(A) = d$ if d is the least integer such that, whenever $1 \leq n \leq d$ the following holds: For every n-tuple $(a_1, \ldots, a_n)$ of elements in A and every $\epsilon > 0$ there exists an n-tuple $(y_1, \ldots, y_n) \subset A$ such that $\sum_{k=1}^{n} y_k^* y_k$ is invertible and $\| \sum_{k=1}^{n} (x_k - y_k)^* (x_k - y_k) \| < \epsilon$. If there is no such d, then we say the stable rank of A is infinite. A comparison of the stable rank of $C(X)$ to the covering dimension of X is given in Exercise 8.7.17.

8.6.7. We should think of AF algebras of "zero-dimensional" objects. Note, however, that there is no definition for "stable rank zero". We get the following theorem for AF algebras. The proof is an exercise; see the exercises at the end of the chapter for a bit of a hint.

Theorem. *Let A be a unital AF algebra. Then A has real rank zero and stable rank one.*

8.6.8. If A is nonunital, we can define its real and stable rank by putting $RR(A) := RR(\tilde{A})$ and $SR(A) := SR(\tilde{A})$. Then the above theorem is also true for nonunital AF algebras.

8.7 Exercises

8.7.1. Let A be a finite-dimensional C*-algebra. Then A is simple if and only if $A \cong M_n$ for some $n \in \mathbb{N}$. (Hint: See Exercises 2.5.10 and 4.4.16.)

8.7.2. Let A be a C*-algebra.

(i) Let $n \in \mathbb{N} \setminus \{0\}$. Suppose A contains a set of elements $\{e_{ij}\}_{1 \leq i,j \leq n} \subset A$ satisfying

$$e_{il} e_{lj} = e_{ij} \quad \text{and} \quad (e_{il})^* = e_{li}.$$

Show that $C^*(\{e_{ij}\}_{1 \leq i,j \leq n_k}) \cong M_n$. (Hint: use the previous exercise.)

(ii) Now suppose $m, n_1, \ldots, n_m \in \mathbb{N} \setminus \{0\}$ and $\{e_{ij}^{(k)}\}_{1 \leq i,j \leq n_k, 1 \leq k \leq m} \subset A$ is a system of matrix units. Show that the C*-subalgebra of A generated by this system of matrix units is a finite-dimensional C*-algebra isomorphic to $M_{n_1} \oplus M_{n_2} \oplus \cdots \oplus M_{n_m}$.

8.7.3. Let $F = \bigoplus_{k=1}^{m} M_{n_k}$ be a finite-dimensional C*-algebra. If $p \in F$ is a projection, show that there are projections $p_k \in M_{n_k}$, $1 \leq k \leq m$ such that $p = \sum_{k=1}^{m} p_k$.

8.7.4. Let $A = \varinjlim(A_n, \varphi_n)$ be the inductive limit of a sequence of C*-algebras and let B be a C*-algebra. For every $n \in \mathbb{N}$ let $\psi^{(n)} : A_n \to B$ be a *-homomorphism satisfying $\psi^{(n+1)} \circ \varphi_n = \psi^{(n)}$ and let $\psi : A \to B$ be the induced *-homomorphism. Then

(i) ψ is injective if and only if $\ker(\psi^{(n)}) \subset \ker(\varphi^{(n)})$ for every $n \in \mathbb{N}$, and

(ii) ψ is surjective if and only if $B = \overline{\cup_{j=1}^{\infty} \psi^{(n)}(A_n)}$.

8.7.5. Let $\mathcal{U}$ be a UHF algebra. Show that the tracial state constructed in Proposition 8.3.4 is unique.

8.7.6. Let (A_n, φ_n) be an inductive sequence of C*-algebras. Let $(n_k)_{k \in \mathbb{N}} \subset \mathbb{N}$ be an increasing sequence of natural numbers. Show that

$$\varinjlim(A_n, \varphi_n) \cong \varinjlim(A_{n_k}, \varphi_{n_k, n_{k+1}}).$$

8.7.7. Show that the definition of a UHF algebra of type $\mathfrak{p}$ is independent of the choice of UHF sequence $(n_k)_{k \in \mathbb{N}}$ for $\mathfrak{p}$. Thus any UHF algebra is uniquely identified with a supernatural number $\mathfrak{p}$.

8.7.8. Let $A \cong \varinjlim(A_n, \varphi_n)$. For $k \in \mathbb{N}$, by abuse of notation, also denote the induced map on $k \times k$-matrices over A by $\varphi_n : M_k(A_n) \to M_k(A_{n+1})$. Show that $\varinjlim(M_k(A_n), \varphi_n) = M_n(A)$.

8.7.9. Let $\mathcal{U}$ be a UHF algebra. Show that $A \otimes_{\min} \mathcal{U} \cong A \otimes_{\max} \mathcal{U}$ for any C*-algebra A, that is, that UHF algebras are nuclear (see Exercise 5.4).

8.7.10. Let $\mathcal{U}_\mathfrak{q}$ be a UHF algebra of infinite type $\mathfrak{p}$. Show that $\mathcal{U}_\mathfrak{p} \otimes \mathcal{U}_\mathfrak{p} \cong \mathcal{U}_\mathfrak{p}$. Suppose that $\mathfrak{p}$ divides $\mathfrak{q}$. Show that $\mathcal{U}_\mathfrak{p}$ absorbs $\mathcal{U}_\mathfrak{q}$ in the sense that $\mathcal{U}_\mathfrak{q} \otimes \mathcal{U}_\mathfrak{p} \cong \mathcal{U}_\mathfrak{q}$. The *universal* UHF algebra, denoted $\mathcal{Q}$ is the UHF algebra associated to the supernatural number $\prod_{p \text{ prime}} p^\infty$. Then the above shows that the universal UHF algebra absorbs all other UHF algebras (including all matrix algebras!).

8.7.11. Show that if A is an AF algebra, then there exists an inductive sequence $(A_n, \varphi_n)_{n \in \mathbb{N}}$ with φ_n injective and $A = \varinjlim(A_n, \varphi_n)$, and if A is moreover unital, the φ_n can be chosen to be unital.

8.7.12. Show that an inductive limit of AF algebras with injective connecting maps is again AF.

8.7.13. Show that an AF algebra is always nuclear.

8.7.14. Let A and B be AF algebras and let $n \in \mathbb{N} \setminus \{0\}$.

(i) Show that $M_n(A)$ is AF.

(ii) Show that $A \otimes B$ is AF.

8.7.15. Let $A = \mathcal{B}(H)$.

(i) Show that two projections $p, q \in A$ are Murray–von Neumann equivalent if and only if $\dim(pH) = \dim(qH)$.

(ii) Show that a projection $p \in A$ is finite if and only if pH is finite-dimensional. Deduce that A is finite if and only if H is finite-dimensional.

(iii) Show that if H is finite-dimensional then every isometry in A is unitary.

(iv) Prove Theorem 8.3.14.

8.7.16. Finish the proof of Theorem 8.6.5 by showing that for any compact metric space X, $\dim X \leq RR(C(X))$.

8.7.17. Let X be a compact metric space. Show that if $\dim X = n$ then

$$SR(C(X)) = [\dim X/2] + 1$$

where $[\dim X/2]$ is $\dim X/2$ if n is even and $(\dim X - 1)/2$ if n is odd.

8.7.18. By Definition 8.6.4 we have that a unital C*-algebra A has *real rank zero* if the invertible self-adjoint elements are dense in A_{sa}.

(a) Let X be the Cantor set. Show that X has covering dimension 0. Show that $[0,1]$ has covering dimension at most 1. Show that $C(X)$ has real rank zero but $C([0,1])$ does not. (See Exercise 4.5.)

(b) Let $a \in M_n = M_{n \times n}(\mathbb{C})$, $b \in M_{n \times 1}(\mathbb{C})$, $c \in M_{1 \times n}(\mathbb{C})$ and $d \in \mathbb{C}$. Let $\epsilon > 0$. Suppose that d is invertible and there is $a' \in M_n$ that is invertible which satisfies $\|a' - (a - bd^{-1}c)\| < \epsilon$. Show that

$$\begin{pmatrix} a' + bd^{-1}c & b \\ c & d \end{pmatrix}^{-1} = \begin{pmatrix} I_n & 0 \\ -d^{-1}c & 1 \end{pmatrix},$$

as elements in M_{n+1} (where I_n is the $n \times n$ identity matrix) and

$$\left\| \begin{pmatrix} a & b \\ c & d \end{pmatrix} - \begin{pmatrix} a' + bd^{-1}c & b \\ c & d \end{pmatrix} \right\| < \epsilon.$$

(c) Let A be a unital C*-algebra. Suppose $b \in \mathrm{Inv}(A)$ and there is $a \in A_{sa}$ with $\|a - b\| < \epsilon$. Show that there is $b' \in A_{sa} \cap \mathrm{Inv}(A)$ with $\|a - b'\| < \epsilon$.

(d) Prove that any AF algebra has real rank zero.

8.7.19. Show that an AF algebra has stable rank one.

8.7.20. Show that if $A = \varinjlim A_n$ is a simple AF algebra with

$$F_n := M_{k(n,1)} \oplus \cdots \oplus M_{k(n,m_n)}$$

then

$$\lim_{n \to \infty} \min\{k(n)_i \mid 1 \leq i \leq m(n)\} \to \infty.$$

8.7.21. Let A be a unital C*-algebra with stable rank one. Show that A is stably finite.

8.7.22. Let $\epsilon > 0$ and $n \in \mathbb{N}$. Show that there exists a $\delta = \delta(\epsilon, n) > 0$, depending only on ϵ and n, such that the following holds: Suppose A is a C*-algebra and $a_0, \ldots, a_n$ are positive elements of norm at most one and $\|a_i a_j\| < \delta$ for every $0 \leq i \neq j \leq n$. Then there are positive elements $b_0, \ldots, b_n \in A$ with $b_i b_j = 0$ for every $0 \leq i \neq j \leq n$ and $\|a_i - b_i\| < \epsilon$, $0 \leq i \leq n$.

8.7.23. Let A be a C*-algebra and $h \in A$ a nonzero positive element. Let $\epsilon > 0$ satisfy $\epsilon < 1/4$. Suppose that $\|h - h^2\| < \epsilon$. Show that there is a projection p in the C*-subalgebra generated by h such that $\|p - h\| < 2\epsilon$ and that php is invertible in $p\,C^*(h)p$ and $\|p - (php)^{-1/2}\| < 4\epsilon$. (Hint: show that there is a gap in the spectrum of h of distance δ on either side of $1/2$ where $\delta = (1 - 4\epsilon)^{1/2}/2$.)

9 Group C*-algebras and crossed products

The group C*-algebra and crossed product construction provide a plethora of interesting examples of C*-algebras. They provide links to harmonic analysis, topological dynamics, quantum groups, and beyond. As such, it is unsurprising that they continue to receive a large amount of attention in the literature. In this chapter, we mainly focus on the constructions themselves. We begin with topological groups and the various ways one may associate an algebra to a group. To construct a group C*-algebra, one must consider first unitary representations, which are interesting objects to study in their own right. To avoid too many technicalities and the chapter growing too large, we focus mainly on discrete groups, particularly once we move to crossed products. Crossed product C*-algebras can be seen as a generalisation of group C*-algebras, where now we take into account further information of how the group acts on a C*-algebra. A group acting on a commutative C*-algebra $C(X)$ corresponds to the group acting by homeomorphisms on X which links the study of such crossed products to the study of topological dynamical systems. Of particular interest are dynamical systems consisting of a space X and a single homeomorphisms. These give rise to $\mathbb{Z}$-actions on $C(X)$, and the corresponding crossed products are always unital and nuclear.

Following the construction of group C*-algebras in the first section, we introduce the notion of *amenability* of a discrete group. Amenable groups are particularly nice because the reduced and full group C*-algebras coincide and are nuclear. In the third section, we show that the reduced group C*-algebras of a discrete abelian group is isomorphic to the C*-algebra of continuous functions on its Pontryagin dual. This partially motivates the study of quantum groups. In the fourth section, we consider the construction of crossed products by group actions. In the final section, we look more closely at the structural properties of crossed products by $\mathbb{Z}$, their link to topological dynamical systems, and characterise simplicity for crossed products by homeomorphisms.

9.1 Group C*-algebras

A *topological group* is a group together with a topology which makes group operations continuous. In particular, any group is a topological group with the discrete topology. We will assume throughout this chapter and the next that the groups are Hausdorff.

9.1.1. A Borel measure on a group G is left-translation-invariant if, for any Borel set $E \subset G$ and any $s \in G$, we have $\mu_G(sE) = \mu_G(E)$.

Theorem (Haar [60]). *Let G be a locally compact group. Then there is a left-translation-invariant Borel measure on G, denoted μ_G, which is unique up to scalar multiple.*

© The Author(s), under exclusive license to Springer Nature Switzerland AG 2021
K. R. Strung, *An Introduction to C*-Algebras and the Classification Program*, Advanced
Courses in Mathematics - CRM Barcelona, https://doi.org/10.1007/978-3-030-47465-2_9

This measure is called a (left) *Haar measure* of G. When G is compact, $\mu(G)$ is finite and so we normalise with $\mu(G) = 1$. If G is infinite and discrete, then we normalise so that $\mu(\{e\}) = 1$.

9.1.2. In general, a left-translation-invariant measure need not be right-translation-invariant. However, if this is the case, then we call G *unimodular*. Unimodular groups include the cases that G is abelian, discrete, or compact. For the sake of brevity, we will stick to the unimodular case, though most of what we will do can be generalised.

9.1.3. There are a number of noncommutative algebras that we can associate with a group. The first is the *group algebra* of G, which is the algebra of formal finite $\mathbb{C}$-linear combinations of elements of G, and is denoted $\mathbb{C}G$. The multiplication in $\mathbb{C}G$ extends the group multiplication. (Here we do not topologise G.)

9.1.4. We also have the function algebra of compactly supported functions on G, denoted $C_c(G)$, which comes equipped with convolution as multiplication,

$$f * g(t) = \int_G f(s)g(s^{-1}t)d\mu(s),$$

and inversion for involution,

$$f^*(s) = \overline{f^*(s^{-1})}.$$

There is a norm on $C_c(G)$ given by $\|f\|_1 = \int_G |f(t)|d\mu(t)$. One must of course check that this multiplication and involution are well defined (that is, that convolution and inversion in fact define elements in $C_c(G)$, as well as that the multiplication is associative). This requires some facility with vector-valued integration. We will use some of this below, but will quickly switch to discrete groups and later on, the integers. We leave it as an exercise to check this, at least in the case of discrete groups where the integrals become sums.

9.1.5. Given $s \in G$, define $\delta_s : G \to \mathbb{C}$ by

$$\delta_s(t) = \begin{cases} 1 & \text{if } t = s, \\ 0 & \text{otherwise.} \end{cases}$$

For every $s \in G$, the function δ_s is continuous if and only if G is discrete. Thus, by identifying s with δ_s, we see that $\mathbb{C}G = C_c(G)$ if and only if G is discrete. When G is locally compact, $C_c(G)$ will not be complete with respect to the $\| \cdot \|_1$ norm. Completing with respect to $\| \cdot \|_1$ gives us yet another group algebra, $L^1(G, \mu)$.

9.1.6. The space $L^1(G, \mu)$ consists of functions $f : G \to \mathbb{C}$ such that

$$\int_G |f(t)|d\mu(t) < \infty.$$

This is a Banach *-algebra when equipped with $\| \cdot \|_1$, convolution and inversion as in the case of $C_c(G)$.

9.1.7. Observe that $C_c(G)$ is a dense *-subalgebra of $L^1(G)$. If G is finite, then $L^1(G, \mu) = \mathbb{C}G$. Since a left Haar measure is unique up to a scalar multiple, for any two left Haar measures μ, μ' we have $L^1(G, \mu) \cong L^1(G, \mu')$ (exercise). Thus henceforth we will write $L^1(G)$ for any $L^1(G, \mu)$ defined with respect to a left Haar measure.

When G is a discrete group, then $L^1(G)$ is denoted by $\ell^1(G)$ and simply consists of those functions $f : G \to \mathbb{C}$ satisfying

$$\sum_{t \in G} |f(t)| < \infty.$$

The algebra of compactly supported functions, $C_c(G)$, which densely spans $\ell^1(G)$, is just the group algebra.

9.1.8. Proposition. *Let G be a locally compact group. Then $L^1(G)$ is unital if and only if G is discrete. In any case, $L^1(G)$ always has a norm one approximate unit.*

Proof. If G is discrete, then a unit $1_{L^1(G)}$ is given by the function δ_e, where e is the identity in G and δ_e is the function defined in 9.1.5. Conversely, if $L^1(G)$ is unital then

$$1_{L^1(G)} * f(e) = \int_G 1_{L^1(G)}(s) f(s^{-1}) d\mu(s) = f(e),$$

for every $f \in L^1(G, \mu)$ only if $1_{L^1(G)}(s) = 0$ for every $s \neq e$. But then if G is not discrete $1_{L^1(G)} = 0$ a.e.μ, thus is not a unit in $L^1(G, \mu)$.

For the second statement, let G be an arbitrary locally compact group and let $\mathcal{O}$ be the collection of open neighbourhoods E of e. For $E \in \mathcal{O}$, let f_E be a function in $L^1(G)$ with $f(e) = 1$, $\mathrm{supp}(f_E) \subset E$, $f_E^* = f^*$ and $\|f\|_1 = 1$. Since the set of all such neighbourhoods is upwards directed with respect to reverse containment, $(f_E)_E$ is an approximate unit for $L^1(G)$. $\qquad\square$

The $\| \cdot \|_1$ norm is not a C*-norm in general, so $L^1(G)$ is not a C*-algebra (see for example 2.1.4). Thus we would like to find a *-homomorphism from $L^1(G)$ into a C*-algebra so that we can complete the image of $L^1(G)$ to a C*-algebra.

9.1.9. Definition. Let G be a locally compact group. A *unitary representation* of G is given by a pair (H, u) consisting of a Hilbert space H and a strongly continuous homomorphism $u : G \to \mathcal{U}(H)$, where $\mathcal{U}(H)$ is the group of unitary operators on H. Here, strongly continuous means that $s \mapsto u(s)\xi$ is continuous for every $\xi \in H$. To simplify the notation, we often write u_s for $u(s)$, $s \in G$.

The condition of strong continuity can be equivalently stated as requiring the continuity of the function

$$G \times H \to H, \quad (s, \xi) \mapsto u(s)\xi.$$

We leave the proof of this equivalence as an exercise.

9.1.10. Definition. We say that a unitary representation is *irreducible* if $u(G)$ does not commute with any proper projections in $\mathcal{B}(H)$.

9.1.11. As with Theorem 5.3.6, the definition for irreducibility of a unitary representation can be equivalently given by there being no proper closed linear subspace $K \subset H$ that is stable under $u(G)$. We leave it as an exercise to check the details.

Proposition. *Let G be a locally compact group and $u : G \to \mathcal{U}(H)$ a unitary representation. Then the C*-subalgebra of $\mathcal{B}(H)$ generated by $u(G)$, written $C^*(u(G))$, is irreducible (with respect to the identity representation on H) if and only if $u(G)$ is irreducible.*

Proof. Exercise.   □

9.1.12. If $u : G \to \mathcal{U}(H)$ is a unitary representation then $\pi : L^1(G) \to \mathcal{B}(H)$ given by

$$\pi(f)\xi = \int_G f(t)u(t)(\xi)d\mu(t), \quad f \in L^1(G), \xi \in H,$$

is a representation of $L^1(G)$.

To see that this is a representation, we must check that it preserves multiplication and adjoints, and that it is norm-decreasing. It is a easy to see, since $u(t)$ is unitary for any $t \in G$, that

$$\|\pi(f)\| \le \int_G |f(t)|dt = \|f\|_1,$$

so the map π is norm-decreasing. Let $f, g \in L^1(G)$. Then

$$\pi(f * g) = \int_G \left(\int_G f(s)g(s^{-1}t)ds \right) u(t)dt$$

$$= \int_G f(s)u(s) \int_G g(s^{-1}t)u(s^{-1}t)dtds$$

$$= \int_G f(s)u(s)ds \int_G g(r)u(r)dr$$

$$= \pi(f)\pi(g).$$

Note that the third line uses the fact that the Haar measure is left-invariant. We were also able to use Fubini's theorem since $\pi(f * g)$ is bounded.

To see that π is a *-representation, we calculate

$$\langle \pi(f)^*\xi, \eta \rangle = \langle \xi, \pi(f)\eta \rangle = \left\langle \xi, \int_G f(t)u(t)\eta \, dt \right\rangle = \int_G \overline{f(t)} \langle \xi, u(t)\eta \rangle dt$$

$$= \int_G \overline{f(t)} \langle u(t)^*\xi, \eta \rangle dt = \int_G \overline{f(t)} \langle u(t^{-1})\xi, \eta \rangle dt = \int_G \overline{f(r^{-1})} \langle u(r)\xi, \eta \rangle dr$$

$$= \left\langle \int_G \overline{f(r)}u(r)\xi dr, \eta \right\rangle = \left\langle \int_G f^*(r)u(r)\xi dr, \eta \right\rangle = \langle \pi(f^*)\xi, \eta \rangle.$$

Conversely, if we have a nondegenerate representation $\pi : L^1(G) \to \mathcal{B}(H)$ then we can find a unique unitary representation of G as follows: Let $(f_E)_{E \in \mathcal{O}}$ be the norm one approximate unit as given in the proof of Proposition 9.1.8. We have

$$\lim_{E \in \mathcal{O}} \pi(f_E)\pi(g)\xi = \pi(g)\xi,$$

for every $g \in L^1(G)$ and every $\xi \in H$. It follows that $\pi(f_E) \xrightarrow{\text{SOT}} 1_{\mathcal{B}(H)}$. Now define $u : G \to \mathcal{B}(H)$ by

$$u(s)\pi(g)\xi = \pi(g_s)\xi \text{ for } s \in G, \xi \in H,$$

where $g_s(t) = g(s^{-1}t)$. In this case we have that $u(s) = \text{SOT-}\lim_{E \in \mathcal{O}} \pi((f_E)_s)$, which in turn implies that u is contractive. Furthermore, it is not hard to check that $u(s)$ is unitary for every $s \in G$ (this uses the assumption of nondegeneracy). Note that the construction ensures that this u is unique.

9.1.13. Definition. Let G be a locally compact group with Haar measure μ. The *left regular representation* of G on the Hilbert space $L^2(G)$, $\lambda : G \to \mathcal{U}(L^2(G))$, is given by

$$\lambda(s)f(t) = f(s^{-1}t).$$

We leave it as an exercise to check that this is indeed a unitary representation of G.

9.1.14. In the case of a discrete group G, we write $\ell^2(G)$ for $L^2(G)$. The standard orthonormal basis will be denoted $\{\delta_s\}_{s \in G}$, where δ_s is the function defined in 9.1.5. In that case, we get $\lambda(s)\delta_t = \delta_{st}$.

9.1.15. Definition. The *reduced group* C*-*algebra* of G, written $C_r^*(G)$ is the closure of $\lambda(L^1(G))$ in $\mathcal{B}(L^2(G))$. The *full group* C*-*algebra*, denoted $C^*(G)$ is the closure of $L^1(G)$ under the direct sum of all unitary equivalence classes of irreducible representations.

As was the case for the maximal and minimal tensor product, the full and reduced C*-algebras are not in general isomorphic. This is often the case for C*-algebraic constructions that involve taking the closure of a *-algebra.

9.1.16. We denote $\lambda(s)$ by u_s for any $s \in G$. Then, for a discrete group G, elements of the form $\sum_{s \in G} b_s u_s$ where $b_s \in \mathbb{C}$ and $b_s = 0$ for all but finitely many $s \in G$, are dense in $C_r^*(G)$.

The norm on the full group C*-algebra is given by

$$\|f\| := \{\|\pi(f)\|_{\mathcal{B}(H)} \mid \pi : L^1(G) \to \mathcal{B}(H) \text{ is a *-representation}\}.$$

That this defines a norm follows from the fact that $\|\pi(f)\| \leq \|f\|_1$ for any *-representation π. Note that the left regular representation extends to a surjective *-homomorphism, which we call the *canonical surjection*

$$\pi_\lambda : C^*(G) \to C_r^*(G).$$

Thus we get $C_r^*(G) \cong C^*(G)/\ker(\pi_\lambda)$.

9.1.17. Recall that a state τ is faithful if $\tau(a^*a) = 0$ implies $a = 0$ (4.1.1).

Theorem. *Let G be a discrete group. Then $C_r^*(G)$ has a tracial state τ which is faithful and satisfies $\tau(u_e) = 1$ and $\tau(u_g) = 0$ for every $g \in G \setminus \{e\}$. Moreover τ is the unique tracial state with these properties.*

Proof. Elements of the form $\sum_{g \in G} b_g u_g$ where $b_g \in \mathbb{C}$ and $b_g = 0$ for all but finitely many $g \in G$, are dense in $C_r^*(G)$. Thus if such a map τ exists, it must be unique.

Let $g \in G$ and let $\delta_g \in \ell^2(G)$ be the standard basis unit associated to g. Define $\tau : C_r^*(G) \to \mathbb{C}$ by

$$\tau(a) := \langle a\delta_e, \delta_e \rangle, \quad a \in C_r^*(G).$$

Then $\tau(u_e) = \langle u_e \delta_e, \delta_e \rangle = \langle \delta_e, \delta_e \rangle = 1$ while if $g \in G \setminus \{e\}$ we have $\tau(u_g) = \langle u_g \delta_e, \delta_e \rangle = \langle \delta_g, \delta_e \rangle = 0$. It is clear that τ is a linear functional. Now let $g, h \in G$. Then

$$\tau(u_g u_h) = \tau(u_{gh}) = \begin{cases} 1 & \text{if } g = h^{-1} \\ 0 & \text{otherwise.} \end{cases}$$

Since $g = h^{-1}$ if and only if $h = g^{-1}$ we have $\tau(u_g u_h) = \tau(u_h u_g)$. Extending by linearity and continuity, we have $\tau(ab) = \tau(ba)$ for every $a, b \in C_r^*(G)$, which is to say, τ is tracial.

Finally, we must show that τ is faithful. Suppose that $\{b_g\}_{g \in G} \subset \mathbb{C}$ and $b_g = 0$ for all but finitely many $g \in G$. Let $b = \sum_g b_g u_g \in C_r^*(G)$. Then

$$b\delta_1 = \sum_{g \in G} b_g u_g \delta_e = \sum_{g \in G} b\delta_g.$$

Since such elements are dense in $\ell^2(G)$, it follows that $\mathbb{C}G\delta_1$ is dense in $\ell^2(G)$. Suppose that $a \in C_r^*(G)$ satisfies $\tau(a^*a) = 0$. Then, for any $c, b \in \mathbb{C}G$ we have

$$|\langle ab\delta_e, c\delta_e \rangle| = |\langle c^*ab\delta_e, \delta_e \rangle| = |\tau(c^*ab)|$$
$$= |\tau(abc^*)| \le \tau(a^*)^{1/2}\tau((bc^*)^*(bc^*))^{1/2} = 0,$$

where we have applied the Cauchy–Schwarz inequality of Proposition 4.1.6. Thus $|\langle a\xi, \eta \rangle| = 0$ for every $\xi, \eta \in \ell^2(G)$, and so $a = 0$. Thus τ is faithful. $\qquad\square$

In general, the reduced group C*-algebra of an arbitrary locally compact group might not have a trace. A characterisation for when $C_r^*(G)$ admits a trace is given by Kennedy and Raum [64], which generalises results of Forrest, Spronk and Wiersma [46]. One is also interested in when a given C*-algebra has a *unique* tracial state, as is the case, for example, for a UHF algebra. Breuillard, Kalantar, Kennedy and Ozawa characterise the case of a unique tracial state for reduced group C*-algebras of discrete groups. In particular, if $C_r^*(G)$ is simple, then it has a unique tracial state [15].

9.2 Amenability

We observed above that, in general, the canonical surjection $\pi_\lambda : C^*(G) \to C^*_r(G)$ is not injective. It is, however, when the group is *amenable*. Amenability of a group also implies other nice properties for its group C*-algebras, such as nuclearity (Theorem 9.2.7).

9.2.1. Let G be a discrete group. For $f \in \ell^\infty(G)$ and $s \in G$, define $s.f \in \ell^\infty(G)$ to be the function $s.f(t) = f(s^{-1}t)$, $t \in G$. A *mean* on $\ell^\infty(G)$ is a linear functional $m : G \to \mathbb{C}$ satisfying

(i) $m(\bar{f}) = \overline{m(f)}$,

(ii) $m(e) = 1$, and

(iii) if $f \geq 0$ then $m(f) \geq 0$.

If $m(s.f) = m(f)$ for every $f \in \ell^\infty(G)$ and every $s \in G$, then m is *left-invariant*.

9.2.2. Definition. Let G be a discrete group. We say that G is *amenable* if $\ell^\infty(G)$ admits a left-invariant mean.

9.2.3. A discrete group G satisfies the *Følner condition* if for any finite subset $E \subset G$ and every $\epsilon > 0$, there exists a finite subset $F \subset G$ such that

$$\max_{s \in E} \frac{|sF \cap F|}{|F|} > 1 - \epsilon.$$

Proposition. *Let G be a countable discrete group that satisfies the Følner condition. Then G is amenable.*

Proof. Since G is countable, we can find, for every $n \in \mathbb{N}$, subsets $E_n \subset G$, such that $E_n \subset E_{n+1}$ and exhaust G. Then, since G satisfies the Følner condition, there are finite subsets $F_n \subset G$ such that

$$\max_{s \in E_n} \frac{|sF_n \cap F_n|}{|F_n|} > 1 - 1/n.$$

In other words

$$\lim_{n \to \infty} \frac{|sF_n \cap F_n|}{|F_n|} = 1.$$

(The sequence $(F_n)_{n \in \mathbb{N}}$ is called a *Følner sequence*.) For $s \in G$, let

$$\mathrm{ev}_s : \ell^\infty(G) \to \{0,1\}$$

denote the evaluation map at s, that is, $\mathrm{ev}_s(f) = f(s)$ for $f \in \ell^\infty(G)$. Let $m_n : \ell^\infty(G) \to [0,1]$ be defined by

$$m_n(f) = |F_n|^{-1} \sum_{s \in F_n} \mathrm{ev}_s(f), \quad f \in \ell^\infty(G).$$

Then one checks that m_n is a mean for every $n \in \mathbb{N}$. Let $E \subset G$ be a finite subset and let $\epsilon > 0$. Observe that for every $f \in \ell^\infty(G)$ with $\|f\| \leq 1$ and every $n \in \mathbb{N}$ we have $|m_n(f)| \leq \|f\| \leq 1$. Thus $(m_n)_{n \in \mathbb{N}}$ is a bounded sequence and hence has a weak-* convergent subsequence. Let $m \in \ell^\infty(G)^*$ be a limit point of $(m_n)_{n \in \mathbb{N}}$. Note that m is also a mean.

Let $f \in \ell^\infty(G)$. Choose $n \in \mathbb{N}$ to be sufficiently large that $2\|f\|/n < \epsilon$. By choosing larger n if necessary, we may assume that $E \cup E^{-1} \subset E_n$. Then, for $t \in E$ and we have

$$m_n(t.f) = |F_n|^{-1} \sum_{s \in F_n} \mathrm{ev}_s(t.f) = |F_n|^{-1} \sum_{s \in F_n} \mathrm{ev}_{t^{-1}s}(f) = |F_n|^{-1} \sum_{r \in t^{-1}F_n} \mathrm{ev}_r(f).$$

We then estimate

$$|m_n(t.f) - m_n(f)| = |F_n|^{-1} \left| \sum_{s \in F_n \setminus (t^{-1}F_n)} \mathrm{ev}_s(f) - \sum_{r \in t^{-1}F_n \setminus F_n} \mathrm{ev}_r(f) \right|$$

$$\leq \frac{\|f\| |F_n \triangle t^{-1} F_n|}{|F_n|}$$

$$\leq 2\|f\|/n$$

$$< \epsilon.$$

Since m is a limit point of $(m_n)_{n \in \mathbb{N}}$, it follows that m is left-invariant. $\qquad\square$

9.2.4. There is also a notion of amenability for locally compact groups which are not necessarily discrete, as well as a Følner condition in this more general context. In both the discrete case and the more general case, the existence of the Følner condition is in fact equivalent to amenability. The converse to Proposition 9.2.3 is a little more difficult to prove, even in the case of a discrete group. A proof of the equivalence, as well as further properties equivalent to amenability, can be found in various places, see for example [18] for the discrete case, or [89, 130] for the general case.

9.2.5. Theorem. *Let G be a locally compact group. Then G is amenable if and only if G satisfies the Følner condition.*

9.2.6. Examples. It is a good exercise to check that the following examples are amenable.

(i) Finite groups are amenable.

(ii) $\mathbb{Z}$ is amenable.

(iii) Subgroups of amenable groups are amenable.

We have the following two important consequences of the Følner condition.

9.2.7. Theorem. *Let G be a discrete group. Then $C_r^*(G)$ has the completely positive approximation property if and only if G is amenable. In particular, $C_r^*(G)$ is nuclear.*

Proof. Again we will only prove one direction: we will show that if G has the Følner condition, then $C_r^*(G)$ has the c.p.a.p. The converse is not so difficult, but it does rely on some von Neumann theory which, for the sake of brevity, we have not introduced.

Let $\lambda : G \to \mathcal{B}(\ell^2(G))$ denote the left regular representation, and let $\{\delta_s\}_{s\in G}$ denote the standard orthonormal basis for $\ell^2(G)$ as in 9.1.14. Let $(F_n)_{n\in\mathbb{N}}$ be a Følner sequence in G. For $n \in \mathbb{N}$, let $p_n \in \mathcal{B}(\ell^2(G))$ be the projection onto the linear subspace of $\ell^2(G)$ spanned by $\{\delta_s \mid s \in F_n\}$. Then $p_n\mathcal{B}(\ell^2(G))p_n \cong M_{|F_n|}$. Let $e_{s,t}^{(n)}$ denote the matrix with 1 in the (s,t)th entry and 0 elsewhere, that is, let $(e_{s,t}^{(n)})_{s,t\in F_n}$ be the canonical set of matrix units for $p_n\mathcal{B}(\ell^2(G))p_n$ after identifying it with $M_{|F_n|}$. Notice that $e_{s,s}^{(n)}$, $s \in F_n$, is just the projection onto the subspace spanned by δ_s and that $\sum_{s,\in F_n} e_{s,s}^{(n)} = p_n$. For $r \in G$ and $f = \sum_{g\in G} \mu_g\delta_g$, $\mu_g \in \mathbb{C}$, we have

$$e_{s,s}^{(n)}\lambda(r)e_{t,t}^{(n)}(f) = \mu_t e_{s,s}^{(n)}\lambda(r)(\delta_t) = \mu_t e_{s,s}^{(n)}\delta_{rt} = \begin{cases} \mu_t\delta_s & \text{if } s = rt, \\ 0 & \text{otherwise.} \end{cases}$$

Also,

$$e_{s,t}^{(n)}(f) = \mu_s e_{s,t}^{(n)}(\delta_t) = \mu_t\delta_s.$$

Thus,

$$e_{s,s}^{(n)}\lambda(r)e_{t,t}^{(n)} = e_{s,r^{-1}s}^{(n)},$$

and from this we get

$$p_n\lambda(r)p_n = \sum_{s,t\in F_n} e_{s,s}^{(n)}\lambda(r)e_{t,t}^{(n)} = \sum_{s\in F_n\cap rF_n} e_{s,r^{-1}s}^{(n)}.$$

Define

$$\varphi_n : C_r^*(G) \to p_n\mathcal{B}(\ell^2(G))p_n, \quad a \mapsto pap.$$

It is straightforward to see that this is a u.c.p. map. Now we define

$$\psi_n : M_{|F_n|} \to C_r^*(G)$$

on the generators of $p_n\mathcal{B}(\ell^2(G))p_n$ as follows:

$$\psi_n(e_{s,t}^{(n)}) = \frac{1}{|F_n|}\lambda(s)\lambda(t^{-1}).$$

It follows from Proposition 7.1.7 that ψ_n is a completely positive map, which is easily seen to be unital. We have

$$\psi_n \circ \varphi_n(\lambda(r)) = \psi_n\left(\sum_{s\in F_n\cap rF_n} e_{s,r^{-1}s}^{(n)}\right) = \frac{1}{|F_n|}\left(\sum_{s\in F_n\cap rF_n}\lambda(r)\right) = \frac{|F_n\cap rF_n|}{|F_n|}\lambda(r).$$

Thus the sequence $\psi_n \circ \varphi_n$ converges pointwise to the identity. It follows that $C_r^*(G)$ has the c.p.a.p. $\square$

9.2.8. Remark. In the case that G is locally compact but not discrete, then if G is amenable again $C_r(G)$ is nuclear, but the converse need not hold. An example is $G = SL_2(\mathbb{R})$.

9.2.9. The proof of the next theorem is omitted for the sake of brevity. For the discrete case, a proof of the following can be found in [18, Section 2.6]. The "if" direction when G is discrete is also covered in [32, Chapter VII]. The general theorem can be found in [89, Chapter 7] or [130, Appendix A.2].

Theorem. *Let G be a locally compact group. Then the canonical surjection $\pi_\lambda : C^*(G) \to C_r^*(G)$ is injective if and only if G is amenable.*

9.3 Group C*-algebras of abelian groups and duality

When a locally compact group is abelian, we can determine its reduced group C*-algebra completely using *Pontryagin duality*. In fact, abelian groups are always amenable (the proof uses the Markov–Kakutani fixed point theorem, see for example [32, Theorem VII.2.2]). Thus in light of Theorem 9.2.9, we need not make the distinction between the full and reduced group C*-algebras.

9.3.1. Definition. Given a locally compact abelian group G, a *character* of G is a continuous group homomorphism $G \to \mathbb{T}$. The set of all characters of G has the structure of a compact abelian group, which we call the *Pontryagin dual* of G and denote by $\hat{G}$.

9.3.2. Let $f \in L^1(G)$. The *Fourier–Plancherel transform* (or sometimes simply *Fourier transform*) $\hat{f}$ on $\hat{G}$ is given by

$$\hat{f}(\gamma) = \int_G \overline{\gamma(t)} f(t) d\mu(t), \quad \gamma \in G.$$

For $f, g \in L^1(G)$ one can show that $\widehat{(f * g)} = \hat{f}\hat{g}$ and $\widehat{(f^*)} = \overline{\hat{f}}$ (exercise).

9.3.3. We will require the following result from abstract harmonic analysis, which we use without proof in Theorem 9.3.4.

Theorem (Plancherel Theorem). *The Fourier–Plancherel transform extends from a map $L^1(G) \to L^1(\hat{G})$ to a unitary operator from $L^2(G) \to L^2(\hat{G})$.*

9.3.4. Theorem. *Let G be an abelian group. Then $C^*(G) \cong C_r^*(G) \cong C_0(\hat{G})$.*

Proof. Let $f, g \in L^1(G)$. Then $f * g(t) = \int_G f(s)g(s^{-1}t)d\mu(s)$ and putting $x = s^{-1}t$, we get $s = tx^{-1} = x^{-1}t$ so

$$\int_G f(s)g(s^{-1}t)d\mu(s) = \int_G f(x^{-1}t)g(x)d\mu(x) = \int_G g(x)f(x^{-1}t)d\mu(x),$$

and we have $f * g = g * f$, that is, $L^1(G)$ is commutative.

Let $\Gamma : L^1(G) \to C_0(\Omega(L^1(G))$ be the Gelfand transform. Recall that $\Omega(L^1(G))$ is the character space of $L^1(G)$ and notice that a character is exactly a one-dimensional representation. As we saw above, every representation of $L^1(G)$ corresponds to a unitary representation of G on the same Hilbert space; here the Hilbert space is $\mathbb{C}$. The one-dimensional unitary representations of G are just the characters of G, that is, $\hat{G}$. Thus the Gelfand transform maps $L^1(G) \to C_0(\hat{G})$. Moreover, we have

$$f \mapsto \hat{f}$$

where $\hat{f}$ is the Fourier–Plancherel transform of f. The range of $\Gamma(L^1(G))$ is clearly self-adjoint. Moreover, by definition of the Fourier–Plancherel transform it separates points. Thus $\Gamma(L^1(G))$ is dense in $C_0(\hat{G})$. By Plancherel's Theorem, this extends to a unitary operator $u : L^2(G) \to L^2(\hat{G})$. Then,

$$u(\lambda(f))u^*(\hat{g}) = u(\lambda(f))g = \widehat{(f * g)} = \hat{f}\hat{g},$$

when $f \in L^1(G)$ and $g \in L^2(G) \cap L^1(G)$. Thus $C_0(G) \ni f \to M_{\hat{f}} \in \mathcal{B}(L^2(\hat{G}))$, where $M_{\hat{f}}$ denotes the operator given by multiplication $\hat{f}$. Since this is isometric, λ is an isometric isomorphism. Thus $C^*(G) \cong C_r^*(G) \cong C_0(\hat{G})$. $\qquad\square$

9.3.5. Let G be a compact group. Since G is in particular a compact Hausdorff space, one can form the unital C*-algebra $C(G)$ of continuous $\mathbb{C}$-valued functions on G. Can we see the group structure in the C*-algebra? Consider group multiplication, $m : G \times G \to G$. Since passing to the function algebra is a contravariant functor, we reverse the arrows and get a *-homomorphism, called the "comultiplication" or "coproduct", which reverses the arrows

$$\Delta : C(G) \to C(G \times G) \cong C(G) \otimes C(G)$$

defined by

$$\Delta(f)(s,t) = f(st), \quad s, t \in G.$$

(The identification of $C(G \times G)$ with the tensor product $C(G) \otimes C(G)$ comes from Exercise 6.5.6, and we need not specify which C*-tensor product here, since commutative C*-algebras are nuclear.) Since m is associative, we have a commutative diagram

$$
\begin{array}{ccc}
G \times G \times G & \xrightarrow{\ m \times \mathrm{id}\ } & G \times G \\
\downarrow{\scriptstyle \mathrm{id} \times m} & & \downarrow{\scriptstyle m} \\
G \times G & \xrightarrow{\ m\ } & G.
\end{array}
$$

Reversing arrows gives us "coassociativity" of the comultiplication,

$$
\begin{array}{ccc}
C(G) & \xrightarrow{\quad\Delta\quad} & C(G) \otimes C(G) \\
\downarrow{\scriptstyle \Delta} & & \downarrow{\scriptstyle \mathrm{id} \otimes \Delta} \\
C(G) \otimes C(G) & \xrightarrow{\ \Delta \otimes \mathrm{id}\ } & C(G) \otimes C(G) \otimes C(G),
\end{array}
$$

which we leave as an exercise to show commutes.

9.3.6. The observations above lead to the notion of *compact quantum groups* which were introduced by Woronowicz, see [141, 142, 143].

Definition (Woronowicz). A *compact quantum group* is a pair (A, Δ) consisting of a unital C*-algebra A and a unital *-homomorphism $\Delta : A \to A \otimes_{\min} A$, called the coproduct, satisfying the following two conditions.

  (i) $\Delta : A \to A \otimes_{\min} A$ is coassociative, that is $(\Delta \otimes \mathrm{id}) \circ \Delta = (\mathrm{id} \otimes \Delta) \circ \Delta$,
 (ii) the linear spans of the sets

$$\{(a \otimes 1)\Delta(b) \mid a, b \in A\} \quad \text{and} \quad \{(1 \otimes a)\Delta(b) \mid a, b \in A\}$$

  are both dense in $A \otimes_{\min} A$.

9.3.7. Remark. Here (i) just says that, after replacing $C(G)$ with A, the (second) diagram of 9.3.5 commutes, hence the term *coassociativity*. The role of the second condition is not as transparent. A first guess for the definition of a compact quantum group might replace (ii) with the existence of a *counit* map and a *coinverse* map. However, these maps in general are only defined on a dense *-subalgebra (where the *coinverse* is also called the *antipode*), where they are unbounded and hence cannot be extended. To get the definition above, Woronowicz uses the equivalent formulation of a group as a semigroup with right and left cancellation. Thus (i) alone tells us we have a *quantum semigroup*, and (ii) tells us we have the noncommutative version of right and left cancellation, giving us a quantum group.

9.3.8. Remark. In the literature, the Definition 9.3.6 is often stated only for separable C*-algebras.

9.3.9. If G is any second countable compact group, then $C(G)$ can be given the structure of a compact quantum group, which we see from the above after checking the density conditions in Definition 9.3.6. If G is countable, discrete and abelian, then by Theorem 9.3.4, $\mathrm{C}_r^*(G) \cong C(\hat{G})$, so $\mathrm{C}_r^*(G)$ can also be given a compact quantum group structure.

Let G be a discrete group. Recall from 9.1.16 that elements of the form $\sum_{g \in G} b_g u_g$ where $b_g \in \mathbb{C}$ and $b_g = 0$ for all but finitely many $g \in G$ are dense in $\mathrm{C}_r^*(G)$.

Proposition. *Let G a discrete group. Then*

$$\Delta(u_g) = u_g \otimes u_g, \quad g \in G,$$

extends to a coassociative coproduct

$$\Delta : \mathrm{C}_r^*(G) \to \mathrm{C}_r^*(G) \otimes_{\min} \mathrm{C}_r^*(G).$$

Moreover, the linear spans of the sets

$$\{(a \otimes 1)\Delta(b) \mid a, b \in \mathrm{C}_r^*(G)\} \quad \text{and} \quad \{(1 \otimes a)\Delta(b) \mid a, b \in \mathrm{C}_r^*(G)\}$$

are dense in $\mathrm{C}_r^*(G) \otimes_{\min} \mathrm{C}_r^*(G)$. In particular, $(\mathrm{C}_r^*(G), \Delta)$ is a compact quantum group.

Proof. It is straightforward to see that Δ extends to a *-homomorphism. If e denotes the identity of G, then $u_e = 1_{C^*_r(G)}$. We have $\Delta(u_e) = u_e \otimes u_e$, so Δ is moreover unital. The two density conditions are obvious. Finally, to see that Δ satisfies the coassociativity condition, we appeal to the associativity of the minimal tensor product, which was established in Exercise 6.5.2. $\qquad\square$

Using this proposition, we see that it makes sense to think of the reduced group C*-algebra of a countable discrete group as the C*-algebra of continuous functions over its Pontryagin dual, even though, since the C*-algebra is noncommutative, there is no actual group underlying the construction.

A more detailed and accessible treatment of group C*-algebras, including many examples, can be found in [32]. A good introduction to compact quantum groups is given in [119].

9.4   Crossed products

An important and interesting generalisation of a group C*-algebra is the crossed product of a C*-algebra by a locally compact group G. The construction of the crossed product has a lot of similarities to the construction of group C*-algebras, but now we have to take into account things like the representations of the C*-algebra on which the group is acting. Again we will restrict ourselves to unimodular groups.

By an action α of G on a C*-algebra A we always mean a strongly continuous group homomorphism $\alpha : G \to \mathrm{Aut}\,(A)$, where $\mathrm{Aut}\,(A)$ denotes the group of *-automorphisms of A. The triple (A, G, α), where A is a C*-algebra with an action α of G on A, is called a C*-*dynamical system*.

9.4.1.   Definition. Suppose that G is a locally compact group acting on a C*-algebra A. A *covariant representation* is a triple (H, π, u) where H is a Hilbert space, (H, π) is a representation for A, (H, u) is a unitary representation for G and π and u satisfy the covariance condition

$$u(g)\pi(a)u(g)^* = \pi(\alpha_g(a)),$$

for every $a \in A$ and $g \in G$.

9.4.2. For a C*-dynamical system (A, G, α), the space $L^1(A, G, \alpha)$ is defined as follows.

First, take compactly supported continuous functions $C_c(G, A)$ with twisted convolution as multiplication

$$(f * g)(t) = \int_G f(s)\alpha_s(g(s^{-1}t))ds.$$

and with involution given by

$$f^*(s) = \alpha_s(f(s^{-1})^*).$$

$L^1(G, A, \alpha)$ is the completion of $C_c(A, G, \alpha)$ with respect to the 1-norm

$$\|f\|_1 = \int_G \|f(s)\| ds.$$

Note that the norm inside the integral is the norm of the C*-algebra A.

9.4.3. Similar to the way a unitary representation of locally compact group G gives rise to a representation of $L^1(G)$, we can relate a covariant representation for (A, G, α) to a representation of $L^1(A, G, \alpha)$ on an L^2-space, which can be defined as the completion of $C_c(A, G, \alpha)$ with respect to the norm

$$\|f\|_2 := \left(\int_G \|f(s)\|^2 ds \right)^{1/2}.$$

Given a covariant representation (H, π, u), the *integrated form* of (H, π, u) is defined to be

$$\pi(f)(g) = \int_G \pi(f(s)) u(s) g \, ds,$$

where $f \in L^1(A, G, \alpha)$ and $g \in L^2(A, G, \alpha)$.

Proposition. *Let G be a locally compact group, A a C*-algebra and $\alpha : G \to \mathrm{Aut}(A)$ an action of G on A. For any covariant representation, the associated integrated form is a representation of $L^1(A, G, \alpha)$ on $H = L^2(A, G, \alpha)$.*

In the opposite direction, a representation $\pi : L^1(A, G, \alpha) \to \mathcal{B}(H)$ also gives a covariant representation of (A, G, α) using an approximate unit for $L^1(G, A)$. Since the details are similar to the case for unitary representations, they are left as an exercise.

9.4.4. Given a Hilbert space H_0, the space $L^2(G, H_0)$ consists of square-integrable functions from G to H_0. It is a Hilbert space with respect to the inner product

$$\langle f, g \rangle = \int_G \langle f(s), g(s) \rangle_{H_0} ds.$$

Let $H := L^2(G, H_0)$. The (left) *regular covariant representation* (H, π, u) corresponding to $\pi_0 : A \to H_0$ is given by $\pi : A \to \mathcal{B}(H)$ defined by

$$\pi(a)(f)(s) = \pi_0(\alpha_{s^{-1}}) f(s), \quad f \in L^2(G, H_0), s \in G,$$

and $u : G \to \mathcal{U}(H)$ defined by

$$u(s)(f)(t) := f(s^{-1} t), \quad f \in L^2(G, H_0), t \in G.$$

9.4.5. Definition. Let $\alpha : G \to \mathrm{Aut}(A)$ be an action of a locally compact group G on a C*-algebra A. Let $\lambda : L^1(A, G, \alpha) \to \mathcal{B}(H)$ denote the direct sum of all integrated forms of regular representations. The *reduced crossed product of A by G*, written $A \rtimes_{r,\alpha} G$, is the closure of $\lambda(L^1(A, G, \alpha)) \subset \mathcal{B}(H)$.

The *full crossed product of A by G* is the closure of $\pi_u(L^1(G, A, \alpha)) \subset \mathcal{B}(H_u)$ where (H_u, π_u) denotes the universal representation, that is, the direct sum of all irreducible representations of $L^1(A, G, \alpha)$.

9.4.6. When G is a discrete group, calculations become easier because the Haar measure is simply counting measure. Thus all our formulas containing integrals are simply sums, and it will usually suffice to consider elements in the crossed product of the form

$$\sum_{g \in G} a_g u_g \text{ where } a_g \in A \text{ and } a_g = 0 \text{ for all but finitely many } g \in G,$$

as such elements are dense. Henceforth we will restrict to discrete groups and in particular the special case where $G = \mathbb{Z}$ and A is commutative. A good reference for more general crossed products is the book by Williams [130].

9.4.7 (Universal property of the full crossed product). Let A be a C*-algebra, G a discrete group and $G \to \text{Aut}\,(A)$ an action. The full crossed product is universal for covariant representations in the following sense. Given any covariant representation (H, π, v) there is a *-homomorphism

$$\rho : A \rtimes_\alpha G \to \mathrm{C}^*(\pi(A), v(G)),$$

satisfying

$$\rho\left(\sum_{g \in G} a_g u_g\right) = \sum_{g \in G} \pi(a_g)v_g,$$

and then extended by continuity. When A is unital, this map is surjective. If G is locally compact but not necessarily discrete, one can also define the analogous universal property, though it is slightly more complicated. The interested reader can see, for example, [130, Section 2.6], which is based on work by Raeburn [97].

Note that in particular that when A is unital we get a map $A \rtimes_\alpha G \twoheadrightarrow A \rtimes_{r,\alpha} G$. When G is amenable [144, 115], or more generally, when the *action* is amenable (see [2]), then this map is in fact a *-isomorphism.

9.4.8. In the case of discrete group acting on a unital C*-algebra, the associated reduced crossed product contains a isometric copy of A as well as a isometric copy of $C_r(G)$ via unital embeddings

$$\iota_A : A \to A \rtimes_\alpha G, \quad \iota_G : C_r^*(G) \mapsto A \rtimes_\alpha G.$$

As a vector space, we have that $A \rtimes_{r,\alpha} G$ is linearly isomorphic to the *vector space* tensor product $A \otimes C_r(G)$, and we define the multiplication restricted to either tensor factor as simply the multiplication in those C*-algebras. However, the multiplication of elements $a \otimes 1_G$ with $1_A \otimes b$, $a \in A$ and $b \in C_r^*(G)$ will be twisted by the action α. In particular, unlike in the usual algebraic tensor product, $a \otimes 1_G$ with $1_A \otimes b$ will not commute. For example, for $a \otimes 1_G$ and $1_A \otimes u_g$ the multiplication will be given by

$$(a \otimes 1_G)(1_A \otimes u_g) = (1_A \otimes u_g)(\alpha_{g^{-1}}(a) \otimes 1_G).$$

In this way, we can think of the crossed product of as a type of "twisted tensor product."

9.4.9. Example. When G acts on a locally compact Hausdorff space X, it induces an action on the C*-algebra $C_0(X)$ by $\alpha_g(f)(x) = f(g^{-1}x)$, $x \in X$, $g \in G$, $f \in C_0(X)$. This provides the following examples:

(a) Let G act trivially on a point x. Then the associated crossed products $\mathbb{C} \rtimes_r G$ and $\mathbb{C} \rtimes_f G$ are just $C_r^*(G)$ and $C_f^*(G)$, respectively.

(b) Suppose that α is the action of G on itself by translation: $\alpha_g(s) = gs$. Then $C_0(G) \rtimes_{f,\alpha} G = C_0(G) \rtimes_{r,\alpha} G \cong \mathcal{K}(L^2(G))$.

9.5   Crossed products by $\mathbb{Z}$ and minimal dynamical systems

In this section we will consider the case that $G = \mathbb{Z}$. In this case, G is amenable so the full and reduced crossed products coincide. Thus we will just denote the crossed product by $A \rtimes_\alpha \mathbb{Z}$. Given a C*-algebra A, a crossed product by $\mathbb{Z}$ arises from a single automorphism on A: If $\alpha : A \to A$ is an automorphism, then we define an action

$$\mathbb{Z} \to \mathrm{Aut}\,(A), \qquad n \mapsto \alpha^n.$$

We have already seen that there is a map $A \hookrightarrow A \rtimes_\alpha \mathbb{Z}$. There is also a map in the other direction $A \rtimes_\alpha \mathbb{Z} \to A$ which is not a *-homomorphism but nevertheless has some very good properties.

9.5.1. Definition. Let A be a C*-algebra and $B \subset A$ a C*-subalgebra. A *conditional expectation* is a c.p.c map $\Phi : A \to B$ satisfying $\Phi(b) = b$ for every $b \in B$ and $\Phi(b_1 a b_2) = b_1 \Phi(a) b_2$ for every $b_1, b_2 \in B$ and $a \in A$. (This last requirement is equivalent to saying that Φ is a left and right B module map, or just B-linear for short.) A conditional expectation Φ is faithful when $\Phi(a^*a) = 0$ if and only if $a = 0$.

Proposition. *Suppose that A and B are C*-algebras and $\Phi : A \to B$ is a positive B-linear idempotent map. Then Φ is a conditional expectation.*

Proof. We just need to show that Φ is contractive and completely positive. Since Φ is positive, we have that $\Phi((a - \Phi(a))^*(a - \Phi(a))) \geq 0$. Expanding using B-linearity together with the fact that since Φ is positive it is also *-preserving, this gives us

$$\Phi((a - \Phi(a))^*(a - \Phi(a))) = \Phi(a^*a - \Phi(a)^*a - a^*\Phi(a) + \Phi(a)^*\Phi(a))$$
$$= \Phi(a^*a) - \Phi(a)^*\Phi(a) - \Phi(a^*)\Phi(a) + \Phi(a)^*\Phi(a))$$
$$= \Phi(a^*a) - \Phi(a^*)\Phi(a)$$
$$= \Phi(a^*a) - \Phi(a)^*\Phi(a).$$

Thus $\Phi(a)^*\Phi(a) \le \Phi(a^*a)$. Suppose that $\|\Phi\| > 1$. Then there exists $a \in A$ with $\|a\| = 1$ and $1 < \|\Phi(a)\|^2$. But if $\|a\| = 1$ then also $\|a^*a\| = 1$ and so

$$1 < \|\Phi(a)\|^2 = \|\Phi(a)^*\Phi(a)\| \le \|\Phi(a^*a)\| \le \|\Phi\|,$$

a contradiction. Thus Φ is contractive.

Now let us show that Φ is completely positive. Let $n > 1$ and observe that for $a_1, \ldots, a_n, x_1, \ldots, x_n \in A$ we have

$$\sum_{i,j=1}^{n} \Phi(x_i)^* a_i^* a_j \Phi(x_j) = \left(\sum_{i=1}^{n} a_i \Phi(x_i)\right)^* \left(\sum_{i=1}^{n} a_i \Phi(x_i)\right) \ge 0.$$

Assume that $B \subset B(\mathcal{H})$ is a concrete C*-algebra and let $\xi \in H$. Then

$$\sum_{i,j=1}^{n} \langle \Phi(x_i)^* \Phi(a_i^* a_j) \Phi(x_j)\xi, \xi\rangle = \sum_{i,j=1}^{n} \langle \Phi(\Phi(x_i)^* a_i^* a_j \Phi(x_j))\xi, \xi\rangle$$
$$= \left\langle \Phi\left(\sum_{i,j=1}^{n} \Phi(x_i)^* a_i^* a_j \Phi(x_j)\right)\xi, \xi\right\rangle$$
$$\ge 0,$$

where the final inequality follows from the positivity of Φ. Thus by Proposition 7.1.5, Φ is completely positive contractive. It follows that Φ is a conditional expectation.   □

9.5.2. Theorem. *Let A be a unital C*-algebra and let $\alpha \in \mathrm{Aut}\,(A)$. Then there is a conditional expectation $\Phi : A \rtimes_\alpha \mathbb{Z} \to A$.*

Proof. Let $u := u_1 \in A \rtimes_\alpha \mathbb{Z}$ be the unitary implementing α in $A \rtimes_\alpha \mathbb{Z}$, that is, u is the unitary satisfying $uau^* = \alpha(a)$ for every $a \in A$ (where we regard A as a C*-subalgebra of $A \rtimes_\alpha \mathbb{Z}$). For $t \in \mathbb{R}$ let $\lambda = e^{2\pi i t}$. Assume that $A \rtimes_\alpha \mathbb{Z}$ is represented on a Hilbert space H. Then

$$u_\lambda : \mathbb{Z} \to \mathcal{U}(A \rtimes_\alpha \mathbb{Z}), \quad n \mapsto (\lambda u)^n$$

determines a unitary representation of $\mathbb{Z}$, and $(H, \mathrm{id}, u_\lambda)$ is a covariant representation. Thus, using the universal property of the crossed product (9.4.7), there is a surjective map

$$\pi_\lambda : A \rtimes_\alpha \mathbb{Z} \to \mathrm{C}^*(A, u_\lambda(\mathbb{Z})) \subset A \rtimes_\alpha \mathbb{Z},$$

such that $\pi_\lambda(a) = a$ for every $a \in A$ and $\pi_\lambda(u) = \lambda u$. It is easy to check that π_λ is injective, hence is an automorphism of $A \rtimes_\alpha \mathbb{Z}$.

Suppose that $x_0 = \sum_{k=-N}^N a_k u^k$ for some $N \in \mathbb{N}$ and $a_k \in A$. For $\lambda = e^{2\pi i t}$, define a function from $\mathbb{R}$ to $A \rtimes_\alpha \mathbb{Z}$ by

$$f_{x_0}(t) = \pi_{e^{2\pi i t}}(x_0) = \sum_{k=-N}^N e^{2\pi i t k} a_k u^k.$$

Observe that f_{x_0} is continuous. Now if $x \in A \rtimes_\alpha \mathbb{Z}$, we can also define the function $f_x(t) = \pi_{e^{2\pi i t}}(x)$. By approximating x by an element of the same form as x_0 (which is possible since these are dense), it follows that $f_x(t)$ is continuous for every $x \in A \rtimes_\alpha \mathbb{Z}$.

Define $\Phi : A \rtimes_\alpha \mathbb{Z} \to A \rtimes_\alpha \mathbb{Z}$ by

$$\Phi(x) = \int_0^1 \pi_{e^{2\pi i t}}(x) dt.$$

Since $\pi_{e^{2\pi i t}}(x) = f_x(t)$, this makes sense. Now, $\pi_{e^{2\pi i t}}$ is a *-isomorphism so in particular is positive and faithful. It follows that Φ is positive and faithful (to convince yourself of this, check what happens on elements of the form $\sum_{k=-N}^N a_k u^k$). It is also straightforward to check that $\Phi(1_{A \rtimes_\alpha \mathbb{Z}}) = 1_A$.

Let $a, b \in A$. Then

$$\Phi(axb) = \int_0^1 \pi_{e^{2\pi i t}}(a) \pi_{e^{2\pi i t}}(x) \pi_{e^{2\pi i t}}(b) dt = \int_0^1 a \pi_{e^{2\pi i t}}(x) b \, dt = a\Phi(x)b,$$

so Φ is A-linear and $\Phi(a) = a\Phi(1_{A \rtimes_\alpha \mathbb{Z}}) = a$.

We now show that $\Phi(A \rtimes_\alpha \mathbb{Z}) \subset A$. First, note that, for $k \neq 0$, we have

$$\Phi(u^k) = \int_0^1 e^{2\pi i t k} u^k \, dt = (2\pi i k)^{-1} (e^{2\pi i k} - 1_{A \rtimes_\alpha \mathbb{Z}}) u^k = 0.$$

Thus, for $x = \sum_{k=-N}^N a_k u^k$ we have $\Phi(x) = a_0 \in A$. Since elements of this form are dense, we see that $\Phi(A \rtimes_\alpha \mathbb{Z}) \subset A$. Finally, since Φ is the identity on A, we have $\Phi^2 = \Phi$. Thus Φ is a conditional expectation by Proposition 9.5.1. $\square$

9.5.3. Let X be an infinite compact metric space and $\alpha : X \to X$ a homeomorphism. Then α induces a $\mathbb{Z}$-action on the C*-algebra $C(X)$ via

$$\mathbb{Z} \to \mathrm{Aut}\,(C(X)), \qquad n \mapsto (f \mapsto f \circ \alpha^{-n}).$$

By slight abuse of notation, since we will rarely write down the action specifically, we write $C(X) \rtimes_\alpha \mathbb{Z}$ for the corresponding crossed product. Note that $C(X) \rtimes_\alpha \mathbb{Z}$ is generated by $C(X)$ and a unitary u satisfying $u f u^* = f \circ \alpha^{-1}$ for every $f \in C(X)$.

9.5.4. Proposition. *Let X be an infinite compact metric space, $\alpha : X \to X$ a homeomorphism, and E a nonempty proper closed α-invariant subset, that is, a nonempty closed subset $E \neq X$ such that $\alpha(E) \subset E$. Then the ideal*

$$I_E := \{f \in C(X) \mid f|_E = 0\} \subset C(X)$$

generates a proper ideal in $C(X) \rtimes_\alpha \mathbb{Z}$.

Proof. Let

$$J_E := \left\{ a \in C(X) \rtimes_\alpha \mathbb{Z} \mid \text{ there is } N \in \mathbb{N} \text{ such that } a = \sum_{k=-N}^{N} f_k u^k, \, f_k \in I_E \right\}.$$

Note that J_E is closed under addition. Suppose that $a = \sum_{k=-N}^{N} f_k u^k \in J_E$ and $b = \sum_{k=-M}^{M} g_k u^k$ for some $M \in \mathbb{N}$ and $g_k \in C(X)$. Then

$$ab = \sum_{k,l} f_k u^k g_l u^{-k} u^{k+l} = \sum_{k,l} f_k (g_l \circ \alpha^{-k}) u^{k+l}.$$

For every k and every l, we have $f_k(g_l \circ \alpha^{-k})|_E = 0$, so $ab \in J_E$. Also,

$$ba = \sum_{k,l} g_k u^k f_l u^{-k} u^{k+l} = \sum_{k,l} g_k (f_l \circ \alpha^{-k}) u^{k+l}.$$

Since E is α-invariant, we have $(f_l \circ \alpha^{-k}) \in I_E$ for every k, and hence for every k and every l we also have $g_k(f_l \circ \alpha^{-k}) \in I_E$. It follows that $ba \in J_E$. Thus $\overline{J_E}$ is an ideal in $C(X) \rtimes_\alpha \mathbb{Z}$.

To see that $\overline{J_E}$ is a proper ideal, consider its image under the conditional expectation $\Phi : C(X) \rtimes_\alpha \mathbb{Z} \to C(X)$ from Theorem 9.5.2. We have $\Phi(J_E) = I_E$ so $\Phi(\overline{J_E}) = I_E$. Thus $\overline{J_E} \neq A$, that is, J_E is proper. $\qquad \square$

9.5.5. Given a homeomorphism $\alpha : X \to X$ of a locally compact Hausdorff space and a point $x \in X$, the *orbit* of x is the set $\{\alpha^n(x) \mid n \in \mathbb{Z}\}$. We also define the *forward orbit* of x to be $\{\alpha^n(x) \mid n \in \mathbb{N}\}$ and the *backward orbit* of x to be $\{\alpha^{-n}(x) \mid n \in \mathbb{N}\}$.

Definition. Let X be a locally compact Hausdorff space. A homeomorphism $\alpha : X \to X$ is *minimal* if X has no proper closed α-invariant subsets. We also say that the topological dynamical system (X, α) is a *minimal dynamical system*.

9.5.6. Proposition. *Let X be a compact metric space and let $\alpha : X \to X$ be a homeomorphism. The following are equivalent:*

  (i) *The homeomorphism α is minimal.*
 (ii) *For every $x \in X$, the orbit of x is dense in X.*
(iii) *For every $x \in X$, the forward orbit of x is dense in X.*
(iv) *For every $x \in X$, the backward orbit of x is dense in X.*

Proof. For (i) implies (ii), suppose that α is minimal and let $x \in X$. Suppose there is some open set $U \subset X$ such that $\alpha^n(x) \notin U$ for every $n \in \mathbb{Z}$. Let $E = X \setminus \cup_{n \in \mathbb{Z}} \alpha^n(U)$ and note that $\alpha(E) \subset E$. Thus minimality implies $E = X$, or $E = \emptyset$. If the latter holds, then $x \in \alpha^n(U)$ for some $n \in \mathbb{Z}$. But then $\alpha^{-n}(x) \in U$, a contradiction. Thus we must $U = \emptyset$, so the orbit of x is dense.

For (ii) implies (iii), we will show that for every $x \in X$ and for any open set $U \subset X$ there is a $n \in \mathbb{N}$ such that $\alpha^n(x) \in U$, which implies that the forward orbit of x is dense. By assumption, every orbit is dense and so $\{\alpha^n(U) \mid n \in \mathbb{Z}\}$ is an open cover of X. By compactness of X there is $N \in \mathbb{N}$ such that

$$\{\alpha^n(U) \mid -N \le n \le N\}$$

is also an open cover of X. Since α is a homeomorphism, we have that $\alpha^{-N}(X) = X$. Thus $\{\alpha^n(U) \mid -2N \le n \le 0\}$ is an open cover of X and therefore there is some m, $-N \le m \le 0$ such that $x \in \alpha^m(U)$. So $\alpha^{-m}(x) \in U$.

Suppose $x \in X$ and $U \subset X$ is an open set. If forward orbits are dense, then $\{\alpha^n(U) \mid n \in \mathbb{N}\}$ is an open cover of X. Thus there is some $m \in \mathbb{N}$ such that $x \in \alpha^m(U)$. So $\alpha^{-m}(x) \in U$. Since x and U were arbitrary, this shows (iii) implies (iv).

Finally, suppose that (iv) holds. Suppose that $E \subset X$ is a closed subset with $\alpha(E) \subset E$. Then $\alpha^n(E) \subset E$ for every $n \in \mathbb{N}$. Let $U = X \setminus E$. If $U \ne \emptyset$, then $\{\alpha^{-n}(U) \mid n \in \mathbb{N}\}$ is an open cover for X. Let $x \in E$. Then there is some $n \in \mathbb{N}$ such that $x \in \alpha^{-n}(U)$ and so $\alpha^n(x) \in U$, contradicting the fact that $E \cap U = \emptyset$. Thus $U = \emptyset$ and $E = X$, which shows that α is minimal. $\qquad\square$

9.5.7. Lemma. *Let X be an infinite compact metric space and $\alpha : X \to X$ a minimal homeomorphism. Suppose that $J \subset C(X) \rtimes_\alpha \mathbb{Z}$ is a nonzero ideal. Then $J \cap C(X) \ne 0$.*

Proof. For a contradiction, suppose $J \cap C(X) = 0$. Let $a \in J$ be nonzero. Then since Φ is faithful, $\Phi(a^*a) \in C(X)$ is also nonzero. We can approximate a^*a by an element $b = \sum_{j=-N}^{N} f_j u^j$ for some $N \in \mathbb{N}$ and $f_j \in C(X)$ with $\|b - a^*a\| < \|\Phi(a^*a)\|/4$. Let $x \in X$ satisfy $|\Phi(a^*a)(x)| \ge 3\|\Phi(a^*a)\|/4$.

Let $B = C(X) + J$. Note that B is subalgebra of $C(X) \rtimes_\alpha \mathbb{Z}$ and J is ideal in B. Let $\rho_0 : C(X) + J \to \mathbb{C}$ be the map given by the composition of the maps

$$C(X) + J \twoheadrightarrow (C(X) + J)/J \cong C(X)/J \cap C(X) \cong C(X) \to \mathbb{C},$$

where the final map $C(X) \to \mathbb{C}$ is the evaluation at x, that is $\mathrm{ev}_x(f) = f(x)$. Then ρ_0 is a state on $C(X) + J$ and so we can extend it to a state $\rho : C(X) \rtimes_\alpha \mathbb{Z} \to \mathbb{C}$. Note that $\rho(a^*a) = 0$ since $a^*a \in J$.

Since x has dense orbit, there is an open set U containing x such that the sets $\{\alpha^n(U) \mid -N \le n \le N\}$ are pairwise disjoint. Let $f_x \in C(X)$ satisfy $f_x(x) = 1$

and $\mathrm{supp}(f_x) \subset U$. Let $-N \leq j \leq N$ and $j \neq 0$. Since ρ is multiplicative on $C(X)$, by Exercise 7.3.3 and 7.3.5 and the fact that $x \notin \mathrm{supp}(f)$, we have,

$$\rho(f_j u^j) = \rho_0(f_j)\rho(u^j) = \rho_0(f_x f_j)\rho(u^j) = \rho(u^j((f_x f_j) \circ \alpha^j)) = 0,$$

since $x \notin \mathrm{supp}((f_x f_j) \circ \alpha^j)$. It follows that $\rho(b) = \rho(f_0)$. Then

$$\Phi(a^*a)/4 > \|b - a^*a\| \leq \|\rho(b - a^*a)\| = \|\rho(b)\|$$
$$= |f_0(x)| \geq |\Phi(a^*a)(x)| - \|b - a^*a\|$$
$$\geq \Phi(a^*a)/2,$$

a contradiction. Thus $J \cap C(X) \neq 0$. $\qquad\square$

9.5.8. Theorem. *Let X be an infinite compact metric space and let $\alpha : X \to X$ be a homeomorphism. Then $C(X) \rtimes_\alpha \mathbb{Z}$ is simple if and only if α is minimal.*

Proof. The fact that simplicity implies minimality follows from Proposition 9.5.4. Now assume that α is minimal and let $J \subset C(X) \rtimes_\alpha \mathbb{Z}$ be a nonempty ideal. Since $I := J \cap C(X)$ is an ideal in $C(X)$, it follows from Exercise 2.5.12 that there is a closed subset $E \subset X$ such that $I = \{f \in C(X) \mid f|_E = 0\}$, or, equivalently, there is an open $U \subset X$ such that $I = C_0(U)$, where

$$U = \{x \in X \mid \text{ there is } f \in I \text{ such that } f(x) \neq 0\}.$$

By the previous lemma, U must be nonempty. Let $h \in I$ and let $x \in X$ be such that $h(x) \neq 0$. Then there is some open set $U \subset X$ containing x such that $h(y) \neq 0$ for every $y \in U$. Since the orbit of x is dense, $\{\alpha^n(U) \mid n \in \mathbb{Z}\}$ is an open cover of X, and since X is compact, there is $N \in \mathbb{N}$ such that $\{\alpha^n(U) \mid -N \leq n \leq N\}$ is an open cover of X. Let $g = \sum_{k=-N}^{N} u^n h u^{-n}$. Then $g \in J$ and $g(x) > 0$ for every $x \in X$, so g is invertible. Thus $1_{C(X) \rtimes_\alpha \mathbb{Z}} \in J$, and so $C(X) \rtimes_\alpha \mathbb{Z}$ is simple. $\qquad\square$

9.6 Exercises

9.6.1. Let G be a discrete group.

(i) Show that convolution and involution in $C_c(G)$ is well defined.

(ii) Show that convolution is associative.

(iii) Show that if G is countable, then $\mathrm{C}_r^*(G)$ is separable.

9.6.2. Let G be a topological group and suppose that μ, μ' are two left Haar measures on G. Show that $L^1(G, \mu) \cong L^1(G, \mu')$.

9.6.3. Let G be a topological group and H a Hilbert space. Suppose that $u : G \to \mathcal{U}(H)$ is a strongly continuous homomorphism, where $\mathcal{U}(H)$ is the group of unitary operators on H. Show that $s \mapsto u(s)\xi$ is continuous for every $\xi \in H$ if and only if the function $G \times H \to H$, $(s, \xi) \mapsto u(s)\xi$ is continuous.

9.6.4. Let G be a locally compact group and $u : G \to \mathcal{U}(H)$ a unitary representation.

(i) Show that u is irreducible if and only if there are no proper closed linear subspaces $K \subset H$ that are stable under $u(G)$.

(ii) Show that C*-subalgebra of $\mathcal{B}(H)$ generated by $u(G)$ is irreducible if and only if u is irreducible.

9.6.5. Let G be a locally compact group. Show that the left regular representation $\lambda : G \to \mathcal{B}(L^2(G))$ is a unitary representation.

9.6.6. For $f \in L^1(G)$, let $\hat{f}$ denote its Fourier–Plancherel transform. Show that when $f, g \in L^1(G)$ we have $(f * g)\widehat{\ } = \hat{f}\hat{g}$ and $(f^*)\widehat{\ } = \overline{\hat{f}}$.

9.6.7. Show that the following discrete groups are amenable.

(i) Finite groups are amenable.

(ii) $\mathbb{Z}$ is amenable.

(iii) Subgroups of amenable groups are amenable.

9.6.8. Let $k \in \mathbb{N} \setminus \{0\}$. What is $C_r^*(\mathbb{Z}^k)$? What is $C^*(\mathbb{Z}^k)$?

9.6.9. Let $H = \ell^2(\mathbb{Z} \times \mathbb{N})$ and let $\{\xi_{n,k} \mid n \in \mathbb{N}, k \in \mathbb{Z}\}$ be an orthonormal basis. Set $0 < q < 1$. Define $a, c \in \mathcal{B}(H)$ as follows:

$$a\,\xi_{n,k} = \sqrt{1_{\mathcal{B}(H)} - q^{2n}}\,\xi_{n-1,k}, \quad c\,\xi_{n,k} = q^n \xi_{n,k+1}.$$

Let $A = C^*(a, c, 1_{\mathcal{B}(H)})$.

(i) Show that $a^*a + c^*c = 1_{\mathcal{B}(H)}$, $aa^* + q^2 c^*c = 1_{\mathcal{B}(H)}$, $c^*c = cc^*$, $ac = qca$ and $ac^* = qc^*a$. Deduce that

$$\begin{pmatrix} a & -qc^* \\ c & a^* \end{pmatrix} \in M_2(A)$$

is a unitary.

(ii) Define $\Delta : A \to A \otimes_{\min} A \subset \mathcal{B}(H \otimes H)$ by

$$\Delta(a) = a \otimes a - qc^* \otimes c, \quad \Delta(c) = c \otimes a + a^* \otimes c.$$

Show that (A, Δ) is a compact quantum group.

(iii) Suppose that H is a Hilbert space and $a, c \in \mathcal{B}(H)$ satisfy the relations in (i) with $q = 1$. Show that $C^*(a, b, 1) \cong C(SU(2))$ (equivalently, $C^*(a, b, 1) \cong C(S_3)$).
(Hint: Use that fact that the Hilbert Nullstellensatz says that $\mathbb{C}[a, b, c, d]/ < ad - bc - 1 >$ is the ring of polynomials on $SU(2)$.)

9.6.10. Let G be a discrete group acting trivially on a point x. Show that the associated crossed products $\mathbb{C} \rtimes_r G$ and $\mathbb{C} \rtimes_f G$ are just $C_r^*(G)$ and $C_f^*(G)$, respectively.

9.6.11. Let $\theta \in \mathbb{R}$. Define $\varphi_\theta : \mathbb{T} \to \mathbb{T}$ by $\varphi_\theta(z) = e^{2\pi i\theta}z$, which is a homeomorphism. Show that $(\mathbb{T}, \varphi_\theta)$ is minimal if and only if θ is irrational. Thus $C(\mathbb{T}) \rtimes_{\varphi_\theta} \mathbb{Z}$ is simple if and only if θ is irrational. If θ is irrational, then $C(\mathbb{T}) \rtimes_{\varphi_\theta} \mathbb{Z}$ is called an *irrational rotation algebra*. What happens when θ is a rational number? An integer?

9.6.12. Let $X = \{0,1\}^{\mathbb{N}}$. Then X is a Cantor set and we can define a homeomorphism $\varphi : X \to X$ as follows. If $x \in X$ we think of x as a sequence $(x_n)_{n\in\mathbb{N}}$ of zeros and ones. If $(x_n)_{n\in\mathbb{N}}$ is the sequence consisting only of ones, then $\varphi(x)$ is the sequence consisting of all zeros. Otherwise, let $m > 0$ be the least integer such that $x_n = 1$ for every $n < m$. In that case, $\varphi(x) = y$ where $y_n = 0$ for $n < m$, $y_m = 1$ and $y_n = x_n$ for every $n > m$. (In other words φ is addition modulo 2 with carry-over.) The dynamical system (X, φ) is called the *2-odometer*.

(i) Show that (X, φ) is minimal and hence $C(X) \rtimes_\varphi \mathbb{Z}$ is simple.
(ii) For $n \geq 2$, define the *n-odometer* (X, φ_n) where φ_n is a homeomorphism on the Cantor set $X = \{0, 1, \ldots, n-1\}^{\mathbb{N}}$ and show that $C(X) \rtimes_{\varphi_n} \mathbb{Z}$ is again simple.

9.6.13. Let (X, φ) be a minimal dynamical system where X is an infinite compact metric space.

(i) Let $Y \subset X$ be a nonempty closed subset. For $x \in Y$, define the *first return time* of x to Y to be

$$r_Y(x) := \min\{n > 0 \mid \varphi^n(x) \in Y\}.$$

Show that for any set Y, there are only finitely many different first return times, that is,

$$\{r_Y(x) \mid x \in Y\}$$

is finite.
(ii) Show that if $Y_1 \supset Y_2$ then $r_{Y_1}(x) \leq r_{Y_2}(x)$ for every $x \in Y_2$.
(iii) For the rest of the exercise let (X, φ) be the Cantor 2-odometer of Exercise 9.6.12. Suppose $Y \subset X$ is a nonempty clopen subset. Show that there is an $n \in \mathbb{N}$ such that any $x \in Y$ has return time 2^n.
(iv) Show that $\alpha^k(Y) \cap \alpha^j(Y) = \emptyset$ when $j \neq k \in \{1, \ldots, 2^n\}$ and $\cup_{k=1}^{2^n} \alpha^k(Y) = X$.
(v) For $1 \leq k \leq 2^n$, let $\chi_{\alpha^k}(Y)$ denote the indicator function on $\alpha^k(Y) \subset X$, which is in $C(X)$ since Y is clopen. Let $u \in C(X) \rtimes_\alpha \mathbb{Z}$ be the unitary satisfying $ufu^* = f \circ \alpha^{-1}$, $f \in C(X)$. Show that the elements

$$u^{j-k}\chi_{\alpha^k(Y)}, \qquad 1 \leq j,k \leq 2^n$$

are a system of matrix units in the C*-subalgebra $C^*(C(X), uC(X \setminus Y)) \subset C(X) \rtimes_\alpha \mathbb{Z} \cong C^*(C(X), u)$ and use this to show that

$$C^*(C(X), uC(X \setminus Y)) \cong C(Y) \otimes M_{2^n}$$

is an AF algebra.

(vi) For a clopen subset $Y \subset X$, let $A_Y := \mathrm{C}^* C(X, uC(X \setminus Y))$. Let $y \in X$ and let $Y_1 \supset Y_2 \supset \cdots$ be a nested decreasing sequence of clopen subsets with $\cap_{n \in \mathbb{N}} Y_n = \{y\}$. Show that $A_{\{y\}} = \varinjlim A_{Y_n}$ and $A_{\{y\}}$ is isomorphic to the UHF algebra M_{2^∞}.

9.6.14. Let (X, φ) be a minimal dynamical system where X is an infinite compact metric space. Let μ be a φ-invariant Borel probability measure on X, that is, $\mu(X) = 1$ and $\mu(\varphi^{-1}(E)) = \mu(E)$ for every Borel set $E \subset X$. Show that

$$\tau(a) = \int_X \Phi(a) d\mu$$

defines a tracial state on $C(X) \rtimes_\varphi \mathbb{Z}$.

10   Cuntz algebras

The Cuntz algebras are a well-studied class of unital simple C*-algebras, named after Joachim Cuntz who constructed them in [28]. Cuntz constructed them to show the existence of a simple separable C*-algebra A which is *infinite*, which is to say, contains an element $x \in A$ such that $x^*x = 1_A$ and $xx^* \neq 1_A$. Previous examples of infinite C*-algebras included the *Calkin algebra* $\mathcal{B}(H)/\mathcal{K}(H)$ or came from the theory of von Neumann algebras – the so-called type III factors – and were never separable. Thus it was unclear if this phenomenon could occur unless the C*-algebra was big enough. We will also see that the Cuntz algebras realise the property of being infinite in a particularly strong way: they are *purely infinite*. Purely infinite simple C*-algebras have many peculiar properties. For example, projections display a strange Banach–Tarski-type paradoxical behaviour in the sense that any given projection can be chopped up into subprojections the same size as the original projection. Following the construction of the Cuntz algebras, many generalisations emerged which could also be used to construct purely infinite C*-algebras. Of particular note are Cuntz–Krieger algebras, graph algebras and Cuntz–Pimsner algebras, which are not always purely infinite, but for which conditions are known that imply pure infiniteness. In fact, it turns out that there are many purely infinite C*-algebras. One of the first major results of the classification program was the Kirchberg–Phillips classification of purely infinite simple C*-algebras. Two of the Cuntz algebras we meet in this section, $\mathcal{O}_2$ and $\mathcal{O}_\infty$ play a particularly interesting role in the classification. We won't have too much to say about that in this book, however there is some discussion in Chapters 16 and 18.

In this chapter, we construct the Cuntz C*-algebras as universal objects, and so the first section focuses on the notion of a universal C*-algebra. In the next section, we collect some facts about purely infinite simple C*-algebras. The final section contains the construction of the Cuntz algebras and the proof that they are purely infinite and simple.

10.1   Universal C*-algebras

Interesting examples of C*-algebras are often described as universal objects given by generators and relations. This has to be done with some care, because some generators and relations cannot be used to construct C*-algebras. This is because the generators and relations have to be realisable as bounded operators on a Hilbert space. Thus relations, which will usually be algebraic relations among the generators and their adjoints, will require a norm condition. A good discussion on the nuances of universal C*-algebras can be found in [81].

10.1.1. Definition. Given a set of generators $\mathcal{G}$ and relations $\mathcal{R}$, a *representation* of $(\mathcal{G}, \mathcal{R})$ on a Hilbert space H is a map $\pi : \mathcal{G} \to \mathcal{B}(H)$ such that $\pi(\mathcal{G})$ are operators satisfying the relations $\mathcal{R}$.

© The Author(s), under exclusive license to Springer Nature Switzerland AG 2021
K. R. Strung, *An Introduction to C*-Algebras and the Classification Program*, Advanced
Courses in Mathematics - CRM Barcelona, https://doi.org/10.1007/978-3-030-47465-2_10

10.1.2. If $\mathcal{A}$ is the free *-algebra on the generators $\mathcal{G}$ subject to the relations $\mathcal{R}$ then this induces a representation $(H, \pi_{\mathcal{G}})$ of $\mathcal{A}$ on H. Let $(\mathcal{G}, \mathcal{R})$ be a set of generators and relations and let $\mathcal{A}$ denote the free algebra on generators $\mathcal{G}$. Suppose that, for every $a \in A$,

$$p(a) := \sup\{\|\pi_{\mathcal{G}}(a)\| \mid \pi_{\mathcal{G}} \text{ is a representation of } (\mathcal{G}, \mathcal{R})\}$$

is finite. Then $p(a)$ is a seminorm on $\mathcal{A}$.

10.1.3. Definition. Given a set of generators and relations $(\mathcal{G}, \mathcal{R})$ such that the $p(a) < \infty$ for every $a \in A$, where $\mathcal{A}$ and p are defined as above. The *universal* C*-algebra of $(\mathcal{G}, \mathcal{R})$, written $\mathrm{C}^*(\mathcal{G}, \mathcal{R})$, is the enveloping C*-algebra of $(\mathcal{A}, p)$.

By the *universal unital* C*-*algebra* on $(\mathcal{G}, \mathcal{R})$ we mean the universal C*-algebra generated by $\mathcal{F} \cup \{1\}$ with relations $\mathcal{R} \cup \{1g = g1 = g \text{ for every } g \in G\}$.

10.1.4. The universal C*-algebra $A = \mathrm{C}^*(\mathcal{G}, \mathcal{R})$ has the following universal property: If a C*-algebra B contains a set of elements X in one-to-one correspondence with $\mathcal{G}$ which also satisfies the relations $\mathcal{R}$, then there is a surjective *-homomorphism $A \to \mathrm{C}^*(X)$ where $\mathrm{C}^*(X)$ is the C*-subalgebra of B generated by X.

10.1.5. Examples. 1. The first example is a nonexample. There is no universal C*-algebra generated by a self-adjoint element. Throughout the book we have repeatedly made use of the fact that a self-adjoint element a in a C*-algebra generates a commutative C*-subalgebra, and so there are plenty of representations $(\mathcal{G}, \mathcal{R})$ where $\mathcal{G} = \{x\}$ and $\mathcal{R} = \{x = x^*\}$. However if $\mathcal{A}$ is the free *-algebra generated by x subject to the relation $x = x^*$, then for any $a \in \mathcal{A}$ we can find a representation of $(\mathcal{G} = \{x\}, \mathcal{R} = \{x = x^*\})$ where the norm of a is arbitrarily large. Thus $p(a)$ will not be finite, and the universal C*-algebra cannot exist.

2. Let $\mathcal{G} = \{a, 1\}$ and $\mathcal{R} = \{\|a\| \leq 1, a^* = a, 1 = 1^* = 1^2, 1a = a1 = a\}$. Then $\mathrm{C}^*(\mathcal{G}, \mathcal{R}) \cong C([-1, 1])$.

3. Let $\mathcal{G} = \{u\}$ and $\mathcal{R} = \{u^*u = uu^* = 1\}$. The universal unital C*-algebra – the unital universal C*-algebra generated by a unitary – is then isomorphic to $C(\mathbb{T})$. Note how this is related to the group C*-algebra construction for $\hat{\mathbb{T}} \cong \mathbb{Z}$. Here there is no explicit mention of a norm bound, however this is implied by the relation $u^*u = 1$ since any operator u on any Hilbert space H satisfying this relation will have to also satisfy $\|u\| = 1$.

10.1.6. An important example is the following. Let $n \in \mathbb{N}$ and let $\mathcal{G} = \{e_{ij} \mid 1 \leq i, j \leq n\} \cup \{1\}$ and $\mathcal{R}$ be the relations $\{e_{ij}e_{kl} = \delta_{jk}e_{il}, e_{ii}^* = e_{ii}^2 = e_{ii}, \sum_{i=1}^n e_{ii} = 1\}$. The universal C*-algebra generated by these generators and relations is of course just M_n. Thus, whenever we find elements satisfying these relations (nontrivially) in a given C*-algebra A, we get a copy of M_n sitting inside of A, as we observed in 8.4.6.

10.2 Purely infinite C*-algebras

This definition for purely infinite, simple, unital C*-algebras appeared as [29, Definition 2.1]. The definition was later generalised to include simple C*-algebras which are not necessarily unital as well as C*-algebras which are not necessarily simple, see for example [68].

10.2.1. Definition. Let A be a unital simple infinite-dimensional C*-algebra. We say A is *purely infinite* if for every $x \in A \setminus \{0\}$ there are $a, b \in A$ such that $axb = 1_A$.

10.2.2. Lemma. *Suppose that A is a unital simple C*-algebra. Then, for any nonzero positive element $a \in A$ there are $n \in \mathbb{N} \setminus \{0\}$ and elements $x_1, \ldots, x_n \in A$ such that $\sum_{k=1}^{n} x_k a x_k^* = 1_A$.*

Proof. Since A is simple and unital, it has no nontrivial algebraic ideals (Exercise 2.5.11). Thus there are $m \in \mathbb{N} \setminus \{0\}$ and $y_1, \ldots, y_m, z_1, \ldots, z_m \in A$ such that $\sum_{k=1}^{m} y_k a z_k = 1_A$. Note that for each $i \leq k \leq m$ we have $(y_k - z_k^*) a (y_k^* - z_k) \geq 0$. Thus

$$0 \leq \sum_{k=1}^{m} (y_k - z_k^*) a (y_k^* - z_k)) = \sum_{k=1}^{m} (y_k a y_k^* + z_k^* a z_k) - \sum_{k=1}^{m} (y_k a z_k + z_k^* a y_k^*).$$

It follows that

$$2 = \sum_{k=1}^{m} (y_k a z_k + z_k^* a y_k^*) \leq \sum_{k=1}^{m} (y_k a y_k^* + z_k^* a z_k).$$

Thus, for $n = 2m$, by scaling the y_k and z_k by $1/\sqrt{2}$, there are $w_1, \ldots, w_n \in A$ such that

$$1_A \leq \sum_{k=1}^{n} w_k a w_k^*.$$

Let $c := \sum_{k=1}^{n} w_k a w_k^*$ and for $1 \leq k \leq n$ and $j \in \mathbb{N} \setminus \{0\}$ set $x_k = c^{-1/2} w_k$. Then

$$\sum_{k=1}^{n} x_k a x_k^* = c^{-1/2} w_k a w_k^* c^{-1/2} = 1_A,$$

which completes the proof. $\square$

10.2.3. Recall that a projection p in a C*-algebra is called infinite if it is not finite (Definition 8.3.11), that is, there exists $v \in A$ such that $v^* v = p$ and $vv^* < p$. We will see that a simple unital purely infinite C*-algebra has many infinite projections.

Theorem. *Let A be a unital simple infinite-dimensional C*-algebra. Then the following conditions are equivalent.*

(i) *A is purely infinite.*

(ii) *Every nonzero hereditary* C*-*subalgebra contains a projection that is Murray–von Neumann equivalent to* 1_A.

(iii) *Every nonzero hereditary* C*-*subalgebra contains an infinite projection.*

Proof. First assume that A is purely infinite. Let $B \subset A$ be a hereditary C*-subalgebra and choose a nonzero element $x \in B$. First we claim that there is an element $c \in A$ such that $cx^*xc^* = 1_A$. Since A is purely infinite, there are $a, b \in A$ such that $axb = 1_A$. Then $1_A = b^*x^*a^*axb \leq \|a\|^2 b^*x^*xb$, so b^*x^*xb is invertible. Let $c = (b^*x^*xb)^{-1/2}b^*$. A simple calculation shows that $cx^*xc^* = 1_A$, which proves the claim. Let $v = xc^*$. Then $v^*v = 1_A$, so vv^* is a projection (Exercise 3.3.1) and vv^* is Murray–von Neumann equivalent to 1_A. Since $vv^* = xc^*c^*x^* \in xAx^* \subset B$, this shows that (i) implies (ii).

Now suppose that (ii) holds. Let $B \subset A$ be a hereditary C*-subalgebra. There are a nonzero projection $p \in B$ and $v \in A$ such that $v^*v = p$ and $vv^* = 1_A$. The hereditary C*-subalgebra pAp is nonzero, so there is a projection $q \in pAp$, hence $q \leq p$ and $w \in A$ such that $w^*w = q$ and $ww^* = 1_A$. Then $(v^*w)^*(v^*w) = w^*w = q$ while $(v^*w)(v^*w)^* = w^*w = p$, so $p \in B$ is infinite, showing (ii) implies (iii).

We show (iii) implies (ii). Suppose (iii) holds and $B \subset A$ is a hereditary C*-subalgebra containing an infinite projection p. Let $q \leq p$ and $v \in A$ satisfy $v^*v = p$ and $vv^* = q$. By the previous lemma, there are $x_1, \ldots, x_n$ such that

$$\sum_{k=1}^{m} x_k(p - q)x_k^* = 1_A.$$

Let $z := \sum_{k=1}^{m} v^k(p - q)x_k^*$. Note that $v^*(p - q)v = v^*vv^*(v^*v)vv^*v - v^*(vv^*)v = v^*qpqv = v^*qv = 0$. We have

$$\|(p - q)(v^j)^*v^i(p - q)\|^4$$
$$= \|(p - q)(v^i)^*v^j(p - q)(v^j)^*v^i(p - q)\|^2$$
$$= \|(p - q)(v^i)^*v^j(p - q)(v^j)^*v^i(p - q)(v^i)^*v^j(p - q)(v^j)^*v^i(p - q)\|.$$

If $i > j$ then $(v^i)^*v^j(p - q)(v^j)^*v^i = 0$ and if $j > i$ then $(v^j)^*v^i(p - q)(v^i)^*v^j = 0$, so in either case, if $i \neq j$ then $(p - q)(v^j)^*v^i(p - q) = 0$. Thus

$$z^*z = \sum_{k=1}^{m}\sum_{j=1}^{m} x_k(p - q)(v^k)^*v^j(p - q)x_j^* = \sum_{k=1}^{m} x_k(p - q)x_k^* = 1_A.$$

On the other hand, $zz^* = \sum_{k=1}^{m}\sum_{j=1}^{m} v^k(p-q)x_k^*x_j(p-q)(v^j)^* \in vAv^* = qvAv^*q$, so $zz^* \in B$ as $qAq \subset pAp \subset B$. So we have that (iii) implies (ii).

Finally, we show that (ii) implies (i). Let $x \in A$ be nonzero. There is a projection $p \in \overline{x^*Ax}$ and $v \in A$ such that $v^*v = p$ and $vv^* = 1_A$. Let $0 < \epsilon < 1$ and choose $b \in A$ such that $\|p - x^*cx\| < \epsilon$. Then x^*bx is invertible in the corner pAp. Let $a := vx^*c$ and $b := (x^*cx)^{-1}v^*$. Then $axb = vx^*cx(x^*cx)^{-1}v^* = vpv^* = 1_A$. □

10.3 Cuntz algebras

In this section, we are interested in universal algebras that result in C*-algebras that are, on the one hand, very far removed from the UHF algebras we have already encountered – they are never stably finite, for example – yet on the other hand bear some interesting similarities: they are simple and, like some of our UHF algebras, some of the Cuntz algebras have a certain self-absorbing property, which we will see in Chapter 16. They were originally introduced by Cuntz in [28], whence the name. Most of this section follows that paper.

10.3.1. Definition. Let $n \in \mathbb{N} \setminus \{0\}$ and let $\mathcal{G} = \{s_1, \ldots, s_n\}$. Define relations on $\mathcal{G}$ by

$$\mathcal{R}_n = \left\{ \sum_{j=1}^{n} s_j s_j^* = 1, s_i^* s_i = 1, 1 \leq i \leq n \right\}.$$

Then the universal unital C*-algebra on $\mathcal{G}$ subject to $\mathcal{R}$ is well defined and we call $C^*(\mathcal{G}, \mathcal{R})$ the *Cuntz algebra* of type n and denote it by $\mathcal{O}_n$.

10.3.2. We can also define a Cuntz algebra of type ∞, denoted $\mathcal{O}_\infty$, in the obvious way:

Definition. The Cuntz algebra $\mathcal{O}_\infty$ is the universal unital C*-algebra generated by a sequence of isometries $(s_i)_{i \in \mathbb{N}}$, $s_i^* s_i = 1$, such that $\sum_{j=1}^{n} s_j s_j^* \leq 1$ for every $n \in \mathbb{N} \setminus \{0\}$.

10.3.3. Remark. The relations described above give norm bounds on the generators, so these universal C*-algebras are well defined (exercise).

10.3.4. Let H be a separable Hilbert space. For $n \in \mathbb{N} \cup \{\infty\}$, fix a sequence of isometries $s_1, \ldots, s_n$ satisfying $\mathcal{R}_n$. Let $B := C^*(s_1, \ldots, s_n)$ be the C*-algebra generated by the isometries $s_1, \ldots, s_n$. Since $\mathcal{O}_n$ is a universal C*-algebra, we have a surjective map $\mathcal{O}_n \twoheadrightarrow B$.

The next lemma implies that the range projections $s_j s_j^*$ of the defining isometries are pairwise orthogonal.

Lemma. *Suppose that $K \in \mathbb{N}$ and $q_0, \ldots, q_K$ are projections in a C*-algebra A satisfying $\|q_0 + \cdots + q_K\| \leq 1$. Then $q_0, \ldots, q_K$ are pairwise orthogonal.*

Proof. Passing to the unitisation if necessary, we may suppose that A is unital. The proof is by induction. We have $\|q_0 + q_1\| \leq 1$, so $1 - (q_0 + q_1) \geq 0$. Thus $q_0(1 - q_0 - q_1)q_0 \geq 0$. But also $q_0(1 - q_0 - q_1)q_0 = -q_0 q_1 q_0 \leq 0$, so $q_0 q_1 q_0 = 0$. Thus $\|q_1 q_0\|^2 = \|q_0 q_1 q_1 q_0\| = \|q_0 q_1 q_0\| = 0$. So q_0 and q_1 are orthogonal projections.

Suppose that $\|q_0 + \cdots + q_k\| \leq 1$ for some $k \geq 1$ implies $q_0, \ldots, q_k$ are pairwise orthogonal. Then $q_0 + \cdots + q_k$ is a projection so if $\|q_0 + \cdots + q_k + q_{k+1}\| \leq 1$, we have that $q_0 + \cdots + q_k$ and q_{k+1} are mutually orthogonal. Then, for any $0 \leq i \leq k$ $0 = q_i(q_0 + \cdots + q_k)q_{k+1} = q_i q_{k+1}$. Thus $q_0, \ldots, q_{k+1}$ are mutually orthogonal. $\square$

10.3.5. For $k \in \mathbb{N}$, we denote by W_k^n the set of k-tuples $(j_1, \ldots, j_k)$ where, for $i = 1, \ldots, k$, we have $j_i \in \{1, \ldots, n\}$ if $n < \infty$ and $j_i \in \mathbb{N}$ if $n = \infty$. Let $W_\infty^n := \bigcup_{k \in \mathbb{N}} W_k^n$.

For $\mu = (j_1, \ldots, j_k) \in W_k^n$ we denote the element $s_{j_1} s_{j_2} \cdots s_{j_k} \in \mathcal{O}_n$ by s_μ. Out of notational convenience, we will also sometimes denote $1_{\mathcal{O}_n}$ by s_0. If $\mu \in W_k^n$ then the length of μ, written $\ell(\mu)$, is k. If $\mu = 0$ then set $\ell(\mu) := 0$.

Lemma. Let $\mu, \nu \in W_\infty^n$. If $\ell(\mu) = \ell(\nu)$ then $s_\mu^* s_\nu = \delta_{\mu,\nu} 1_{\mathcal{O}_n}$.

Proof. If $\ell(\mu) = \ell(\nu) = 0$, then this is obvious. Suppose that (i) holds for any $\mu, \nu \in W_\infty^n$ with $\ell(\mu) = \ell(\nu) = k \geq 0$. Let $\ell(\mu) = \ell(\nu) = k+1$. Then $\mu = (i_1, \ldots, i_k, i_{k+1})$ and $\nu = (j_1, \ldots, j_k, j_{k+1})$. Put $\mu' := (i_1, \ldots, i_k)$ and $\nu' := (j_1, \ldots, j_k)$ if $k \geq 1$ and $\nu' = \mu' = 0$ otherwise. By the previous lemma,

$$s_{i_{k+1}}^* s_{j_{k+1}} = s_{i_{k+1}}^* (s_{i_{k+1}} s_{i_{k+1}}^*)(s_{j_{k+1}} s_{j_{k+1}}^*) s_{j_{k+1}} = s_{i_{k+1}}^* s_{j_{k+1}}$$
$$= \delta_{i_{k+1}, j_{k+1}} s_{i_{k+1}}^* s_{j_{k+1}} = \delta_{i_{k+1}, j_{k+1}} 1_{\mathcal{O}_n}.$$

Thus

$$s_\mu^* s_\nu = s_{i_{k+1}}^* (s_{\mu'}^* s_{\nu'}) s_{j_{k+1}}$$
$$= \delta_{\mu', \nu'} s_{i_{k+1}}^* s_{j_{k+1}}$$
$$= \delta_{\mu, \nu} 1_{\mathcal{O}_n},$$

since $\delta_{\mu', \nu'} \delta_{i_{k+1}, j_{k+1}} = \delta_{\mu, \nu}$. $\square$

10.3.6. Lemma. Let $\mu, \nu \in W_\infty^n$ and let $p = s_\mu s_\mu^*$ and $q = s_\nu s_\nu^*$. Suppose $s_\mu^* s_\nu \neq 0$. It holds that

(i) if $\ell(\mu) = \ell(\nu)$ then $s_\mu = s_\nu$ and $p = q$,

(ii) if $\ell(\mu) < \ell(\nu)$ then $s_\nu = s_\mu s_{\mu'}$ with $\mu' \in W_{\ell(\nu)-\ell(\mu)}^n$ and $q < p$,

(iii) if $\ell(\mu) > \ell(\nu)$ then $s_\mu = s_\nu s_{\nu'}$ with $\nu' \in W_{\ell(\mu)-\ell(\nu)}^n$ and $p < q$.

Proof. For (i) the proof is again by induction on the length of $\mu = \nu$. If $\ell(\mu) = \ell(\nu) = 0$ then $s_\nu = s_\mu = 1_{\mathcal{O}_n} = p = q$. Suppose that (i) holds for any $\mu, \nu \in W_\infty^n$ with $\ell(\mu) = \ell(\nu) = k \geq 0$. Suppose $\mu', \nu' \in W_\infty^n$ and $\ell(\mu') = \ell(\nu') = k+1$. Then $\mu' = (i_1, \ldots, i_k, i_{k+1})$ and $\nu' = (j_1, \ldots, j_k, j_{k+1})$ for some $1 \leq i_1, \ldots, i_{k+1}, j_1, \ldots, j_{k+1} \leq n$ (where we allow $n = \infty$). Let $\mu = (i_1, \ldots, i_k)$ and $\nu = (j_1, \ldots, j_k)$. By the previous lemma,

$$1_{\mathcal{O}_n} = s_{\mu'}^* s_{\nu'} = (s_{i_{k+1}})^* s_\mu^* s_\nu s_{j_{k+1}}$$
$$= (s_{i_{k+1}})^* s_{j_{k+1}} = \delta_{i_{k+1}, j_{k+1}}.$$

Thus $i_{k+1} = j_{k+1}$ and, applying the induction hypothesis to μ and ν implies that $s_{\mu'} = s_{\nu'}$ and hence $p = q$.

Now suppose that $\ell(\mu) < \ell(\nu)$. Then $\nu = \alpha\mu'$ where $\ell(\alpha) = \ell(\nu)$ and $\ell(\mu') = \ell(\nu) - \ell(\mu)$. Since $s_\mu^* s_\nu \neq 0$ and $s_\mu^* s_\nu = s_\mu^* s_\alpha s_{\mu'}$, we clearly have $s_\mu^* s_\alpha \neq 0$. Thus by (i), we have $\alpha = \mu$ and

$$q = s_\nu s_\nu^* = s_\mu(s_{\mu'} s_{\mu'}^*) s_\mu^* < s_\mu 1_{\mathcal{O}_n} s_\mu^* = p.$$

The proof of (iii) is similar and left as an exercise.   □

10.3.7. Lemma. *If $w \neq 0$ is a word in $\{s_i \mid i \in \{1, \overset{.}{\ldots}, n\}\} \cup \{s_i^* \mid i \in \{1, \ldots, n\}\}$ then there are unique elements $\mu, \nu \in W_\infty^n$ such that $w = s_\mu s_\nu^*$.*

Proof. Let $w = x_1 \cdots x_r$ with $x_j \in \{s_i \mid i \in \{1, \ldots, n\}\} \cup \{s_i^* \mid i \in \{1, \ldots, n\}\}$. Since $s_i^* s_j = \delta_{i,j} 1_{\mathcal{O}_n}$, and $w \neq 0$, we can reduce w by cancelling any instance of $s_i^* s_i$. Thus we can write $w = y_1 \cdots y_t$ with $t \leq r$ and such that for some i_0, if $y_j = s_{i_0}$ then $y_{j-1} = s_i$ for some $i \in \{1, \ldots, n\}$. Let $0 \leq j_0 \leq t$ be the integer such that $y_{j_0} = s_{i_0}$ and satisfies $y_j = s_{i(j)}$ for every j with $j \geq 1$ and $j \leq j_0$ and $y_j \in s_{i(j)}^*$ for any j with $j_0 < j$ and $j \leq t$. Let $\mu \in W_{j_0}^n$ be given by $(i(1), i(2), \ldots, i(j_0)) = i_0$ and $\nu \in W_{t-j_0}^n$ be given by $(i(j_0 + 1), i(j_0 + 2), \ldots, i(t))$. Then $w = s_\mu s_\nu^*$, which shows existence.

Suppose that there are $\alpha, \beta \in W_\infty^n$ with $w = s_\alpha s_\beta^*$. Since $w^* w \neq 0$, we must also have $s_\alpha^* s_\mu \neq 0$. Then, by the previous lemma, $\alpha = \mu$. Applying the same argument to ww^*, we also have $\beta = \nu$. This shows uniqueness.   □

10.3.8. Let $F_0^n = \mathbb{C}$ and for $k > 0$ let

$$F_k^n := \mathrm{C}^*(s_\mu s_\nu^* \mid \mu, \nu \in W_k^n) \quad \text{and} \quad F^n := \cup_{k \in \mathbb{N}} F_k^n.$$

Theorem. *If $n < \infty$ then $F_k^n \cong M_{n^k}$. Moreover $F_k^n \subset F_{k+1}^n$ and $F^n \cong \mathcal{U}_{n^\infty}$, the UHF algebra of type n^∞. If $n = \infty$ the $F_k^n \cong \mathcal{K}$ and F^n is an AF algebra.*

Proof. Let $\mu, \mu', \nu, \nu' \in W_k^n$. It follows from above that $(s_\mu s_\nu^*)(s_{\mu'} s_{\nu'}^*) = \delta_{\nu, \mu'} s_\mu s_{\nu'}^*$. Also, $(s_\mu s_\nu^*)^* = s_\nu s_\mu^*$. Thus $s_\mu s_\nu^*$, $\mu, \nu \in W_k^n$ are a system of matrix units. If $n < \infty$, then $|W_k^n| = n^k$, and so $F_k^n \cong M_{n^k}$. Furthermore, if $n < \infty$ we have $\sum_{j=1}^n s_j s_j^* = 1_{\mathcal{O}_n}$ so

$$s_\mu s_\nu^* = \sum_{j=1}^n s_\mu s_j s_j^* s_\nu^* \in F_{k+1}^n.$$

Thus $F_k^n \subset F_{k+1}^n$, and it is not hard to see that under the identification $F_k^n \cong M_{n^k}$, these embeddings are just the connecting maps from 8.3.1. Thus $F^n \cong \mathcal{U}_{n^\infty}$.

Suppose now that $n = \infty$. Then $F_k^n = \bigcup_{m \in \mathbb{N}} F_k^m$ and $F_k^m \subset F_k^{m+1}$ where the embedding is into the top left $m_k \times m_k$ corner of M_{m+1}^k. Thus $F_k^n \cong \mathcal{K}$. The inductive limit of an AF algebra with injective connecting maps is again AF (Exercise 8.7.12), so F^n is an AF algebra.   □

10.3.9. The proof of the following theorem should look familiar: it is quite similar in construction to the conditional expectation on a crossed product by $\mathbb{Z}$. (One can construct the Cuntz algebra as a crossed product by $\mathbb{N}$, a generalisation of crossed products by the integers we saw in 9.5. In fact if, one tensors with the compact operators, $\mathcal{O}_n \otimes \mathcal{K}$ is isomorphic to a crossed product by $\mathbb{Z}$. The interested reader should see [29, Section 2].)

Theorem. *For every $n \in \mathbb{N} \cup \{\infty\}$, there exists a faithful conditional expectation $\Phi : \mathcal{O}_n \to F^n$.*

Proof. Let $t \in \mathbb{R}$. Then $e^{2\pi i t} s_1, e^{2\pi i t} s_2, \ldots$ satisfy the relations $\mathcal{R}_n$ of Definition 10.3.1, so, by the universal property, there is a *-homomorphism

$$\rho_t : \mathcal{O}_n \to C^*(e^{2\pi i t} s_1, e^{2\pi i t} s_2, \ldots, e^{2\pi i t} s_n) \subset \mathcal{O}_n$$

which maps the generators $s_j \mapsto e^{2\pi i t} s_j$. Note that $\rho_t \circ \rho_{-t} = \mathrm{id}_{\mathcal{O}_n}$ so in particular, ρ_t is injective. Since $\rho_t(s_j^*) = e^{-2\pi i t} s_j^*$, a straightforward calculation shows

$$\rho_t(s_\mu s_\nu^*) = e^{2\pi i t(\ell(\mu) - \ell(\nu))} s_\mu s_\nu^*.$$

It follows that for arbitrary $x \in \mathcal{O}_n$ the function $f : \mathbb{R} \to \mathcal{O}_n$ defined by $f_x(t) = \rho_t(x)$ is continuous. Thus we may define

$$\Phi(x) = \int_0^1 \rho_t(x) dt = \int_0^1 f_x(t) dt.$$

Since ρ_t is an injective *-homomorphism, Φ is positive and faithful. Now

$$\Phi(s_\mu s_\nu^*) = \left(\int_0^1 e^{2\pi t(\ell(\mu) - \ell(\nu))} dt \right) s_\mu s_\nu^* = \begin{cases} 0 & \text{if } \ell(\mu) \neq \ell(\nu), \\ s_\mu s_\nu^* & \text{if } \ell(\mu) = \ell(\nu), \end{cases}$$

so by extending to $\mathcal{O}_n$ we see that $\Phi(\mathcal{O}_n) \subset F^n$ and that $\Phi(x) = x$ for every $x \in F^n$. Finally, we get that for $a, b \in F^n$ and $x \in \mathcal{O}_n$

$$\Phi(axb) = \int_0^1 \rho_t(axb) dt = \int_0^1 \rho_t(a) \rho_t(x) \rho_t(b) dt = a \left(\int_0^1 \rho_t(x) dt \right) b = a\Phi(x)b,$$

so Φ is a conditional expectation by Proposition 9.5.1. $\square$

10.3.10. Lemma. *Suppose that $k \in \mathbb{N}$ and $\mu, \nu \in W_\infty^n$ satisfy $\ell(\mu), \ell(\nu) \leq k$ with $\ell(\mu) \neq \ell(\nu)$. Let $s_\gamma := s_1^k s_2$. Then $s_\gamma^*(s_\mu s_\nu^*) s_\gamma = 0$.*

Proof. We have that $(s_1^k)^* s_\mu = 0$ unless $s_\mu = s_1^{\ell(\mu)}$. Thus

$$s_\gamma^*(s_\mu s_\nu^*) s_\gamma = s_2^*(s_1^{k-\ell(\mu)})^* s_\nu^* s_\gamma.$$

Similarly, $s_\nu^* s_\gamma = 0$ unless $s_\nu = (s_1^{l(\nu)})$. Thus

$$s_2^*(s^{k-\ell(\mu)})^* s_1^{k-\ell(\nu)} s_2 = \begin{cases} s_2^* s_1^{\ell(\mu)-\ell(\nu)} s_2 & \text{if } \ell(\mu) > \ell(\nu), \\ s_2^*(s_1^{\ell(\nu)-\ell(\mu)})^* s_2 & \text{if } \ell(\mu) < \ell(\nu). \end{cases}$$

However, $s_2^* s_1 = 0 = s_1^* s_2$. Thus $s_\gamma^*(s_\mu s_\nu^*)s_\gamma$ can only be nonzero if $\ell(\mu) = \ell(\nu)$, which is not the case by assumption. Thus $s_\gamma^*(s_\mu s_\nu^*)s_\gamma = 0$. $\qquad\square$

10.3.11. Proposition. *For every $k \in \mathbb{N} \setminus \{0\}$ there exists an isometry $w \in \mathcal{O}_n$, commuting with F_k^n, such that*

$$\Phi(x) = wxw^*,$$

for every $x \in \mathrm{span}\{s_\mu, s_\nu^* \mid \ell(\mu), \ell(\nu) \le k\}$.

Proof. Assume first that $n < \infty$. Let $s_\gamma = s_1^{2k} s_2$ and define

$$w := \sum_{\ell(\alpha)=k} s_\alpha s_\gamma s_\alpha^*.$$

Then

$$w^* w = \sum_{\ell(\alpha)=k} \sum_{\ell(\beta)=k} s_\alpha s_\gamma^* s_\alpha^* s_\beta s_\gamma s_\beta^*$$

$$= \sum_{\ell(\alpha)=k} s_\alpha s_\gamma^* s_\gamma s_\alpha^* = \sum_{\ell(\alpha)=k} s_\alpha s_\alpha^* = 1_{\mathcal{O}_n}.$$

Furthermore, when $\ell(\mu) = \ell(\nu) = k$ we have

$$w s_\mu = \sum_{\ell(\alpha)=k} s_\alpha s_\gamma s_\alpha^* s_\mu = s_\mu s_\gamma,$$

while

$$s_\mu^* w = \sum_{\ell(\alpha)=k} s_\mu^* s_\alpha s_\gamma s_\alpha^* = s_\gamma s_\mu^*.$$

Thus

$$w s_\mu s_\nu^* = s_\mu s_\gamma s_\nu^* = s_\mu s_\nu^* w.$$

Since $s_\mu s_\nu^*$ are a system of matrix units generating F_k^n, it follows that $wF_k^n = F_k^n w$. Thus $w^* xw = x$ for every $x \in F_k^n$. On the other hand, if $\ell(\mu) \ne \ell(\nu)$ then

$$w^* s_\mu s_\nu^* w = \sum_{\ell(\alpha),\ell(\beta)=k} s_\alpha s_\gamma^* s_\alpha^* s_\mu s_\nu^* s_\beta s_\gamma s_\beta^* = 0,$$

by the previous lemma. It follows that for every $x \in \mathrm{span}\{s_\mu s_\nu^* \mid \ell(\mu), \ell(\nu) \le k\}$ we have $\Phi(x) = w^* xw$, which proves the lemma in the case that $n < \infty$.

Suppose now $n = \infty$. Let N be the largest integer appearing in μ and ν. Let $\tilde{s} \in \mathcal{B}(H)$ (where $\mathcal{O}_\infty \subset \mathcal{B}(H)$) be an operator satisfying $\tilde{s}^* \tilde{s} = 1_{\mathcal{O}_\infty}$ and

$1_{\mathcal{O}_\infty} - \sum_{j=1}^N s_j s_j^* = \tilde{s}\tilde{s}^*$. Then, with s_γ as above, let

$$w = \sum_{\ell(\alpha)=k} s_\alpha s_\gamma s_\alpha^*,$$

where $\alpha \in \{1, \ldots, N, N+1\}^k$ and we interpret $s_{N+1} = \tilde{s}$. Then one checks that w satisfies the requirements of the lemma.   □

10.3.12. Proposition. *Suppose that $n < \infty$ and $x \in \mathcal{O}_n$ is nonzero. Then there exist $a, b \in \mathcal{O}_n$ such that $axb = 1_{\mathcal{O}_n}$.*

Proof. Let $x \in \mathcal{O}_n$ be nonzero. Then, since Φ is faithful, we have $\Phi(x^*x) > 0$. Multiplying by a scalar if necessary, we may assume that $\|\Phi(x^*x)\| = 1$. Find some $N \in \mathbb{N}$ and $\mu_j, \nu_j \in W_\infty^n$ such that $y = \sum_{j=1}^N s_{\mu_j} s_{\nu_j}^*$ is self-adjoint and satisfies $\|x^*x - y\| < 1/4$. Then

$$\Phi(y) \geq \Phi(x^*x) - 1/4 = 3/4.$$

Let $m := \max\{\ell(\mu_j), \ell(\nu_j) \mid 1 \leq j \leq N\}$. By Proposition 10.3.11, there exists an isometry $w \in \mathcal{O}_n$ which satisfies $\Phi(y) = wyw^* \in F_m^n$. From Theorem 10.3.8, we have $F_m^n \cong M_{n^m}$ if $n < \infty$ and $F_m^n \cong \mathcal{K}$ if $n = \infty$. In either case, $\Phi(y)$ is positive and hence diagonalisable. Thus there exists a minimal projection $q \in F_m^n$ (corresponding to the entry with value $\|\Phi(y)\|$ in the diagonalisation) such that

$$q\Phi(y) = \Phi(y)q = \|\Phi(y)\|q > (3/4)q,$$

and $u \in F_m^n$ satisfying $uqu^* = s_1^m(s_1^*)^m$ (since $s_1^m(s_1^*)^m$ is identified with the e_{11} matrix unit for F_m^n). Define

$$z := \|\Phi(y)\|^{-1/2}(s_1^*)^m uqw^*.$$

Then

$$zyz^* = \|\Phi(y)\|^{-1}(s_1^*)^m uqw^* ywqu^* s_1^m = \|\Phi(y)\|^{-1}(s_1^*)^m uq\Phi(y)qu^* s_1^m$$
$$= \|\Phi(y)\|^{-1}(s_1^*)^m u\|\Phi(y)\|qu^* s_1^m = (s_1^*)^m uqu^* s_1^m = 1_{\mathcal{O}_n}.$$

Thus

$$\|zx^*xz - 1\| = \|zx^*xz - zyz^*\| \leq \|z\|^2/4 \leq \|\Phi(y)\|^{-1}/4 = 1/3.$$

It follows from Theorem 1.2.4 that zx^*xz is invertible. Let $b = z^*(zx^*xz)^{-1/2}$ and let $a = b^*x^*$. Then

$$axb = (zx^*xz)^{-1/2}zx^*xz^*(zx^*xz)^{-1/2} = 1_{\mathcal{O}_n},$$

completing the proof.   □

Putting all the above results together, we arrive at the final theorem of this chapter.

10.3.13. Theorem. *For any $n \in \mathbb{N}_{\geq 2} \cup \{\infty\}$, the Cuntz algebra $\mathcal{O}_n$ is simple and purely infinite.*

10.4 Exercises

10.4.1. Let $\theta \in \mathbb{R}$. Let $\mathcal{G} = \{u, v\}$ and define relations $\mathcal{R}$ by

$$uu^* = u^*u = 1, \quad vv^* = v^*v = 1, \quad uv = e^{2\pi i \theta} vu.$$

If $e^{2\pi i \theta} = 1$, then $C^*(\mathcal{G} \mid \mathcal{R}) \cong C(\mathbb{T}^2)$. For this reason, when $e^{2\pi i \theta} \neq 1$, then $C^*(\mathcal{G} \mid \mathcal{R})$ is called a *noncommutative torus*. Show that if θ is irrational, then $C^*(\mathcal{G} \mid \mathcal{R}) \cong C(\mathbb{T}) \rtimes_{\varphi_\theta} \mathbb{Z}$ is the irrational rotation algebra of Exercise 9.6.11.

10.4.2. Let $q \in (0, 1)$. Consider four generators a, b, c, d and relations $\mathcal{R}$ given by

$$a^*a + c^*c = 1, \quad a^* + q^2 c^*c = 1, \quad c^*c = cc^*, \quad ac = qca, \quad ac^* = qc^*a.$$

(i) Show that the universal C*-algebra $C^*(\{a, b, c, d\}, \mathcal{R})$ exists.
(ii) Show that $C^*(\{a, b, c, d\}, \mathcal{R})$ can be given a comultiplication making it into a compact quantum group. (Hint: see Exercise 9.6.9.)

The universal C*-algebra $C^*(\{a, b, c, d\}, \mathcal{R})$ is denoted $C_q(\mathrm{SU}_2)$ or $C(SU_q(2))$ and is thought of as "continuous functions on *quantum* SU_2".

(iii) Show that as a C*-algebra, $C_q(\mathrm{SU}_2)$ does not depend on the parameter $q \in (0, 1)$. Show that if $q = 1$ then $C^*(\{a, b, c, d\}, \mathcal{R})$ is commutative.
(iv) What is a good definition for the compact quantum group $C_q(\mathrm{SU}_n)$, for $n \in \mathbb{N}_{>2}$?

10.4.3. Let A be a unital C*-algebra. Show that if $T(A) \neq \emptyset$ then A is not purely infinite.

10.4.4. Show that the relations for the Cuntz algebras do allow one to define universal C*-algebras by showing that the generators and relation can be realised as bounded operators on some Hilbert space, and that there is a bound on the norm of the generators in every representation.

10.4.5. Show that there is an injective unital *-homomorphism $\varphi : \mathcal{O}_3 \to \mathcal{O}_2$. (Hint: Consider the elements $\{s_1, s_2 s_1, s_2^2\} \subset \mathcal{O}_2$.) More generally, show that there is an injective unital *-homomorphism $\varphi : \mathcal{O}_{k(n-1)+1} \to \mathcal{O}_n$ for every $k \geq 1$.

10.4.6. Suppose that $n \geq 2$ and k divides n. Show that $M_k(\mathcal{O}_n) \cong \mathcal{O}_n \otimes M_k \cong \mathcal{O}_n$ (Hint: consider the matrix $a_{kj+i} := (a_{l,m})_{l,m=1,\dots,k}$ defined by $a_{l,m} = 0$ if $l \neq i$ and $a_{i,m} = s_{kj+m}$ otherwise.)

10.4.7. Show that

$$\begin{pmatrix} 0 & 0 & 0 \\ 1 & 0 & 0 \\ 0 & s_1 & s_2 \end{pmatrix}, \quad \begin{pmatrix} s_1 & s_2 s_1 & s_2^2 \\ 0 & 0 & 0 \\ 0 & 0 & 0 \end{pmatrix}$$

generate a copy of $M_3(\mathcal{O}_2) = \mathcal{O}_2 \otimes M_3$ inside $\mathcal{O}_2$. Conclude that there are *-isomorphisms $M_3(\mathcal{O}_2) \cong \mathcal{O}_2 \otimes M_3 \cong \mathcal{O}_2$.

10.4.8. More generally, show that

(i) For every $n \geq 1$ we have $\mathcal{O}_2 \otimes M_n \cong \mathcal{O}_2$.

(ii) For every UHF algebra $\mathcal{U}$ we have $\mathcal{O}_2 \otimes \mathcal{U} \cong \mathcal{O}_2$.

10.4.9. A projection p is called *properly infinite* if there exists orthogonal projections q_1, q_2 such that $q_1 \sim p \sim q_2$. A unital (not necessarily simple) C*-algebra A is called *properly infinite* if 1_A is properly infinite.

(i) Show that the Cuntz algebras are properly infinite.

(ii) Show that if A is a properly infinite, then there is a unital embedding $\mathcal{O}_\infty \hookrightarrow A$. (Hint: if s_1 and s_2 are orthogonal isometries, consider the isometries $\{s_2^n s_1 \mid n \geq 0\}$.)

(iii) Show that every projection in a simple unital purely infinite C*-algebra is properly infinite.

10.4.10. Let A, B, C, D be unital C*-algebras and let $\Phi : A \to C$ and $\Psi : B \to D$ be conditional expectations. Show that the tensor product

$$\Phi \otimes \Psi : A \otimes_{\min} B \to C \otimes_{\min} D,$$

is again a conditional expectation.

11 Quasidiagonality and tracial approximation

This chapter is somewhat different from the previous three chapters. Instead of constructing C*-algebras which we then show are simple (or find conditions under which they are simple), here we start with a simple C*-algebra and show it has a particular structure. This is useful if we can also prove things about the C*-algebras with this structure. For example, if we are given some C*-algebra A and are able to show it is isomorphic to an AF algebra by finding an appropriate inductive limit model, then we immediately know that A has real rank zero, stable rank one, admits a tracial state, and so forth. In this chapter, instead of approximating a C*-algebra by building blocks, such as finite-dimensional C*-algebras, in terms of the norm via an inductive limit structure, we will see that C*-algebras can also be approximated in more measure-theoretic way, by asking for building blocks that are arbitrarily large with respect to all traces. This turns out to be much easier than trying to construct an inductive limit, as we won't have to deal with things like connecting maps. The concept of tracial approximation also plays a key role in the classification program (which we will meet in Part III), as it allows a way around complicated intertwining arguments (see Chapter 13).

In this chapter, we begin in Section 11.1 by investigating an approximation property known as *quasidiagonality*. Quasidiagonal C*-algebras are always stably finite, and it remains an open question whether or not all separable, stably finite nuclear C*-algebras are quasidiagonal (this is often called the *Blackadar–Kirchberg problem*, as the question was originally posed in [10]). In Section 11.2 we will see that in the simple unital case, provided a C*-algebra has enough projections, a C*-algebra is quasidiagonal precisely when it is a *Popa algebra*. Popa algebras are simple, unital C*-algebras with a local approximation property. They are the precursor to the *tracially approximately finite* (TAF) C*-algebras, which, along with tracial approximation by more general building blocks, is the subject of the final section. We finish the chapter by showing that certain approximately homogeneous C*-algebras are TAF.

11.1 Quasidiagonality

As we will see in this chapter, quasidiagonal C*-algebras admit certain "external" approximations by matrix algebras. We have already seen some "internal" approximations by matrix algebras when we defined the completely positive approximation property, which plays a fundamental role in the classification program. Indeed, quasidiagonality has strong links to nuclearity, and with an eye on the final section, we will develop some results for simple quasidiagonal C*-algebras. Of particular importance are the Popa algebras (which, for some simple C*-algebras, characterise quasidiagonality) as well as the refined notion of tracial approximation by finite-dimensional C*-algebras.

© The Author(s), under exclusive license to Springer Nature Switzerland AG 2021
K. R. Strung, *An Introduction to C*-Algebras and the Classification Program*, Advanced
Courses in Mathematics - CRM Barcelona, https://doi.org/10.1007/978-3-030-47465-2_11

11.1.1. Definition. Let H be a Hilbert space and let $S \subset \mathcal{B}(H)$ be a collection of operators. We say that S is quasidiagonal if, for every finite subset $\mathcal{F} \subset S$, every finite set $\{\xi_1, \ldots, \xi_n\} \subset H$, and every $\epsilon > 0$, there exists a finite rank projection $p \in \mathcal{B}(H)$ such that

(i) $\|ps - sp\| < \epsilon$ for every $s \in S$, and

(ii) $\|p\xi_i - \xi_i\| < \epsilon$ for $1 \leq i \leq n$.

11.1.2. Definition. A C*-algebra A is said to be *quasidiagonal* if A has a faithful representation $\pi : A \to \mathcal{B}(H)$ such that $\pi(A)$ is a quasidiagonal collection of operators.

Thanks to Voiculescu [127], instead of the definition given above, we can (and usually will) use the characterisation of quasidiagonality given by Theorem 11.1.3 (iii). The theorem also says that if there is one faithful representation which is a quasidiagonal collection of operators, then in fact every faithful unital representation, provided it does not contain a nonzero compact operator, is also a quasidiagonal collection of operators. An accessible proof can be found in [18].

11.1.3. Theorem. *Let A be a separable unital* C*-*algebra. Then the following are equivalent:*

(i) *A has a faithful representation $\pi : A \to \mathcal{B}(H)$ such that $\pi(A)$ is a quasidiagonal collection of operators.*

(ii) *If $\pi : A \to B(H)$ is a faithful unital representation which does not contain a nonzero compact operator, then $\pi(A)$ is a quasidiagonal collection of operators.*

(iii) *There is a net of c.p.c. maps $\varphi_\lambda : A \to M_{n_\lambda}$ such that*

(a) *$\|\varphi_\lambda(ab) - \varphi_\lambda(a)\varphi_\lambda(b)\| \to 0$, for every $a, b \in A$,*

(b) *$\|\varphi_\lambda(a)\| \to \|a\|$, for every $a \in A$.*

11.1.4. Lemma. *Suppose A is a unital quasidiagonal* C*-*algebra. Then the maps in Theorem 11.1.3 (iii) can always be chosen to be unital.*

Proof. Exercise. (For a hint, see the exercises at the end of the chapter.) □

11.1.5. Although Theorem 11.1.3 (iii) provides a definition of quasidiagonality in terms of limits, it is actually a local property. We can give another equivalent definition as follows, the proof of which is an exercise.

Proposition. *A* C*-*algebra A is quasidiagonal if and only if, for every finite subset $\mathcal{F} \subset A$ and every $\epsilon > 0$, there is a c.p.c. map $\varphi : A \to M_n$ such that*

(i) *$\|\varphi(ab) - \varphi(a)\varphi(b)\| < \epsilon$, for every $a, b \in \mathcal{F}$,*

(ii) *$\|\varphi(a)\| > \|a\| - \epsilon$, for every $a \in \mathcal{F}$.*

11.1.6. Lemma. *Let A be a unital* C*-*algebra and let $2 \leq n \in \mathbb{N}$. Then A is quasidiagonal if and only if $M_n(A)$ is quasidiagonal.*

Proof. We have $M_n(A) \cong A \otimes M_n$ (Exercise 4.4.15). Suppose that $A \otimes M_n$ is quasidiagonal. Then there is a net $\varphi_\lambda : A \otimes M_n \to M_{k_\lambda}$ of completely positive contractive maps satisfying

$$\|\varphi_\lambda((a \otimes 1_{M_n})(b \otimes 1_{M_n})) - \varphi_\lambda(a \otimes 1_{M_n})\varphi_\lambda(b \otimes 1_{M_n})\| \to 0,$$

and such that $\|\varphi_\lambda(a \otimes 1_{M_n})\| \to \|a\|$ for every $a \in A$. Then the maps $\tilde{\varphi}_\lambda : A \to M_{k_\lambda}$ defined by $\tilde{\varphi}_\lambda(a) = \varphi_\lambda(a \otimes 1_{M_n})$ give a net of c.p.c maps satisfying the requirements. So A is quasidiagonal.

Now suppose that A is quasidiagonal and let $\varphi_\lambda : A \to M_{k_\lambda}$ be the associated net of completely positive maps. Then $\varphi_\lambda \otimes \mathrm{id} : A \otimes M_n \to M_{k_\lambda} \otimes M_n$ is a net satisfying the requirements for quasidiagonality of $A \otimes M_n \cong M_n(A)$. $\square$

We saw in Theorem 8.3.14 that unital AF algebras are stably finite. The same is also true for quasidiagonal C*-algebra. We show this in the unital case.

11.1.7. Theorem. *Let A be a unital quasidiagonal C*-algebra. Then A is stably finite.*

Proof. We will show that, for every $n \in \mathbb{N}$, every isometry $v \in M_n(A)$ is a unitary. This will show that A is stably finite by Proposition 8.3.13. The proof is by contradiction. Suppose $v \in M_n(A)$ is an isometry which is not unitary, that is, $v^*v = 1_{M_n(A)}$ but $vv^* \neq 1_{M_n(A)}$. Since A is quasidiagonal if and only if $M_n(A)$ is quasidiagonal, we may assume without loss of generality that $v \in A$. Choose a net $\varphi_\lambda : A \to M_{k_\lambda}$ of c.p.c. maps satisfying the requirements for quasidiagonality of A. By Lemma 11.1.4 we may moreover choose the φ_λ to be unital. Then

$$\begin{aligned}
\|\varphi_\lambda(vv^* - 1_A)\| &= \|\varphi_\lambda(vv^*) - \varphi_\lambda(1_A)\| \\
&\leq \|\varphi_\lambda(vv^*) - \varphi_\lambda(v)\varphi_\lambda(v^*)\| + \|\varphi_\lambda(v^*)\varphi_\lambda(v) - \varphi_\lambda(1_A)\| \\
&\leq \|\varphi_\lambda(vv^*) - \varphi_\lambda(v)\varphi_\lambda(v^*)\| + \|\varphi_\lambda(v^*)\varphi_\lambda(v) - \varphi_\lambda(v^*v)\| \\
&\to 0.
\end{aligned}$$

But also $\|\varphi_\lambda(vv^* - 1_A)\| \to \|vv^* - 1_A\| > 0$, a contradiction. So A must be stably finite. $\square$

11.1.8. Remark. This means that purely infinite C*-algebras, such as the Cuntz algebras, are not quasidiagonal.

11.2   Popa algebras

In the case that a C*-algebra is unital, simple, and has "sufficiently many" projections (for example, if A has real rank zero), there is an intrinsic characterisation of quasidiagonality, that is, a characterisation which makes no mention of maps into matrices or projections on an ambient Hilbert space. Popa algebras were introduced in [95].

11.2.1. Definition. Let A be a simple unital C*-algebra. We call A a *Popa algebra* if for any finite subset $\mathcal{F} \subset A$ and any $\epsilon > 0$ there exists a finite-dimensional C*-subalgebra $F \subset A$ and projection $p \in A$ with $1_F = p$ satisfying

(i) $\|pa - ap\| < \epsilon$ for every $a \in \mathcal{F}$, and

(ii) $\operatorname{dist}(F, pap) < \epsilon$ for every $a \in \mathcal{F}$.

11.2.2. Lemma. *Let A be a simple unital C*-algebra. Suppose that A is a Popa algebra. Then A is quasidiagonal.*

Proof. Exercise. □

11.2.3. Recall from Definition 8.6.4 that a C*-algebra has real rank zero if the invertible self-adjoint elements of A are dense in the self-adjoint elements of A. The next proposition is due to L. Brown and Pedersen [16, Theorem 2.6].

Proposition. *Let A be a C*-algebra. The following are equivalent:*

(i) *A has real rank zero,*

(ii) *every self-adjoint element $a \in A$ can be approximated by elements with finite spectrum,*

(iii) *every hereditary C*-subalgebra of A has a (not necessarily increasing) approximate unit consisting of projections.*

Proof. Suppose A has real rank zero. By definition, if A in nonunital then $RR(A) = RR(\tilde{A})$ (8.6.8), so we may assume that A is unital. Let $a \in A_{sa}$. Normalising if necessary we may assume that $\operatorname{sp}(a) \subset [-1, 1]$. Let $\epsilon > 0$ be given. Choose $t_0 := -1 < t_1 < \cdots < t_n := 1$ satisfying $|t_i - t_{i+1}| < \epsilon/2$. Since $a - t_0$ is self-adjoint, there exists an invertible self-adjoint element $\tilde{a}_0 \in A$ approximating $a - t_0$ up to $\epsilon_0 := \epsilon/4$.

Let $a_0 = \tilde{a}_0 + t_0$ and observe that $a - t_0$ is invertible and $\|a - a_0\| < \epsilon_0$. The element $a_0 - t_0$ is invertible, so t_0 is not contained in the spectrum of a_0. Since $\operatorname{sp}(a_0) \cup \{0\}$ is closed, there exists $\epsilon_1 > 0$ with $\epsilon_1 < \epsilon/8$ such that $(t_0 - \epsilon_1, t_0 + \epsilon_1) \cap \operatorname{sp}(a_0) = \emptyset$.

Now we repeat this with t_1 and a_0 to get a_1 such that $a_1 - t_1$ is invertible and $\|a_1 - a_0\| < \epsilon_1 < \epsilon/8$. Since a_1 is within ϵ_1 of a_0, the choice of ϵ implies that $t_0 \notin \operatorname{sp}(a_1)$. Thus $\{t_0, t_1\} \cap \operatorname{sp}(a_1) = \emptyset$.

Repeating this for each t_i we arrive at an element $a_n \in A$ with $a_n - t_n$ invertible, $\|a_n - a_{n-1}\| < \epsilon_n < \epsilon/2^{n+2}$ and $\{t_1, \ldots, t_n\} \cap \operatorname{sp}(a_n) = \emptyset$. Furthermore,

$$\|a - a_n\| \le \sum_{i=0}^{n-1} \|a - a_i\| < \sum_{i=0}^{n-1} \epsilon/2^{n+2} < \epsilon/2.$$

Let $\chi_{(t,t']}$ denote the indicator function on the interval $(t, t']$ (see 8.6.1). Since $t_i \notin \operatorname{sp}(a_n)$, $f(t) = \sum_{i=0}^{n-1} t_{i+1} \chi_{(t_i, t_{i+1}]}(t)$ is continuous on $\operatorname{sp}(a_n)$ and approximates the identity function $g(t) = t$ within $\epsilon/2$. Applying the functional calculus we then have that $\|a_n - f(a_n)\| < \epsilon/2$. Since $f(t)$ takes finitely many values, $f(a_n)$ has finite spectrum. Let $b := f(a_n)$. Then $\|b - a\| < \|b - a_n\| + \|a_n - a\| < \epsilon$. This proves (i) implies (ii).

Assume that (ii) holds. Let $B \subset A$ be a hereditary C*-subalgebra. Let $b_1, \ldots, b_n \in B$. Without loss of generality, we may assume that each b_i is positive. Let $b = \sum_{i=1}^{n} b_i$. Rescaling if necessary, assume that b has norm one.

Let $0 < \epsilon < 1/4$, and define the continuous function $f_\epsilon : [0, 1] \to [0, 1]$ by

$$f_\epsilon(t) := \begin{cases} 0 & t \in [0, \epsilon/2], \\ \text{linear} & t \in (\epsilon/2, \epsilon), \\ 1 & t \in [\epsilon, 1]. \end{cases}$$

Let $a, b \in A$, with $\|a\| \leq 1$. Approximating f_ϵ by polynomials, we can find a $\delta > 0$ sufficiently small so that $\|f_\epsilon(a) - f_\epsilon(b)\| < \epsilon$ provided $\|a - b\| < \delta$. Shrinking δ if necessary, we may assume that $\delta < \epsilon$. Since b is positive, we may choose such an a to be self-adjoint, and since self-adjoint elements of finite spectrum are dense, we can moreover assume that a has finite spectrum, say $\mathrm{sp}(a) = \{\lambda_1, \ldots, \lambda_m\} \subset \mathbb{R} \cap [-1, 1]$ for $m \in \mathbb{N} \setminus \{0\}$. It follows that for every i, $1 \leq i \leq m$, the indicator function on $\{\lambda_i\}$, which we denote by χ_i, is continuous on the spectrum of a. Note that $\chi_i(a)$, $1 \leq i \leq m$ are mutually orthogonal projections, and $a = \sum_{i=1}^{m} \lambda_i \chi_i(a)$. Let $I := \{i \mid \lambda_i \geq \epsilon\} \subset \{1, \ldots, m\}$ and define a projection by $p := \sum_{i \in I} \chi_i(a)$. Then p satisfies

$$\|pa - a\| < \epsilon, \quad f_\epsilon(a)p = p.$$

Thus

$$\|pb - b\| < \|pb - pa\| + \|pa - a\| < 2\epsilon.$$

We estimate

$$\begin{aligned} \|f_\epsilon(b)pf_\epsilon(b) - p\| &\leq \|f_\epsilon(b)pf_\epsilon(b) - f_\epsilon(b)pf_\epsilon(a)\| + \|f_\epsilon(b)pf_\epsilon(a) - p\| \\ &< \epsilon + \|f_\epsilon(b)p - p\| \\ &= \epsilon + \|f_\epsilon(b)p - f_\epsilon(a)p\| \\ &< 2\epsilon. \end{aligned}$$

Since $\epsilon < 1/4$ and $f_\epsilon(b)pf_\epsilon(b)$ is evidently self-adjoint, we may apply Lemma 8.4.1 to find a projection q in the C*-subalgebra generated by $f_\epsilon(b)pf_\epsilon(b)$ satisfying

$$\|p - q\| < 4\epsilon.$$

Now $f_\epsilon(b)pf_\epsilon(b) \in B$, so also $q \in B$. Furthermore,

$$\|qb - b\| \leq \|qb - pb\| + \|pb - b\| < 4\epsilon + 2\epsilon = 6\epsilon.$$

Thus for every $1 \leq i \leq n$ we have $\|qb_i - b_i\| \leq \|qb - b\|$. Since $b_1, \ldots, b_n$ and ϵ were arbitrary, it follows that B has an approximate unit of projections.

Finally, we show that (iii) implies (i). Suppose that every hereditary C*-sub-algebra of A has an approximate unit consisting of projections. Let $a \in A$ be self-adjoint and put $a = b - c$ where $b, c \geq 0$ and $bc = cb = 0$. As usual, we may assume that a has norm 1. Let $0 < \epsilon < 1/2$ and find a projection p in the hereditary C*-subalgebra generated by b such that $\|pb - b\| < \epsilon$. Since $bc = cb = 0$, we have that $cp = 0$. Define

$$a' := pbp + 2\epsilon p + (1_A - p)a(1_A - p) - 2\epsilon(1_A - p).$$

Then $\|a - a'\| \leq 3\epsilon$. We have

$$pa'p = pbp + 2\epsilon p \geq 2\epsilon p.$$

Thus

$$\|p - pa'p\| \leq |1 - 2\epsilon| \|p\| < 1,$$

so $pa'p$ is invertible in pAp, that is, there exists $x \in pAp$ with $pa'px = xpa'p = p$. Also

$$(1_A - p)a'(1_A - p) = (1_A - p)b(1_A - p) - (1_A - p)c(1_A - p) - 2\epsilon(1_A - p)$$
$$\leq 0 - \epsilon(1_A - p) - 2\epsilon(1_A - p)$$
$$\leq -\epsilon(1_A - p).$$

Thus

$$\|(1_A - p) + (1_A - p)a'(1_A - p)\| \leq |1 - \epsilon| < 1,$$

so $-(1_A - p)a'(1_A - p)$ and hence $(1_A - p)a'(1_A - p)$ is invertible in $(1_A - p)$, that is, there is $y \in (1_A - p)A(1_A - p)$ such that $(1_A - p)a'(1_A - p)y = y(1_A - p)a'(1_A - p) = 1_A - p$.

Notice that a' commutes with p. Then,

$$a'(x + y) = (a'p + a'(1_A - p))(x + y)$$
$$= (pa'p + (1_A - p)a'(1_A - p))(x + y)$$
$$= p + 1_A - p$$
$$= 1_A.$$

Thus a' is invertible.   □

To prove the reverse implication of Lemma 11.2.2, we will follow the strategy of proof due to N. Brown [17], which allows us to avoid the von Neumann algebra theory in Popa's original proof. First, we will require some background.

11.2.4. Definition. Let A be a C*-algebra. A state $\phi : A \to \mathbb{C}$ can be *excised* if there exists a net $(h_\lambda)_\Lambda$ of positive norm one elements in A such that, for every $a \in A$,

$$\lim_\lambda \|h_\lambda^{1/2} a h_\lambda^{1/2} - \phi(a)h_\lambda\| = 0.$$

If the h_λ can be taken to be projections, then we say that ϕ can be *excised by projections*.

11.2.5. Recall that a state ϕ on a C*-algebra A is *pure* if, whenever $\psi : A \to \mathbb{C}$ is a positive linear functional with $\phi \leq \psi$ then $\phi = t\psi$ for some $t \in [0,1]$. We will require the following theorem of Glimm [52], which, for the sake of brevity, we will use without proof.

Theorem (Glimm). *Let A be a C*-algebra acting irreducibly on a Hilbert space H. Suppose that $A \cap \mathcal{K}(H) = \{0\}$. Then the pure states of A are weak-* dense in the state space of A.*

11.2.6. Corollary. *Let A be a simple unital infinite-dimensional C*-algebra. Then the pure states of A are weak-* dense in the state space of A.*

Proof. Let $\pi : A \to \mathcal{B}(H)$ be an irreducible representation of A. Then, since A is simple, π is faithful. If $\pi(a) \in \pi(A)$ is a compact operator, then so is the operator defined by $\pi(xay) = \pi(x)\pi(a)\pi(y)$ for any $x, y \in A$, since $\mathcal{K}(H)$ is an ideal in $\mathcal{B}(H)$. But then also $\pi(A) \cap \mathcal{K}(H)$ is an ideal in $\pi(A)$. The faithfulness of π implies $\pi(A) \cap \mathcal{K}(H)$ is either 0 or $\pi(A)$. Since A is unital, $\pi(1_A)$ is a projection in $\mathcal{B}(H)$ which is a unit for $\pi(A)$. If $\pi(A) \cap \mathcal{K}(H) = \pi(A)$ then $\pi(1_A)$ is a finite-dimensional projection. But then A would have to be finite-dimensional. So A does not contain a compact operator. Thus A satisfies Glimm's theorem and the pure states of A are indeed weak-* dense in the state space of A. $\square$

11.2.7. Again we do not include the proof of the next proposition to keep this chapter from growing too long. For a proof of the next proposition, the reader should consult Section 2 of [1].

Proposition (Akemann–Anderson–Pedersen). *Let $\phi : A \to \mathbb{C}$ be a state on a C*-algebra A. Suppose that ϕ is a weak-* limit of pure states. Then ϕ can be excised.*

11.2.8. Proposition. *Let A be a simple unital infinite-dimensional C*-algebra with real rank zero. Then any state on A can be excised by projections.*

Proof. Let ϕ be a state on A. Then, combining Corollary 11.2.6 and Proposition 11.2.7, ϕ can be excised. Let $(h_\lambda)_\Lambda$ be a net of positive norm one elements in A such that

$$\lim_\lambda \|h_\lambda^{1/2} a h_\lambda^{1/2} - \phi(a)h_\lambda\| = 0 \text{ for every } a \in A.$$

Since A has real rank zero, by Proposition 11.2.3 we may assume that each h_λ has finite spectrum. In that case, there are finitely many mutually orthogonal projections $p_i^{(\lambda)}$, $1 \leq i \leq k(\lambda)$, and $1 = \alpha_1^{(\lambda)} > \alpha_2^{(\lambda)} > \cdots > a_{k(\lambda)}^{(\lambda)} > 0$ such that

$$h_\lambda = \sum_{i=1}^{k(\lambda)} \alpha_i^{(\lambda)} p_i^{(\lambda)}.$$

(To see this, look at the proof of Proposition 11.2.3 where we did something similar using functional calculus.) Observe that $p_1^\lambda h_\lambda = p_1^\lambda$. Thus we calculate

$$\|p_1^{(\lambda)} a p_1^{(\lambda)} - \phi(a) p_1^\lambda\| = \|p_1^{(\lambda)}(h_\lambda^{1/2} a h_\lambda^{1/2} - \phi(a) h_\lambda) p_1^{(\lambda)}\|$$
$$\leq \|h_\lambda^{1/2} a h_\lambda^{1/2} - \phi(a) h_{(\lambda)}\|,$$

which implies that the projections p_1^λ excise ϕ. $\qquad\square$

11.2.9. Let ϕ be a state on an infinite-dimensional C*-algebra A. For $y \in A$ let $\hat{y}$ denote the image of y in $A/N_\phi \subset H_\phi$ corresponding to the GNS construction (4.2.1, 4.2.2). Note that we can choose, for any $n \in \mathbb{N}$ elements $y_1, \ldots, y_n \in A$ such that $\phi(y_j^* y_i) = \delta_{ij}$; this simply amounts to finding n many orthonormal elements in H_ϕ (we leave it as an exercise to show such elements always exist). If p is the orthogonal projection in $\mathcal{B}(H_\phi)$ onto the span of $\hat{y}_1, \ldots, \hat{y}_n$, then

$$\Phi : A \to p\mathcal{B}(H_\phi)p, \quad a \mapsto p\pi_\phi(a)p$$

defines a unital completely positive map. We will use this notation in the next proposition.

Proposition. *Let A be a unital C*-algebra and suppose that ϕ is a state on A which can be excised by projections. Let $y_1, \ldots, y_m \in A$ be elements satisfying $\phi(y_j^* y_i) = \delta_{ij}$ and let $p \in \mathcal{B}(H_\phi)$ denote the orthogonal projection onto the span of the vectors $\hat{y}_1, \ldots, \hat{y}_n$. Then, for any finite subset $\mathcal{F} \subset A$ and any $\epsilon > 0$ there exists a *-monomorphism*

$$\psi : p\pi_\phi(A)p \to A$$

such that

$$\|\psi(p)a\psi(p) - \psi(\Phi(a))\| < \epsilon, \text{ for every } a \in \mathcal{F}.$$

Moreover, for every unitary element $u \in A$ we have

$$\|u\psi(p) - \psi(p)u\|^2 \leq \|p\pi_\phi(u) - \pi_\phi(u)p\|^2 + 2\|\psi(p)u\psi(p) - \psi(\Phi(u))\|.$$

Proof. Let $\mathcal{F} \subset A$ and $\epsilon > 0$ be given. We may assume that $\mathcal{F}$ is composed of norm one elements and that $1_A \in \mathcal{F}$. Choose $\delta > 0$ satisfying

$$\delta < \min\{\delta(\epsilon/4m^2, m), \epsilon/2m^2\},$$

where $\delta(\epsilon/4m^2, m)$ is as given in Lemma 8.4.5. Since ϕ can be excised by projections, there is a projection $p \in A$ such that, for every $1 \leq i, j \leq m$, we have

$$\|p(y_j^* a y_i)p - \phi(y_j^* a y_i)p\| < \delta, \text{ for every } a \in \mathcal{F}.$$

Put $w_i := y_i p$. Then by Lemma 8.4.5 we can perturb the w_i to partial isometries $v_i \in A$, $1 \leq i \leq m$ satisfying

$$v_j^* v_i = \delta_{ij} p \text{ and } \|w_i - v_i\| < \epsilon/4m^2.$$

Set $e_{ij} := v_i v_j^*$. Then $e_{ik}e_{kj} = e_{ij}$ and $e_{ij}^* = e_{ji}$, $1 \leq i, j, k \leq m$, so the e_{ij} are a system of matrix units (8.4.6) generating a C*-subalgebra of A isomorphic to M_m. Let $q := \sum_{i=1}^m e_{ii}$, which is just the unit of this matrix algebra. Now

$$\left\| qaq - \sum_{i,j=1}^m \phi(y_i^* ay_j)e_{ij} \right\| = \left\| \sum_{i,j=1}^m (v_i v_i^* av_j v_j^* - \phi(y_i^* ay_j)v_i v_j^*) \right\|$$

$$\leq \sum_{i,j=1}^m \left(\|v_i w_i^* av_j v_j^* - \phi(y_i^* ay_j)v_i v_j^*\| + \|v_i v_i^* av_j v_j^* - v_i w_i^* av_j v_j^*\| \right)$$

$$\leq \epsilon/4 + \sum_{i,j=1}^m \left(\|v_i w_i^* aw_j v_j^* - \phi(y_i^* ay_j)v_i v_j^*\| + \|v_i w_i^* av_j v_j^* - v_i w_i^* aw_j v_j^*\| \right)$$

$$\leq \epsilon/2 + \sum_{i,j=1}^m \|v_i p(y_i^* ay_j)pv_j^* - \phi(y_i^* ay_j)v_i pv_j^*\|$$

$$\leq \epsilon/2 + \epsilon/2 < \epsilon.$$

Observe that, with respect to the orthonormal basis $\{\hat{y}_1, \ldots, \hat{y}_m\}$, we have

$$p\pi_\phi(a)p = \sum_{i,j=1}^m \langle \pi_\phi(a)\hat{y}_j, \hat{y}_i \rangle = \sum_{i,j=1}^m \phi(y_i ay_j).$$

It follows that

$$\psi : p\pi_\phi(A)p \to \mathrm{C}^*(\{e_{ij} \mid 1 \leq i, j \leq m\}) \subset A, \quad p\pi_\phi(a)p \mapsto \sum_{i,j=1}^m \phi(y_i ay_j)e_{i,j},$$

is a well-defined *-monomorphism, and

$$\|\psi(p)a\psi(p) - \psi(\Phi(a))\| = \left\| qaq - \sum_{i,j=1}^m \phi(y_i ay_j)e_{i,j} \right\| < \epsilon.$$

Now let $u \in A$ be a unitary. Then,

$$\|uq - qu\|^2 = \|uq - quq + quq - qu\|^2$$

$$= \|qu(1_A - q) - (1_A - q)uq\|^2$$

Since $qu(1_A - q)$ and $(1_A - q)uq$ are orthogonal, the above gives

$$\|qu(1_A - q) - (1_A - q)uq\|^2 = \max\{\|qu(1_A - q)\|^2, \|(1_A - q)uq\|^2\}$$

$$= \max\{\|(1_A - q)u^*q\|^2, \|(1_A - q)uq\|^2\}.$$

Applying the C*-equality and approximating quq and qu^*q by

$$\psi(\Phi(u)) \quad \text{and} \quad \psi(\Phi(u^*))$$

respectively, we get

$$\max\{\|(1_A - q)u^*q\|^2, \|(1_A - q)uq\|^2\}$$
$$= \max\{\|qu(1_A - q)u^*q\|, \|qu^*(1_A - q)uq\|\}$$
$$\leq \max\{\|q - \psi(\Phi(u)\phi(u^*))\|, \|q - \psi(\Phi(a^*)\phi(u))\|\} + 2\|quq - \psi(\Phi(u))\|$$
$$= \max\{\|p - \pi_\phi(u^*)p\pi_\phi(u)p\|\|p - \pi_\phi(u)p\pi_\phi(u^*)p\| + 2\|quq - \psi(\Phi(u))\|$$
$$= \|p\pi_\phi(u) - \pi_\phi(u)p\|^2 + 2\|qaq - \psi(\Phi(u))\|.$$

Thus,

$$\|u\,\psi(p) - \psi(p)\,u\|^2 \leq \|p\,\pi_\phi(u) - \pi_\phi(u)\,p\|^2 + 2\|\psi(p)\,u\,\psi(p) - \psi(\Phi(u))\|,$$

which proves the lemma. $\square$

11.2.10. Corollary. *Let A be a unital C^*-algebra and suppose that ϕ is a state on A which can be excised by projections. Let $p \in \mathcal{B}(H_\phi)$ be a finite rank projection. Then, for any finite subset $\mathcal{F} \subset A$ and any $\epsilon > 0$, there exists a *-monomorphism*

$$\psi : p\pi_\phi(A)p \to A$$

such that

$$\|\psi(p)a\psi(p) - \psi(\Phi(a))\| < \epsilon, \text{ for every } a \in \mathcal{F}.$$

Moreover, for every unitary element $u \in A$ we have

$$\|u\psi(p) - \psi(p)u\|^2 \leq \|p\pi_\phi(u) - \pi_\phi(u)p\|^2 + 2\|\psi(p)u\psi(p) - \psi(\Phi(u))\|.$$

Proof. Let (H_ϕ, π_ϕ) be the GNS representation corresponding to ϕ. Then H_ϕ is the completion of A/N_ϕ, so if p is finite rank it can be approximated by a finite rank projection onto the span of vectors $\hat{y}_1, \ldots, \hat{y}_m$ where $y_1, \ldots, y_m \in A$. The result then follows from Proposition 11.2.9. $\square$

11.2.11. Theorem. *Let A be a simple separable unital infinite-dimensional C^*-algebra which is quasidiagonal and has real rank zero. Then A is a Popa algebra.*

Proof. Let ϕ be a state on A. Since A is simple and unital, the GNS construction applied to ϕ yields a faithful representation containing no compact operators. Since A is unital, separable and quasidiagonal, $\pi(A)$ is a quasidiagonal collection of operators. Let $\mathcal{F} \subset \mathcal{B}(H_\phi)$ and $\epsilon > 0$. Since A is unital, it is linearly spanned by its unitaries (Proposition 3.1.6), so without loss of generality, we may assume $\mathcal{F}$ is composed of unitaries. Let p be a finite rank projection satisfying

$$\|p\pi_\phi(a) - \pi_\phi(a)p\| < \epsilon/2 \text{ for every } a \in \mathcal{F}.$$

Since ϕ can be excised by projections, there is a *-monomorphism $\psi : p\pi_\phi(A)p \to A$ with

$$\|\psi(p)a\psi(p) - \psi(p\pi_\phi(a)\,p)\| < \epsilon/4 \text{ for every } a \in \mathcal{F},$$

and

$$\|u\,\psi(p) - \psi(p)\,u\|^2 \le \|p\,\pi_\phi(u) - \pi_\phi(u)\,p\|^2 + 2\|\psi(p)\,u\,\psi(p) - \psi(\Phi(u))\|,$$

for every unitary $u \in A$. In particular, this is satisfied for every element in the finite subset $\mathcal{F}$.

Let $F := \psi(p\,\pi_\phi(A)p) \subset A$, which is a finite-dimensional C*-subalgebra of A with $1_F = \psi(p)$. Then,

$$\begin{aligned}
\|1_F a - a1_F\|^2 &= \|\psi(p)a - a\psi(p)\|^2 \\
&\le \|p\,\pi_\phi(a) - \pi_\phi(a)\,p\|^2 + 2\|\psi(p)\,a\,\psi(p) - \psi(p\pi_\phi(a)p)\| \\
&< \epsilon/2 + 2\epsilon/4 \\
&= \epsilon,
\end{aligned}$$

for every $a \in A$, showing that A satisfies (i) of Definition 11.2.1. Also,

$$\|1_F a 1_F - \psi(p\pi_\phi(a)p)\| < \epsilon/4 < \epsilon,$$

for every $a \in F$, so $\mathrm{dist}(1_F a 1_F, F) < \epsilon$, showing (ii) of Definition 11.2.1. Thus A is a Popa algebra. □

11.2.12. It is useful to know when a C*-algebra has arbitrarily "small" projections, which is made precise by the following definition.

Definition. A C*-algebra A has property (SP) if every nonzero hereditary C*-algebra has a nonzero projection.

Note that if A has real rank zero, then by Proposition 11.2.3, A has property (SP), however (SP) does not imply real rank zero in general.

11.2.13. We will show that Popa algebras have property (SP), but first we need a perturbation lemma.

Lemma. *For any $\epsilon > 0$ there exists a $\delta > 0$ such that for any C*-algebra A and any $a \in A_+$ with $0 \le a \le 1$, if there exists a projection $p \in A$ with*

$$\|ap - p\| < \delta,$$

then there exists a projection q in the hereditary C-algebra generated by a satisfying*

$$\|p - q\| < \epsilon.$$

Proof. Let $\delta < \min\{\epsilon/4, 1/4\}$. Then

$$\|apa - p\| = \|apa - ap + ap - p\| < 2\|ap - p\| < 2\delta.$$

So $\|apa\| \ge 1 - 2\delta \ge 1/2$ and therefore by Lemma 8.3.7 there is a projection q in the hereditary C*-subalgebra generated by a satisfying

$$\|apa - q\| \le 2\|apa - p\| < 4\delta < \epsilon.$$
□

11.2.14. Lemma. *Let $f \in C([-1,1])$. For any $\epsilon > 0$ there exists a $\delta = \delta(f, \epsilon) > 0$ satisfying the following: For any C^*-algebra A and any self-adjoint element $a \in A$ with $\|a\| \leq 1$ and projection $p \in A$ such that*

$$\|pa - ap\| < \delta,$$

we have

$$\|f(a)p - f(pap)\| < \epsilon.$$

Proof. We leave the details as an exercise. Hint: use the functional calculus and the fact that any such f can be approximated by polynomials. $\square$

11.2.15. The next lemma is easy to prove, but will come in useful. As with the previous lemma, the details are an exercise.

Lemma. *Let A be a finite-dimensional C^*-algebra. Suppose a, b are nonzero positive elements satisfying $ab = b$. Then a is a projection.*

11.2.16. We will make use of the functional calculus with respect to the following function. Let $0 < r < 1$ and define $f_r : [0, 1] \to [0, 1]$ by

$$f_r(t) := \begin{cases} 0 & t \in [0, r/4), \\ \frac{4}{r}t - 1 & t \in [r/4, r/2), \\ 1 & t \in (r/2, 1]. \end{cases}$$

11.2.17. Proposition. *Let A be a Popa algebra. Then A has property (SP).*

Proof. Let B be a hereditary C^*-subalgebra of A. If B is finite-dimensional, then B contains a nonzero projection. So assume that B is infinite-dimensional. Then B contains a nonzero positive element a with infinite spectrum and norm 1 (Exercise 3.4.10). Choose some r, $0 < r < 1/4$ and define $a_1 := f_r(a)$, $a_2 := f_{2r}(a)$ and $a_3 := f_{4r}(a)$ where f_r is the function defined in 11.2.16. Then $0 \leq a_i \leq 1_A$, $i = 1, 2, 3$ and satisfy

$$a_1 a_2 = a_2, \quad a_2 a_3 = a_3.$$

Let $\mathcal{F} := \{a_i, a_i^{1/2} \mid i = 1, 2, 3\}$ and let $\epsilon_0 < \min\{1/8, \delta/3\}$ where $\delta = \delta(1/8, 2)$ is as given by Exercise 8.7.22. Then, since A is a Popa algebra, there is a projection $p \in A$ and a finite-dimensional C^*-subalgebra $F \subset pAp$ with $1_F = p$ satisfying

(i) $\|pf - fp\| < \epsilon_0$ for every $f \in \mathcal{F}$, and

(ii) $\operatorname{dist}(pfp, F) < \epsilon_0$ for every $f \in \mathcal{F}$.

Let $b_i \in F$ be elements satisfying $\|pa_ip - b_i\| < \epsilon_0$, $0 \leq i \leq 3$. Then,

$$\|b_1 b_2 - b_2\| < \|pa_1 pa_2 p - pa_2 p\| + 2\epsilon_0 < \|pa_1 a_2 p - pa_1 p\| + 3\epsilon_0 = 3\epsilon_0 < \delta.$$

Similarly, $\|b_2 b_3 - b_3\| < \delta$. Thus $\|(p - b_1)b_2\| = \|b_2 - b_1 b_2\| < \delta$. Then, by Exercise 8.7.22 and the choice of δ, there exists $z, c_2 \in F$ such that $c_2 z = 0$ and $\|(p - b_1) - z\|$,

$\|c_2 - b_2\| < 1/8$. Let $c_1 := p - z$. Then $c_1 c_2 = c_2$. It follows from Lemma 11.2.15 that c_1 is a projection. Moreover,

$$
\begin{aligned}
\|a_1^{1/2} p a_1^{1/2} - c_1\| &< 2\epsilon_0 + \|p a_1 p - c_1\| \\
&< 2\epsilon_0 + \|p a_1 p - b_1\| + \|b_1 - c_1\| \\
&< 3\epsilon_0 + 1/8 < 1/2.
\end{aligned}
$$

Thus, by Lemma 8.4.1 there exists a projection in the hereditary C*-subalgebra $a_1^{1/2} p a_1^{1/2}$ which is contained in $\overline{aAa} \subset B$. □

11.3 Tracial approximation by building blocks

We already saw, in the chapter on inductive limits, that we can often approximate a C*-algebra in norm by tractable "building block" algebras, for example the finite-dimensional C*-algebras approximating an AF algebra. Here we will see that we can also approximate a C*-algebra in a tracial way: instead of asking that our building blocks are large in terms of a norm estimate, as happens in inductive limits, we simply ask that they are "large" when measured by tracial states.

Let $\mathcal{S}$ be a class of separable unital C*-algebras. The next definition, due to Lin, was originally defined for the class $\mathcal{S} = F$ of finite-dimensional C*-algebras. In that case, the definition asks that A is a Popa algebra and that we can always choose the finite-dimensional subalgebra $F \subset A$ to be "large", by asking that $1_A - 1_F$ can be made small enough to be twisted under any prescribed positive element.

11.3.1. Definition (cf. [74]). A simple separable unital C*-algebra is said to be *tracially approximately* $\mathcal{S}$, or TA$\mathcal{S}$ for short, if the following holds. For every finite subset $\mathcal{F} \subset A$, every $\epsilon > 0$, and every nonzero positive element $c \in A$ there is a projection $p \in A$ and a C*-subalgebra $B \subset pAp$ with $1_B = p$ and $B \in \mathcal{S}$ such that

(i) $\|pa - ap\| < \epsilon$ for every $a \in \mathcal{F}$;

(ii) $\text{dist}(pap, B) < \epsilon$ for every $a \in \mathcal{F}$;

(iii) $1_A - p$ is Murray–von Neumann equivalent to a projection in $\overline{cAc}$.

Despite the name, the definition above seems to make no mention of traces. However, (iii) implies that $\tau(1_A - p) \leq \tau(c)$ for every $\tau \in T(A)$, which implies that $1_A - p$ can be made arbitrarily small in trace. In other words, the C*-subalgebra B is large inside A, when measured by tracial states.

11.3.2. Remark. If $\mathcal{S} = F$ is the class of finite-dimensional C*-algebras, then a C*-algebra which is TAF is often said to have *tracial rank zero*. It is not hard to see that if A is a simple, unital AF algebra, then A is TAF. However, TAF algebras are more general; we will see an example in Theorem 11.3.12. More generally, if $\mathcal{I}^k$ denotes the class of C*-algebras which are finite direct sums of C*-algebras of the form $p(M_n(C(X)))p$, where X is a finite CW complex with dimension k and p is a projection in $M_n(C(X))$, then a C*-algebra that is $TA\mathcal{I}^k$ is said to have *tracial rank no more than* k. Tracial rank was introduced by Lin in [73]. If $\mathcal{I}$ denotes the class of *interval algebras*, that is, C*-algebras which are finite direct sums of C*-algebras of the form $M_n(C([0,1]))$, then a simple unital C*-algebra is TAI if and only if it has tracial rank no more than one [73, Theorem 7.1].

11.3.3. Lemma. *Let A be a C*-algebra. Suppose $p, q, r \in A$ are projections and $a \in A_+$ a positive element. Suppose that $r \in \overline{aAa}$, q is Murray–von Neumann equivalent to r, and p is Murray–von Neumann equivalent to a projection in rAr. Then p is Murray–von Neumann equivalent to a projection in $\overline{aAa}$.*

Proof. Exercise.   □

11.3.4. Theorem. *Suppose $\mathcal{S}$ is a class of unital C*-algebras which is closed under passing to unital hereditary C*-subalgebras. Then, if A is a simple separable unital C*-algebra which is TA$\mathcal{S}$, so is any unital hereditary C*-subalgebra of A.*

Proof. Let A be tracially approximately $\mathcal{S}$ and suppose that $e \in A$ is a projection. We must show that eAe is also a tracially approximately $\mathcal{S}$. Let $\mathcal{F} \subset eAe$ be a finite subset, $\epsilon > 0$ and $c \in eAe_+$. Without loss of generality we may assume that the elements in $\mathcal{F}$ have norm at most one. Let $\epsilon' < \min\{1/11, \epsilon/65\}$. Since A is TA$\mathcal{S}$ there is a projection $p \in A$ and a C*-subalgebra $B \subset pAp$ with $1_B = p$ such that

(i) $\|pa - ap\| < \epsilon'$ for every $a \in \mathcal{F} \cup \{e\}$,

(ii) $\mathrm{dist}(pap, B) < \epsilon'$ for every $a \in \mathcal{F} \cup \{e\}$, and

(iii) $1_A - p$ is Murray–von Neumann equivalent to a projection in $\overline{cAc}$.

From (i), we have $\|(epe)^2 - epe\| = \|epepe - epe\| < \epsilon'$, so by Lemma 8.7.23 there is a projection $q_1 \in eAe$ such that $\|q_1 - epe\| < 2\epsilon'$. Then

$$\begin{aligned} \|q_1 - pep\| &< \|q_1 - ep\| + \epsilon' \\ &< \|q_1 - epe\| + 2\epsilon' \\ &< 4\epsilon'. \end{aligned}$$

From (ii), there is an $x \in B$, which we may assume is self-adjoint, such that $\|x - pep\| < \epsilon'$. Then

$$\begin{aligned} \|x^2 - x\| &< \|xpep - pep\| + \epsilon' \\ &< \|pepep - pep\| + 2\epsilon' \\ &< 3\epsilon'. \end{aligned}$$

Thus, again by Lemma 8.7.23 there is a projection $q_2 \in B$ with $\|q_2 - x\| < 6\epsilon'$. Then

$$\|q_1 - q_2\| \leq \|q_1 - pep\| + \|pep - x\| + \|x - q_2\| < 11\epsilon'.$$

Since $11\epsilon' < 1$, by Lemma 8.3.6 there exists a unitary $u \in A$ satisfying $q_1 = uq_2u^*$ and $\|1_A - u\| \leq 11\sqrt{2}\epsilon' < 16\epsilon'$.

Let $B' = uBu^*$, which is evidently a finite-dimensional C*-subalgebra. Define $C := q_1 B' q_1$. This is finite-dimensional since B' is, and moreover, since $q_1 \in eAe$, it is a finite-dimensional C*-subalgebra of eAe.

If $a \in \mathcal{F}$, there exists $x_a \in C$ satisfying $\|pap - px_a p\| < \epsilon'$. Define $a' := q_1 ux_a u^* q_1$ and observe that $a' \in C$. We compute

$$\|q_1 a q_1 - a'\| = \|q_1 a q_1 - uq_2 x_a q_2 u^*\| = \|q_1 a q_1 - uq_2 px_a p q_2 u^*\|$$
$$< \|q_1 a q_1 - uq_2 pap q_2 u^*\| + \epsilon' = \|q_1 a q_1 - uq_2 a q_2 u^*\| + \epsilon'$$
$$= \|q_1 a q_1 - q_1 uau^* q_1\| + \epsilon' < \|a - uau^*\| + \epsilon'.$$

Moreover,

$$\|a - uau^*\| \leq \|1_A - u\|\|a\| + \|ua - au^*\| + \|a\|\|1_A - u^*\|$$
$$\leq 32\epsilon' + \|u - 1_A\|\|a\| + \|a\|\|1_A - u^*\|$$
$$= 64\epsilon'.$$

so $\mathrm{dist}(q_1 a q_1, C) < 65\epsilon' < \epsilon$, showing (ii) of Definition 11.3.1 holds. To show (i) of Definition 11.3.1, we have

$$\|q_1 a - a q_1\| < \|pepa - apep\| + 2\epsilon'$$
$$< \|ppea - aepp\| + 4\epsilon'$$
$$= \|pa - ap\| + 4\epsilon'$$
$$< 5\epsilon' < \epsilon,$$

for every $a \in F$.

Finally, we have

$$\|(e - q_1) - (1_A - p)e(1_A - p)\| = \|e - q_1 - e + pe + ep - pep\|$$
$$\leq \|pe - q_1\| + \|ep - pep\|$$
$$< \|pep - q_1\| + \epsilon' + \epsilon'$$
$$< 6\epsilon' < 1,$$

so $e - q_1$ is Murray–von Neumann equivalent to a projection in $(1_A - p)A(1_A - p)$. Since $1_A - p$ is Murray–von Neumann equivalent to a projection in $\overline{cAc}$, it follows from Lemma 11.3.3 that $e - q_1$ is as well, which shows (iii) of Definition 11.3.1. Thus eAe is TA$\mathcal{S}$. $\qquad\square$

11.3.5. We have already seen that AF algebras have real rank zero. The same turns out to be true for TAF C*-algebras.

Theorem. *Let A be a simple unital TAF C*-algebra. Then A has real rank zero.*

Proof. Let $a \in A$ be a self-adjoint element and let $\epsilon > 0$. Without loss of generality, assume that $\epsilon < 1$. We need to show that there is $x \in A$ which is invertible, self-adjoint and such that $\|a - x\| < \epsilon$. Let $f \in C(-\|a\|, \|a\|)$ be a continuous function satisfying $0 \leq f \leq 1$ and $f(t) = 1$ for every $\|t\| < \epsilon/128$ and $f(t) = 0$ for $|t| \geq \epsilon/64$. Observe that if $f(a) = 0$ then 0 cannot be in the spectrum of a, in which case we simply take $b = a$ since a itself is invertible. Thus we may assume that $f(a) \neq 0$.

Let B be the hereditary C*-subalgebra generated by $f(a)$. By Proposition 11.2.17, A has property (SP), so B contains a nonzero projection q. Let $b = f_{\epsilon/2}(a)$, where $f_{\epsilon/2}$ is defined as in 11.2.16. Then $f_{\epsilon/2}(t)f(t) = f(t)f_{\epsilon/2}(t)$ for every $t \in [-\|a\|, \|a\|]$, so $bq = qb = 0$, and

$$\|b - a\| < \epsilon/2.$$

By Theorem 11.3.4, the unital hereditary C*-subalgebra $(1_A - q)A(1_A - q)$ is also TAF. Thus, there is a projection $p \in (1_A - q)A(1_A - q)$ and a finite-dimensional unital C*-subalgebra $C \subset p(1_A - q)A(1_A - q)p$ such that

(i) $\|pb - bp\| < \epsilon/8$,

(ii) $\operatorname{dist}(pap, C) < \epsilon/8$,

(iii) $1_A - q - p$ is Murray–von Neumann equivalent to a projection in qAq.

We have that

$$\|b - pbp - (1_A - q - p)b(1_A - q - p)\| = \| - pbp - (-bp - pb + pbp)\|$$
$$\leq \|pbp - bp\| + \|pb - pbp\|$$
$$< \epsilon/2.$$

Now, C is finite-dimensional, so in particular has real rank zero. Thus, using (i), we can find an invertible self-adjoint element $b' \in C$ such that

$$\|pbp - b'\| < \epsilon/4.$$

Let $v \in A$ be a partial isometry satisfying $v^*v = 1_A - q - p$ and $vv^* \leq q$, which exists by (iii). Set $c := (1_A - q - p)b(1_A - q - p)$. We claim that the element

$$z := b_2 + (\epsilon/16)v + (\epsilon/16)v^* + (\epsilon/4)(q - vv^*)$$

is invertible in $(1 - p)A(1 - p)$. First, note that $y := (c + \epsilon/16)v + (\epsilon/16)v^*$ is contained in $((1_A - q) + vv^*)A((1_A - q) + vv^*)$. In matrix notation, we have

$$y = \begin{pmatrix} c & (\epsilon/16)v^* \\ (\epsilon/16)v & 0 \end{pmatrix}.$$

It is easy to check that this has inverse given by

$$\begin{pmatrix} 0 & (16/\epsilon)v^* \\ (16/\epsilon)v & (16^2/\epsilon^2)vcv^* \end{pmatrix},$$

which we will denote by d. Since $(1_A - q + vv^*)(q - vv^*) = 0$ and d is contained in $((1_A - q) + vv^*)A((1_A - q) + vv^*)$, we have

$$zd = (y + (\epsilon/4))(q - vv^*))d = ((1_A - p) + vv^*),$$

and so

$$(d + (4/\epsilon))(q - vv^*) = (d + (4/\epsilon))(q - vv^*)c = 1_A - p,$$

which shows that c is invertible in $(1_A - p)A(1_A - p)$. Finally, one checks that $x := b' + c$ is a self-adjoint element which is invertible in A and

$$\begin{aligned}
\|a - x\| &= \|a - b' - c\| \\
&= \|a - b' - (1_A - q - p)b(1_A - q - p)\| \\
&= \|a - b' - b + bp + pb - pbp\| \\
&\leq \|a - b\| + \|bp - b'\| + \|pb - pbp\| \\
&< \epsilon/2 + \epsilon/8 + \epsilon/4 + \epsilon/8 < \epsilon,
\end{aligned}$$

as required. $\qquad\square$

It is straightforward to see that a simple unital AF algebra is always TAF. The reader might question whether or not the class of simple unital TAF algebras contains anything else. It contains many approximately homogeneous C*-algebras as well.

11.3.6. Recall from 8.5.3 that an approximately homogeneous (AH) algebra is an inductive limit $A = \varinjlim(A_n, \varphi_n)$ where each A_n is finite direct sum of C*-algebras of the form $p(C(X) \otimes M_n)p$ for some compact Hausdorff space X and projection $p \in C(X) \otimes M_n$.

Note that in the above we allow that X is a single point so that any AF algebra is also AH. (One could also compose connecting maps with point evaluations, but by allowing points we can restrict to the case where the maps are injective.) It is not immediately clear that the class of simple unital AH algebras is strictly larger than the class of simple unital AF algebras. However, this does turn out to be true. In Chapter 9 we saw that if X is an infinite compact metric space and $\alpha : X \to X$ is a minimal homeomorphism, then the crossed product C*-algebra $C(X) \rtimes_\alpha \mathbb{Z}$ is unital and simple (Theorem 9.5.8). If X is a Cantor set and $\alpha : X \to X$ is minimal, then Putnam showed that $C(X) \rtimes_\alpha \mathbb{Z}$ is an inductive limit of circle algebras (an A$\mathbb{T}$ algebra) [96], which is an AH algebra with each $X = \mathbb{T}$ in the definition above. Similarly, the irrational rotation algebra (Exercise 9.6.11) is an A$\mathbb{T}$ algebra ([39, Theorem 4]). None of these examples can be AF algebras. One way to see this is by looking at the unitaries: an AF algebra will always have a connected unitary group (see Exercises 12.6.3 and 12.6.4), while any crossed product by the integers will not [14, Proposition 4.4.1]. This can also be seen by the nontriviality of the K_1-group (which we introduce in Chapter 12), which can be computed via the Pimsner–Voiculescu exact sequence [94].

11.3.7. The definition for covering dimension was given in Definition 8.6.2. Let $d < \infty$. If each compact metric space in the direct sum of each A_n has covering dimension at most d, then we will say that the AH algebra $A = \varinjlim(A_n, \varphi_n)$ is an *AH algebra with bounded dimension*. The class of simple unital AH algebras with bounded dimension is already quite large, note for example that it contains all the crossed products of the previous paragraph.

We will show that many simple unital AH algebras are TAF. However, we will cut a few corners and use a slightly different characterisation of TAF C*-algebras due to Winter. First, we need to introduce comparability of projections. Recall that in a matrix algebra M_n, the canonical trace determines the rank of a projection, and that furthermore rank completely determines the Murray–von Neumann equivalence classes in M_n (Exercise 3.3.1 (c)). So the trace is an accurate measure of the size of projections in M_n. This is not necessarily the case in an arbitrary C*-algebra, so we introduce the definition below.

11.3.8. Definition (cf. [7, 1.3.1]). A C*-algebra A has *comparison of projections* if, whenever two projections $p, q \in A$ satisfy $\tau(p) < \tau(q)$ for every tracial state $\tau \in T(A)$, then p is Murray–von Neumann equivalent to a projection $p' < q$.

In the literature, this property is sometimes called *Blackadar's second fundamental comparability property*. It is implied by a property called *strict comparison of positive element*, which will be introduced in Chapter 14.

11.3.9. We will use the following lemma, which gives an alternative characterisation of simple unital TAF algebras in the case of real rank zero and comparison of projections. For a proof see [132].

Lemma ([132, Lemma 3.2]). *Let A be a simple unital C*-algebra with real rank zero and comparison of projections. Then A is TAF if and only there is $n \in \mathbb{N} \setminus \{0\}$ such that for any finite subset $\mathcal{F} \subset A$ and $\epsilon > 0$ there exists a projection $p \in A$ and finite-dimensional unital C*-subalgebra $B \subset pAp$ satisfying*

(i) *$\|pa - ap\| < \epsilon$ for every $a \in \mathcal{F}$,*
(ii) *$\mathrm{dist}(pap, B) < \epsilon$ for every $a \in \mathcal{F}$,*
(iii) *$\tau(p) > 1/n$ for every $\tau \in T(A)$.*

11.3.10. Lemma. *Let (X, d) be a compact metric space with $\dim(X) \leq m < \infty$. Suppose that μ is a Borel probability measure with full support. Then, for any finite subset $\mathcal{F} \subset C(X)$ and any ϵ, there are $n \in \mathbb{N}$ and pairwise disjoint open subsets $U_1, \ldots, U_n$ such that*

(i) *$|f(x) - f(y)| < \epsilon$ for every $x, y \in U_i$, $1 \leq i \leq n$, and*
(ii) *$\mu(\bigcup_{i=1}^n U_i) \geq 1/m$.*

Proof. Choose $\delta > 0$ so that $|f(x) - f(y)| < \epsilon$ whenever $d(x, y) < \delta$. For $x \in X$, let $B(x, \delta)$ denote the open ball of radius δ and centre x. Then $\{B(x, \delta) \mid x \in X\}$ is an open cover of X. Since X is compact, it has to find a finite subcover, and since

X has covering dimension at most m, we can furthermore assume that the finite open cover is of the form $\{U_{i,j} \mid 0 \le i \le m, 1 \le j \le n(i)\}$ for some $n(i) \in \mathbb{N}$, where $U_{i,j} \cap U_{i,j'} = \emptyset$ for every $0 \le i \le d$, $1 \le j \ne j' \le n(i)$. Since $\mu(\bigcup_{i=0}^{d} \bigcup_{j=1}^{n(i)} U_{i,j}) = \mu(X) = 1$, there must be some i such that $\mu(\bigcup_{j=1}^{n(i)} U_{i,j}) \ge 1/m$. Then the open sets $U_{i,1}, \ldots, U_{i,n(i)}$ satisfy the requirements. $\qquad\square$

11.3.11. Let A be a C*-algebra. Suppose that X is a compact metric space such that $C(X) \subset A$ is a C*-subalgebra. If $\tau \in T(A)$ is a tracial state, then by the Riesz Representation theorem there is a unique Borel measure μ_τ on X satisfying

$$\tau(f) = \int f d\mu_\tau, \quad f \in C(X).$$

For any open set $V \subset X$ and any $\epsilon > 0$, let $U \subset V$ be an open subset such that $U \subset \bar{U} \subset V$ and $\mu(U) \ge \mu(V) - \epsilon$. Let $f : X \to [0,1]$ be a continuous function that is identically 1 for every $x \in U$ and zero for every $x \in X \setminus V$. Then f satisfies $\tau(f) \ge \mu(V) - \epsilon$.

11.3.12. The following is a simpler case of a more general theorem, implicit in [41], rephrased in the language of tracial approximation. See also [76] for a similar statement about AH algebras without real rank zero. The results of Chapter 17 will imply that any AH algebra with a bound on the dimensions of the spaces in the inductive limit will have strict comparison of positive elements. Thus the assumption of comparison of projections below is in fact unnecessary.

Theorem. *Let* $A = (\varinjlim A_n, \varphi_n)$ *be a simple unital AH algebra such that each*

$$A_n \cong \bigoplus_{k=1}^{m_n} C(X_{n,k}) \otimes M_{r(n,k)}$$

and that

$$\max\{\dim(X_{n,k}) \mid n \in \mathbb{N}, 1 \le k \le m_n\} \le d < \infty.$$

Suppose that A has real rank zero, has a unique tracial state, and comparison of projections. Then A is TAF.

Proof. By Lemma 11.3.9, it is enough to show that for any finite subset $\mathcal{F} \subset A$ and $\epsilon > 0$ there exists a projection $p \in A$ and finite-dimensional unital C*-subalgebra $B \subset pAp$ satisfying

(i) $\|pa - ap\| < \epsilon$ for every $a \in \mathcal{F}$,

(ii) $\mathrm{dist}(pap, B) < \epsilon$ for every $a \in \mathcal{F}$,

(iii) $\tau(p) > 1/d$ for every $\tau \in T(A)$.

Let $\mathcal{F} \subset A$ and $\epsilon > 0$ be given. Without loss of generality, we may assume that $\mathcal{F} = \{f_1, \ldots, f_r\} \subset \bigoplus_{k=1}^m C(X_k) \otimes M_{n_k}$ with $\dim(X_k) \leq d$. Going further out in the inductive sequence if necessary, we may assume that

$$\tau\left(1 - \left(\sum_{k=1}^m 1_{C(X_k)} \otimes 1_{n_k}\right)\right) < 1/4,$$

where 1_{n_k} denotes the unit in M_{n_k}.

Let μ_k denote the Borel measure on X_k induced by τ. For every $k \in \mathbb{N}$, use Lemma 11.3.10 to find pairwise disjoint open subsets $W_{k,1}, \ldots, W_{k,s(k)}$ and pairwise disjoint open subsets $U_{k,i} \subset \overline{U_{k,i}} \subset W_{k,i}$, $1 \leq i \leq s(k)$ of X_k such that

$$|f_j(x) - f_j(y)| < \epsilon_1 \text{ for every } x, y \in U_{k,i}, 1 \leq i \leq s(k), 1 \leq j \leq r;$$

and

$$\mu_k\left(\bigcup_{i=1}^{s(k)} U_{k,i}\right) \geq \tau(1_{C(X_k)} \otimes 1_{n_k})/d > 3/(4d).$$

Let $V_{k,i} \subset \overline{V_{k,i}} \subset U_{k,i}$ and

$$\mu_k(U_{k,i} \setminus V_{k,i}) < \left(8d\sum_{k=1}^m s(k)\right)^{-1}.$$

Let $\gamma_{k,i} \in C(X_k)$ be a function supported in $U_{k,i}$ such that $\gamma_{k,i}(x) = 1$ for every $x \in V_{k,i}$, and let $\alpha_{k,i} \in C(X_k)$ be a function supported in $V_{k,i}$ with $0 \leq \alpha_{k,i} \leq 1$. Then, for each $k = 1, \ldots, m$, we have

$$\tau\left(\sum_{k=1}^m \sum_{i=1}^{s(k)} \alpha_{k,i} \otimes 1_{n_k}\right) = \sum_{k=1}^m \mu_k(V_{k,I}) \geq 5/(8d).$$

Let $\epsilon_1 < \epsilon/4$. Extend $U_{k,1}, \ldots, U_{k,s(k)}$ to a finite open cover of X_k, $U_{k,1}, \ldots, U_{k,s(k)}$, $U_{k,s(k)+1}, \ldots, U_{k,t(k)}$ such that for every $1 \leq j \leq r$ and every $s(k) + 1 \leq i \leq t(k)$, we have $\|f_j(x) - f_j(y)\| < \epsilon_1$ for every $x, y \in U_{k,i}$, and so that $\{\gamma_{k,i}\}_{i=1}^{t(k)}$ extends to a partition of unity $\{\gamma_{k,i}\}_{i=1}^{s(k)}$ subordinate to this open cover, where $\mathrm{supp}(\gamma_{k,i}) \subset U_{k,i}$ for every $s(k) + 1 \leq i \leq t(k)$.

For every $k = 1, \ldots, m$ and $i = 1, \ldots, s(k)$, we can find a function $\beta_{k,i} \in C(X_k)$ satisfying $\gamma_{k,i}\beta_{k,i} = \beta_{k,i}$ and $\beta_{k,i}\alpha_{k,i} = \alpha_{k,i}$.

Define, for every $1 \leq k \leq m$ and every $1 \leq i \leq s(k)$,

$$a_{k,i} := \alpha_{k,i} \otimes 1_{n_k}, \quad b_{k,i} = \beta_{k,i} \otimes 1_{n_k}, \quad c_{k,i} = \gamma_{k,i} \otimes 1_{n_k}.$$

Since A has real rank zero, the hereditary C*-subalgebra $\overline{b_{k,i}Ab_{k,i}}$ contains an approximate unit of projections. In particular, there is a projection $p_{k,i}$ such that $\|p_{k,i}b_{k,i} - b_{k,i}\| < \min\{\epsilon_1, 1/(8d)\}$. Note that the $p_{k,i}$ are pairwise orthogonal and, by approximating $p_{k,i}$ by elements in $b_{k,i}Ab_{k,i}$, we have $p_{k,i}c_{k,i} = p_{k,i}$.

Let

$$p_k := \sum_{i=1}^{s(k)} p_{k,i}, \quad \text{and} \quad p := \sum_{k=1}^{m} p_k.$$

Set

$$B := p\left(\bigoplus_{k=1}^{m} 1_{C(X_k)} \otimes M_{n_k}\right)p = \bigoplus_{k=1}^{m} p_k(1_{C(X_k)} \otimes M_{n_k})p_k \subset pAp.$$

We will show that p and B satisfy (i)–(iii) with respect to the finite set $\mathcal{F} = \{f_1, \ldots, f_r\}$, $\epsilon > 0$ and $1/(2d)$. Let $f_j \in \mathcal{F}$. For every $k = 1, \ldots, m$ and $i = 1, \ldots, s(k)$ there exist matrices $m^{(k,i)} \subset M_{n_k}$ such that $\|f_j(x) - m^{(k,i)}\| < \epsilon_1$. Let $g_j := \sum_{k,i} \gamma_{k,i} \otimes m^{(k,i)}$. Then we have $\|f_j - g_j\| < \epsilon_1$ and

$$p_{k,i}g_j = p_{k,i}c_{k,i}1_{C(X_k)} \otimes m^{(k,i)} = p_{k,i}1_{C(X_k)} \otimes m^{(k,i)}.$$

We have

$$\|pf_j - f_jp\| = 2\epsilon_1 + \|pg_j - g_jp\|$$
$$= 2\epsilon_1 + \max_{k,i} \|p_{k,i}g_j - g_jp_{k,i}\|$$
$$\leq 2\epsilon_1 + \max_{k,i} \|p_{k,i}b_{k,i}1_{C(X_k)} \otimes m^{(k,i)} - 1_{C(X_k)} \otimes m^{(k,i)}b_{k,i}p_{k,i}\|$$
$$\leq 2\epsilon_1 + 2\max_{k,i} \|p_{k,i}b_{k,i}1_{C(X_k)} \otimes m^{(k,i)} - b_{k,i}1_{C(X_k)} \otimes m^{(k,i)}\|$$
$$< 4\epsilon_1$$
$$< \epsilon,$$

showing (i).

For (ii),

$$\left\| pg_jp - \sum_{k=1}^{m}\sum_{i=1}^{s(k)} p_{k,i}(1_{C(X_k)} \otimes m^{(k,i)})p_{k,i} \right\|$$
$$= \max_{k,i} \|p_{k,i}g_jp_{k,i} - p_{k,i}(1_{C(X_k)} \otimes m^{(k,i)})p_{k,i}\|$$
$$= \max_{k,i} \|p_{k,i}c_{k,i}1_{C(X_k)} \otimes m^{(k,i)}c_{k,i}p_{k,i} - p_{k,i}(1_{C(X_k)} \otimes m^{(k,i)})p_{k,i}\|$$
$$= 0.$$

Thus

$$\text{dist}(pf_jp, B) \leq \|pf_jp - pg_jp\| \leq \epsilon_1 < \epsilon.$$

Finally,

$$\tau(p) = \tau\left(\sum_{k,i} p_{k,i}c_{k,i}\right) \geq \tau\left(\sum_{k,i} p_{k,i}b_{k,i}\right) = \tau\left(\sum_{k,i} b_{k,i}\right) - 1/(8d)$$

$$> \tau\left(\sum_{k,i} a_{k,i}\right) - 1/(8d) > 5/(8d) - 1/(8d) = 1/(2d).$$

showing (iii). Thus A is TAF. □

11.4 Exercises

11.4.1. Suppose that A is a unital quasidiagonal C*-algebra. Show that there is a net of *unital* completely positive maps $\varphi_\lambda : A \to M_{n_\lambda}$ such that $\|\varphi_\lambda(ab) - \varphi_\lambda(a)\varphi_\lambda(b)\| \to 0$ and $\|\varphi_\lambda(a)\| \to \|a\|$ for every $a, b \in A$. Hint: Use functional calculus on the operators $\varphi_\lambda(1_A)$ to show that there are projections p_λ satisfying $\|p_\lambda - \varphi_\lambda(1_A)\| \to 0$ and that $\varphi(1_A)p_\lambda$ is invertible in $p_\lambda M_{n_\lambda} p_\lambda$.

11.4.2. Prove Proposition 11.1.5, that quasidiagonality can be reframed as a local property.

11.4.3. Let A be a simple unital C*-algebra. Suppose that A is a Popa algebra. Show that A is quasidiagonal. (Hint: see Exercise 7.3.7.)

11.4.4. Let ϕ be a state on an infinite-dimensional C*-algebra A. Show that, for any $n \in \mathbb{N}$ elements, there are $y_1, \ldots, y_n \in A$ such that $\phi(y_j^* y_i) = \delta_{ij}$;

11.4.5. Let A be a unital quasidiagonal C*-algebra. Show that A has a tracial state.

11.4.6. Prove Lemma 11.2.14: Let $f \in C([-1,1])$. For any $\epsilon > 0$ there exists a $\delta = \delta(f, \epsilon) > 0$ satisfying the following: For any C*-algebra A and any self-adjoint element $a \in A$ with $\|a\| \leq 1$ and projection $p \in A$ such that

$$\|pa - ap\| < \delta,$$

we have

$$\|f(a)p - f(pap)\| < \epsilon.$$

11.4.7. Let A be a finite-dimensional C*-algebra. Suppose a, b are nonzero positive elements satisfying $ab = b$. Show that a is a projection.

11.4.8. Let A be a C*-algebra. Suppose $p, q, r \in A$ are projections and $a \in A_+$ a positive element. Suppose that $r \in \overline{aAa}$, q is Murray–von Neumann equivalent to r, and p is Murray–von Neumann equivalent to a projection in rAr. Show that p is Murray–von Neumann equivalent to a projection in $\overline{aAa}$.

11.4.9. Let $\mathcal{C}$ denote any class of unital C*-algebras. Suppose that A is a simple unital inductive limit of C*-algebras in $\mathcal{C}$ and that A has at least one tracial state. Show that A is TA$\mathcal{C}$.

11.4.10. Let A be a simple unital C*-algebra. Suppose that A is TAF. Show that A has stable rank one. (Hint: see the proof that TAF implies real rank zero.)

11.4.11. Let A be a simple unital C*-algebra. Suppose that for every finite subset $\mathcal{F} \subset A$, every $\epsilon > 0$, and every nonzero positive element $c \in A_+$ the following holds: There exist a projection $p \in A$ and a unital AF C*-subalgebra B with $1_B = p$ satisfying

(i) $\|pa - ap\| < \epsilon$ for every $a \in \mathcal{F}$,

(ii) $\mathrm{dist}(pap, B) < \epsilon$ for every $a \in \mathcal{F}$,

(iii) $1_A - p$ is Murray–von Neumann equivalent to a projection in $\overline{cAc}$.

Show that A is TAF.

One can use this to show that the AF algebra of Exercise 9.6.13 is arbitrarily "tracially large" in the crossed product given by the Cantor 2-odometer and hence the crossed product is TAF. More generally, a similar technique was used in various contexts to show that a crossed product by a minimal homeomorphism can be classified by tracial approximation techniques, see for example [80, 114, 124, 113, 79].

Part III

Introduction to Classification Theory

12 K-theory

It would be difficult to talk about the classification of C*-algebras without talking about K-theory. For the reader familiar with topological K-theory, K-theory for C*-algebras can be thought of its noncommutative version. For a commutative C*-algebra $C(X)$, the C*-algebraic K-theory of $C(X)$ turns out to be isomorphic to the topological K-theory of X, although in general (once one moves to the noncommutative setting) there is no ring structure for C*-algebraic K-theory. The K-theory for C*-algebras consists of two functors K_0 and K_1 which take the category of C*-algebras with *-homomorphisms to the category of abelian groups with group homomorphisms. It is an important isomorphism invariant and plays a fundamental role in the classification program for separable nuclear C*-algebras, as we will see. In this chapter, we provide a very basic introduction, focusing only on the constructions as well as on some results we will require later on. More involved introductions to the K-theory for C*-algebras, which should be easily accessible to the reader, can be found in [103] and [128]. A very thorough development is given in [8].

In Section 12.1 we introduce the K_0-group of a unital C*-algebra and see that it has an order structure whenever A is stably finite. In the second section, we prove that K_0 is continuous with respect to inductive limits, a property which often makes computation of K-theory for C*-algebras relatively easy. In Section 12.3 we show how to construct the K_0-group for a nonunital C*-algebras and prove that K_0 is *half exact*, another useful property for computational purposes. Section 12.4 constructs the K_1-group. In the final section, we collect a number of other useful

© The Author(s), under exclusive license to Springer Nature Switzerland AG 2021
K. R. Strung, *An Introduction to C*-Algebras and the Classification Program*, Advanced Courses in Mathematics - CRM Barcelona, https://doi.org/10.1007/978-3-030-47465-2_12

facts and properties about K-theory, such as split exactness, Bott periodicity and the 6-term exact sequence.

12.1   The abelian group $K_0(A)$

Let A be a C*-algebra and let $M_\infty(A) = \bigcup_{n \in \mathbb{N}} M_n(A)$ where $M_n(A)$ is included into $M_{n+1}(A)$ by copying M_n into the top left corner of $M_{n+1}(A)$, that is,

$$a \mapsto \begin{pmatrix} a & 0 \\ 0 & 0 \end{pmatrix}.$$

If $p \in M_n(A)$ and $q \in M_m(A)$ then we define $p \oplus q \in M_{n+m}(A)$ to be

$$p \oplus q = \begin{pmatrix} p & 0 \\ 0 & q \end{pmatrix}.$$

Recall that two projections p, q in a C*-algebra A are said to be Murray–von Neumann equivalent if there exists a partial isometry $v \in A$ such that $vv^* = p$ and $v^*v = q$ (Definition 8.3.11). For projections $p, q \in M_\infty(A)$, set $p \sim q$ if there is some $m, n \in \mathbb{N}$ and some $v \in M_{m,n}(A)$ such that $v^*v = p$ and $vv^* = q$. This is an equivalence relation on the projections in $M_\infty(A)$.

It is easy to see that $p \oplus 0 \sim p$, so we may identify p with $p \oplus 0$. If $p \in M_m(A)$ and $q \in M_n(A)$, with $m < n$ we can add $n - m = k$ zeros to p to consider it as a projection in $M_n(A)$. Then $p \oplus 0_{M_k} \sim q$ is just Murray–von Neumann equivalence in $M_n(A)$.

12.1.1. The next proposition collects some useful facts. The details are left as an exercise, which is not too difficult but should be helpful familiarising the reader with the concepts.

Proposition. *Let $p_1, p_2, q_1, q_2, r \in M_\infty(A)$ be projections.*

  (i) *If $p_1 \sim p_2$ and $q_1 \sim q_2$ then $p_1 \oplus q_1 \sim p_2 \oplus q_2$.*
 (ii) *Up to equivalence, $\oplus$ is commutative: $p_1 \oplus q_1 \sim q_1 \oplus p_1$.*
(iii) *If $p_1, q_1 \in M_n(A)$ and $p_1 q_1 = 0$, then $p_1 + q_1 \sim p_1 \oplus q_1$,*
 (iv) *Up to equivalence, $\oplus$ is associative: $(p_1 \oplus q_1) \oplus r \sim p_1 \oplus (q_1 \oplus r)$.*

12.1.2. Let A be a unital C*-algebra and let $[p]$ denote the $\sim$ equivalence class of the projection $p \in M_\infty(A)$. Denote by $V(A)$ all $\sim$ equivalence classes of projections in $M_\infty(A)$, that is

$$V(A) := M_\infty(A)/ \sim .$$

By Proposition 12.1.1 (i), (ii), and (iv), we can endow $V(A)$ with the structure of an abelian semigroup. We call $V(A)$ the *Murray–von Neumann semigroup* of A. Moreover, $V(A)$ has an identity element given by the equivalence class of 0, so in fact it is a monoid. We put an equivalence relation on the set of formal differences of elements in $V(A)$ by declaring $[p] - [q] \sim_G [p'] - [q']$ if and only if there is $[r] \in V(A)$ such that $[p] + [q'] + [r] = [p'] + [q] + [r] \in V(A)$.

$K_0(A)$ is now defined to be

$$K_0(A) := \{[p] - [q] \mid [p], [q] \in V(A)\} / \sim_G .$$

The construction of a group from a semigroup via this equivalence on formal differences is called the *Grothendieck* construction. Thus $K_0(A)$ is the *Grothendieck group* of the semigroup $V(A)$.

12.1.3. Suppose that A is unital and $p, q \in M_\infty(A)$ are projections such that $p \oplus r \sim q \oplus r$ for some projection $r \in M_n(A)$. Let 1_n denote $\mathrm{diag}(1_A, \ldots, 1_A) \in M_n(A)$. Then

$$p \oplus 1_n \sim p \oplus r \oplus (1_n - r) \sim q \oplus r \oplus (1_n - r) \sim q \oplus 1_n,$$

so $[p] = [q]$ if and only if $p \oplus 1_n \sim q \oplus 1_n$ or some $n \in \mathbb{N}$.

At this point, we move from projections to unitaries. It is easy to see that if A is unital and $p \in A$ is a projection, then for any unitary $u \in A$, we have that $q := upu^*$ is also a projection. If such a unitary exists, we say that p and q are unitarily equivalent. If $q = upu^*$ then define $v := up$ and observe that $q = vv^*$ and $p = v^*v$. In other words, unitary equivalence implies Murray–von Neumann equivalence. In general, the reverse implication does not hold in a C*-algebra A, however, when we consider projections in all of $M_\infty(A)$, we can show that Murray–von Neumann equivalence is enough to establish unitary equivalence. This is part of the content of Proposition 12.1.5, where we show that Murray–von Neumann equivalence classes, unitary equivalence classes and homotopy equivalence classes in $M_\infty(A)$ coincide. First, we require a lemma.

12.1.4. Lemma. *Let A be a unital C*-algebra and let $u \in M_2(A)$ be a unitary. Suppose that $\mathrm{sp}(u) \neq \mathbb{T}$. Then there is a continuous path $w : [0, 1] \to M_2(A)$ such that*

(i) *$w(t)$ is unitary for every $t \in [0, 1]$,*

(ii) *$w(0) = u$,*

(iii) *$w(1) = \begin{pmatrix} 1_A & 0 \\ 0 & 1_A \end{pmatrix}$.*

In other words, u is homotopic to $1_{M_2(A)}$.

Proof. Let $u \in M_2(A)$ be a unitary with $\mathrm{sp}(u) \neq \mathbb{T}$. Since $\mathrm{sp}(u) \neq \mathbb{T}$, there is some $\theta \in \mathbb{R}$ such that $e^{i\theta} \notin \mathrm{sp}(u)$. Thus e^{it} has a well-defined inverse function for $t \in (\theta, 2\pi + \theta)$ (by taking the appropriate branch of the complex logarithm). In particular, $\phi(e^{it}) = t$, for $t \in (\theta, 2\pi + \theta)$ is a real-valued function that is well defined and continuous on $\mathrm{sp}(u)$, which moreover satisfies $\lambda = e^{i\phi(\lambda)}$ for every $\lambda \in \mathrm{sp}(u)$.

Using functional calculus we have $u = (e^{i\phi(u)})$. Define $w(t) = e^{it\phi(u)}$ for $t \in [0,1]$. Since u is unitary, we have $\phi(\lambda) \in \mathbb{R}$ for every $\lambda \in \mathrm{sp}(u)$ and so $\phi(u)$ is self-adjoint. Thus $w(t)$ is a unitary for every $t \in [0,1]$. Since $w(\cdot, \lambda) : [0,1] \to \mathbb{T}$ given by $w(t, \lambda) = e^{it\lambda}$ is continuous for every $\lambda \in \mathrm{sp}(\phi(u))$, it follows that $w(t)$ is a continuous path from $w(0) = 1_{M_2(A)}$ to $w(1) = u$. $\square$

12.1.5. This next proposition tells us that we could have also defined $K_0(A)$ for a unital C*-algebra A using either unitary equivalence or homotopy equivalence of projections in $M_\infty(A)$, in place of Murray–von Neumann equivalence. In what follows, for $n \in \mathbb{N}$ we denote

$$0_n := \underbrace{0 \oplus \cdots \oplus 0}_{n \text{ times}} \in M_n(A).$$

Proposition. *Let A be a unital C*-algebra. Suppose $p, q \in M_\infty(A)$ are projections. The following are equivalent:*

(i) *there are $k, l, m \in \mathbb{N}$ and there is a continuous path of projections*

$$r : [0,1] \to M_m(A),$$

such that $r(0) = p \oplus 0_k$ and $r(1) = q \oplus 0_l$,

(ii) *there are $k, l, m \in \mathbb{N}$ and a unitary $u \in M_m(A)$ such that*

$$u(p \oplus 0_k)u^* = q \oplus 0_l,$$

(iii) *$[p] = [q]$ in $V(A)$.*

Proof. For (i) implies (ii), suppose there are $k, l, m \in \mathbb{N}$ and a path of projections $r : [0,1] \to \bigcup_{n \in \mathbb{N}} M_n(A)$ such that $r(0) = p \oplus 0_k$ and $r(1) = q \oplus 0_l$. Choose $\delta > 0$ such that if $|t - t'| < \delta$ then $\|r(t) - r(t')\| < 1$. By Lemma 8.3.6, if $|t - t'| < \delta$ there exists a unitary $u(t) \in M_m(A)$ such that $u(t)r(t)u(t')^* = r(t')$. Thus (ii) holds by compactness of the interval and the fact that the unitaries in $M_m(A)$ form a group under multiplication.

For (ii) implies (iii), if there are $k, l, m \in \mathbb{N}$ and a unitary $u \in \bigcup_{n \in \mathbb{N}} M_n(A)$ such that $u(p \oplus 0_k)u^* = q \oplus 0_l$, then $v := u(p \oplus 0_k)$ satisfies $v^*v = p \oplus 0_k$ and $vv^* = q \oplus 0_l$. Hence $[p] = [q]$ in $V(A)$.

For (iii) implies (i), if $[p] = [q]$ in $V(A)$, we can choose k, l, m so that both $p \oplus 0_k, q \oplus 0_l \in M_m(A)$ and there is a partial isometry $v \in M_m(A)$ satisfying

$$v^*v = p \oplus 0_k, \qquad vv^* = q \oplus 0_l.$$

Put $p' := p \oplus 0_k$ and $q' := q \oplus 0_l$. It is straightforward to check that

$$u := \begin{pmatrix} v & 1_{M_m(A)} - vv^* \\ v^*v - 1_{M_m(A)} & v^* \end{pmatrix}$$

is a unitary in $M_{2m}(A)$. Let

$$w(t) := \begin{pmatrix} \cos(\frac{\pi}{2})v & 1_{M_m(A)} - (1_{M_m(A)} - \sin(\frac{\pi}{2}))vv^* \\ (1_{M_m(A)} - \sin(\frac{\pi}{2}))v^*v - 1_{M_m(A)} & \cos(\frac{\pi}{2})v^* \end{pmatrix}.$$

Then $w : [0,1] \to M_{2m}(A)$ is a continuous path of unitaries with

$$w(0) = u, \quad w(t) = \begin{pmatrix} 0 & 1_{M_m(A)} \\ -1_{M_m(A)} & 0 \end{pmatrix}.$$

It now follows from Lemma 12.1.4, after rescaling paths, that there is a continuous path of unitaries $w' : [0,1] \to M_{2m}(A)$ with $w(0) = u$ and $w(1) = 1_{M_{2m}(A)}$. Since

$$u(p \oplus 0_k \oplus 0_m)u^* = q \oplus 0_l \oplus 0_m,$$

(i) follows. $\qquad\qquad\qquad\qquad\qquad\qquad\qquad\qquad\qquad\qquad\qquad\qquad\qquad\square$

12.1.6. Recall that a unital C*-algebra is stably finite if $M_n(A)$ is finite for every $n \in \mathbb{N}$, that is, $1_{M_n(A)}$ is a finite projection for every $n \in \mathbb{N}$ (Definition 8.3.12). By Proposition 8.3.13, A is finite if and only if every isometry in A is a unitary. Here, we give another equivalent condition for finiteness of a unital C*-algebra.

12.1.7. Proposition. *Let A be a unital C*-algebra. Then A is finite if and only if every projection $p \in A$ is finite.*

Proof. If every projection is finite, then this is in particular true for 1_A, so A is finite. Conversely, assume that A is finite. Let $p, q \in A$ be projections with $q \leq p$ and suppose that $v^*v = p$ and $vv^* = q$, for some $v \in A$. We need to show that $p = q$. Let

$$s := v + (1_A - p).$$

Since $q \leq p$, by Proposition 3.1.14, we have $pq = qp = q$. Thus

$$\begin{aligned} \|v^* - v^*p\|^2 &= \|(v^* - v^*p)^*(v^* - v^*p)\| \\ &= \|vv^* - vv^*p - pvv^* + pvv^*p\| \\ &= \|q - qp - pq - pqp\| = 0, \end{aligned}$$

so $v^*(1_A - p) = 0$, and also $(1_A - p)v = (v^*(1_A - p))^* = 0$. From this it follows that

$$s^*s = (v^* + (1_A - p))(v + (1_A - p)) = v^*v + (1_A - p) = 1_A,$$

which is to say, s is an isometry. Since A is finite, every isometry is a unitary by Proposition 8.3.13. Thus

$$\begin{aligned} 1_A = ss^* &= (v + (1_A - p))(v + (1_A - p))^* \\ &= vv^* + v(1_A - p) + (1_A - p)v^* + (1_A - p) \\ &= q + v - vv^*v + v^* - v^*vv^* + 1_A - p \\ &= q + 1_A - p, \end{aligned}$$

so $p - q = 0$. $\qquad\qquad\qquad\qquad\qquad\qquad\qquad\qquad\qquad\qquad\qquad\qquad\square$

12.1.8. An *ordered abelian group* is an abelian group G together with a partial order that respects the group structure (in the sense that if $x \leq y$ then $x+z \leq y+z$ for every $x, y, z \in G$), and which satisfies

$$G = \{x \in G \mid 0 \leq x\} - \{x \in G \mid 0 \leq x\}.$$

As for C*-algebras, we denote the positive elements of G by G_+.

Of course, (G, G_+) should properly be called a *partially ordered abelian group*, but in the C*-algebra literature – particularly the classification literature – the adverb "partially" is usually dropped for brevity. Since this is the general convention, it is the terminology that we will use.

12.1.9. A *cone* in an abelian group G is a subset H such that $H + H \subset H$, $G = H - H$ and $H \cap (-H) = \{0\}$. If G has a partial order, then G_+ is a cone. Conversely, if H is a cone in G, then setting $x \leq y$ if and only if $y - x \in H$, defines a partial order on G.

12.1.10. Theorem. *Let A be a unital stably finite C*-algebra. Then*

$$K_0(A)_+ := \{[p] \mid p \in M_\infty(A) \text{ is a projection}\}$$

is a cone and hence $K_0(A)$ is an ordered abelian group with the induced partial order.

Proof. That $K_0(A)_+ + K_0(A)_+ \subset K_0(A)_+$ and $K_0(A) = K_0(A)_+ - K_0(A)_+$ is immediate. So we just need to show that $K_0(A)_+ \cap (-K_0(A)_+) = \{0\}$.

Let $x \in K_0(A)_+ \cap (-K_0(A)_+)$. Then $x = [p] - [0] = [0] - [q]$ for some projections $p, q \in M_\infty(A)$. Thus there is a projection $r \in M_\infty(A)$ with $[p \oplus q \oplus r] = [r]$. Let $n \in \mathbb{N}$ be sufficiently large so that $p \oplus q \oplus r \in M_n(A)$. Then, adding zeros where necessary, we can find projections $p', q', r' \in M_n(A)$ with $p' \sim p$, $q' \sim q$, and $r' \sim r$, with p', q', r' mutually orthogonal. Then $p'+q'+r'$ is Murray–von Neumann equivalent to r' in $M_n(A)$. Since A is stably finite, $p' + q' + r'$ cannot be equivalent to a proper subprojection by Proposition 12.1.7. It follows that $p' + q' = 0$, and hence $p' = q' = 0$. Thus $x = [p] = [q] = 0 \in K_0(A)_+$. $\qquad\square$

12.1.11. Lemma. *Let p and q be projections in a unital C*-algebra A. Then $q = u^*pu$ for some unitary u in A if and only if $p \sim q$ and $1_A - p \sim 1_A - q$.*

Proof. Suppose $q = upu^*$. Set $v = up$ and $w = u(1_A - p)$. Then

$$v^*v = pu^*up = p, \qquad vv^* = uppu^* = q$$

and

$$w^*w = (1_A - p)u^*u(1_A - p) = 1_A - p, \qquad ww^* = u(1_A - p)(1_A - p)u^* = 1_A - q.$$

Hence $p \sim q$ and $1_A - p \sim 1_A - q$.

Conversely, if $p \sim q$ and $1_A - p \sim 1_A - q$ then there are v, w in A with $p = v^*v$, $q = vv^*$, $1_A - p = w^*w$ and $1_A - q = ww^*$. Let $z = v + w$. Then

$$
\begin{aligned}
z^*z &= v^*v + v^*w + w^*v + w^*w \\
&= p + v^*q(1_A - q)w + w^*(1_A - q)qv + (1_A - p) \\
&= 1_A.
\end{aligned}
$$

Similarly, $zz^* = 1_A$, so z is a unitary. Conjugating q by z gives

$$
z^*qz = (v^*q + w^*q)(v + w) = (v^* + w^*(1 - q)q)(v + w) = v^*v + v^*q(1_A - q)w = p,
$$

as required. $\qquad\qquad\qquad\qquad\qquad\qquad\qquad\qquad\qquad\qquad\qquad\qquad\square$

12.1.12. We say that a C*-algebra A has *cancellation of projections* if, for any four projections $p, q, e, f \in A$ satisfying $pe = 0, qf = 0, e \sim f$ and $p + e \sim q + f$, we have $p \sim q$. When A is unital we have $p(1_A - p) = 0$ for any projection p, so cancellation of projections implies that $p \sim q$ if and only if $1_A - p \sim 1_A - q$. By Lemma 12.1.11, this holds if and only if there is a unitary $u \in A$ such that $q = u^*pu$. If $M_n(A)$ has cancellation of projections for all $n \in \mathbb{N}$, then A has *cancellation*.

Recall that a C*-algebra has stable rank one if the invertible elements of A are dense in A (Definition 8.6.6). C*-algebras with stable rank one are particularly well behaved. In particular, Theorem 12.1.14 shows that they have cancellation of projections.

12.1.13. Proposition. *Let A be a unital C*-algebra with stable rank one. Then $M_n(A)$ has stable rank one for every $n \in \mathbb{N}$.*

Proof. The proof is by induction. If $n = 1$ then $M_n(A) = A$ has stable rank one by assumption. Assume that $n > 1$ and that $M_n(A)$ has stable rank one. Then, identifying $M_{n+1}(A)$ with $A \otimes M_{n+1}$ (Exercise 4.4.15), set $p := 1_A \otimes e_n$ where $e_n := 1_n \oplus 0$ denotes the embedding of the identity matrix 1_n in M_n into the top left corner of M_{n+1}. Then

$$
pM_{n+1}(A)p \cong M_n(A),
$$

and

$$
(1_{n+1} - p)M_{n+1}(A)(1_{n+1} - p) \cong A,
$$

so $pM_{n+1}(A)p$ and $(1_{n+1} - p)M_{n+1}(A)(1_{n+1} - p)$ both have stable rank one. Let $\epsilon > 0$. For $x \in M_{n+1}(A)$, we can write

$$
x = \begin{pmatrix} a & b \\ c & d \end{pmatrix},
$$

where

$$
a \in pM_n(A)p, \quad b \in pM_{n+1}(A)(1_{n+1} - p), \quad c \in (1_{n+1} - p)M_{n+1}(A)p,
$$

and
$$d \in (1_{n+1} - p)M_{n+1}(A)(1_{n+1} - p).$$

Since $(1_{n+1} - p)M_{n+1}(A)(1_{n+1} - p)$ has stable rank one, there exists an invertible element $d' \in (1_{n+1} - p)M_{n+1}(A)(1_{n+1} - p)$ with $\|d - d'\| < \epsilon$. Let $y = b(d')^{-1}c$. Then we have $y \in pM_{n+1}(A)p$ and since $pM_{n+1}(A)p$ also has stable rank one, there exists an invertible element $z \in pM_{n+1}(A)p$ such that $\|(a - y) - z\| < \epsilon$. Let $a' := y + z$ and set

$$x' := \begin{pmatrix} a' & b \\ c & d' \end{pmatrix}.$$

Then one checks that

$$\begin{pmatrix} z^{-1} & -z^{-1}b(d')^{-1} \\ -(d')^{-1}cz^{-1} & (d')^{-1}cz^{-1}b(d')^{-1} + (d')^{-1} \end{pmatrix}$$

is an inverse for x' and $\|x - x'\| \le \max\{\|a - a'\|, \|d - d'\|\} < \epsilon$. Thus $M_{n+1}(A)$ has stable rank one.   □

12.1.14. Theorem. *Let A be a unital C*-algebra with stable rank one. Then A has cancellation.*

Proof. Let p and q be projections in $M_n(A)$ with $p \sim q$ and $v \in M_n(A)$ such that $v^*v = p$ and $vv^* = q$. By 12.1.12, it is enough to show that there exists a unitary $s \in M_n(A)$ such that $q = sps^*$. Since A has stable rank one, so does $M_n(A)$ (Proposition 12.1.13). Thus there is an invertible $x \in M_n(A)$ with $\|x - v\| < 1/8$. Since $\|v\| = 1$, we may assume that $\|x\| \le 1$. Let $u = x(x^*x)^{-1/2}$. Then u is a unitary in $M_n(A)$ and we have

$$\|x^*x - p\| \le \|x^*x - x^*v\| + \|x^*v - v^*v\| < 1/4.$$

Similarly,
$$\|xx^* - q\| \le \|xx^* - xv^*\| + \|xv^* - vv^*\| < 1/4.$$

Thus

$$\|upu^* - q\| = \|upu^* - u(x^*x)u^* + u(x^*x)u^* - q\| \le \|\|p - x^*x\| + \|xx^* - q\| < 1/2.$$

Then by Lemma 8.3.6, there is a unitary $w \in M_n(A)$ such that $q = w^*upu^*w$.   □

12.1.15. An *order unit* for an ordered abelian group G is an element $u \in G_+$ such that, for every $x \in G$ there exists an $n \in \mathbb{N}$ such that $-nu \le x \le nu$. Note that in general an order unit is *not* unique.

Proposition. *Let A be a unital stably finite C*-algebra. Then $[1_A]$ is an order unit for $K_0(A)$.*

Proof. Exercise.   □

12.1.16. We will refer to a triple $(K_0(A), K_0(A)_+, [1_A])$ as an *ordered abelian group with distinguished order unit*, or sometimes as a *pointed ordered abelian group*.

12.1.17. Definition. An ordered abelian group (G, G_+) is *simple* if every nonzero positive element is an order unit.

12.1.18. Proposition. *Let A be a simple unital stably finite C*-algebra. Then*

$$(K_0(A), K_0(A)_+)$$

is simple.

Proof. Exercise (see the exercises for a hint).   □

12.1.19. Observe that any *-homomorphism $\varphi : A \to B$ induces a group homomorphism $\varphi_0 : K_0(A) \to K_0(B)$ by simply defining $\varphi_0([p] - [q]) = [\varphi(p)] - [\varphi(q)]$. Here when we apply φ on the right-hand side, inside the brackets, we are actually identifying φ with its inflation to the appropriate matrix algebra over A. One needs to check this is well defined, but that is not difficult.

12.1.20. If (G, G_+, u) and (H, H_+, v) are ordered abelian groups with distinguished order units, then a *unital positive homomorphism* is a homomorphism $\varphi : G \to H$ satisfying $\varphi(G_+) \subset H_+$ and $\varphi(u) = v$. If φ is a group isomorphism and φ^{-1} is also a unital positive homomorphism, then we call φ a *unital order isomorphism*.

Let A and B be C*-algebras and $\varphi : A \to B$ a *-homomorphism. If A and B are stably finite, we also have $\varphi_0(K_0(A)_+) \subset K_0(B)_+$, and if φ is unital, then $\varphi_0([1_A]) = [\varphi(1_A)] = [1_B]$ so φ_0 preservers the order unit. If $\varphi : A \to B$ is a *-isomorphism, then φ_0 is a unital order isomorphism.

12.2 Continuity of K_0

As for C*-algebras, we can define inductive limits of abelian groups.

Let $(G_n, \varphi_n)_{n \in \mathbb{N}}$ be a sequence of abelian groups together with homomorphisms $\varphi_n : G_n \to G_{n+1}$. We call such a sequence an *inductive sequence of abelian groups*. We leave it as an exercise to formulate the proper definition for the limit of such a sequence and to verify that it is indeed an abelian group which satisfies the universal property given in Theorem 12.2.1.

As in the case of inductive limits of C*-algebras, if $G = \varinjlim(G_n, \varphi_n)$ we will denote the homomorphism induced by φ_n into the limit as $\varphi^{(n)} : G_n \to G$. The proofs of the next two theorems and proposition are effectively simpler versions of the analogous statements for inductive limits of C*-algebras (Theorem 8.2.6 and Theorem 8.2.7, respectively), so they are also left as exercises.

12.2.1. Theorem. *Let $(G_n, \varphi_n)_{n \in \mathbb{N}}$ be an inductive sequence of abelian groups with limit $G = \varinjlim G_n$. Suppose there is an abelian group H and for every $n \in \mathbb{N}$ there* *are group homomorphisms $\psi^{(n)} : G_n \to H$ making the diagrams*

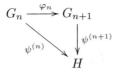

commute. Then there is a unique *-homomorphism $\psi : G \to H$ making the diagrams

commute.

12.2.2. Proposition. *Let $(G_n, \varphi_n)_{n \in \mathbb{N}}$ be an inductive sequence of abelian groups with limit $G = \varinjlim G_n$. Then $G = \bigcup_{n=1}^{\infty} \varphi^{(n)}(G_n)$.*

12.2.3. Theorem. *Let $(G_n, \varphi_n)_{n \in \mathbb{N}}$ be an inductive sequence of abelian groups with limit $G = \varinjlim G_n$. Suppose there is a sequence of homomorphisms $\psi^{(n)} : G_n \to H$ to an abelian group H satisfying $\psi^{(n+1)} \circ \phi_n = \psi^{(n)}$. Then, letting $\psi : A \to B$ be the induced *-homomorphism, we have*

  (i) *ψ is injective if and only if $\ker(\psi^{(n)}) \subset \ker(\phi^{(n)})$ for every $n \in \mathbb{N}$,*

  (ii) *ψ is surjective if and only if $H = \overline{\bigcup_{j=1}^{\infty} \psi^{(n)}(G_n)}$.*

12.2.4. In what follows, if $\varphi : A \to B$ is a *-homomorphism, we will denote the induced map $K_0(A) \to K_0(B)$ by $[\varphi]_0$ to avoid confusion with the *-homomorphism $\varphi_0 : A_0 \to A_1$ given in the inductive sequence.

Theorem. *Let $A = \varinjlim(A_n, \varphi_n)$ be an inductive limit of unital C*-algebras. Then $\varinjlim K_0(A_n) \cong K_0(\varinjlim A_n)$ as abelian groups, and if each $(K_0(A_n), K_0(A_n))_+$, $n \in \mathbb{N}$ and $(K_0(A), K_0(A)_+)$ are ordered abelian groups, then the isomorphism is an isomorphism of ordered abelian groups. Furthermore,*

  (i) *$K_0(A) = \bigcup_{n=1}^{\infty} [\varphi^{(n)}]_0(K_0(A_n))$,*

  (ii) *$K_0(A)_+ = \bigcup_{n=1}^{\infty} [\varphi^{(n)}]_0(K_0(A_n)_+)$,*

  (iii) *$\ker([\varphi^{(n)}]_0) = \bigcup_{m=n+1}^{\infty} \ker([\varphi_{m,n}]_0)$ for every $n \in \mathbb{N}$.*

Proof. We will prove (i), (ii) and (iii) and leave the first statements, which follow, as an exercise. For $k \in \mathbb{N}$, by abuse of notation, we also denote the induced map by $\varphi_n : M_k(A_n) \to M_k(A_{n+1})$. In that case, $\varinjlim(M_k(A_n), \varphi_n) = M_k(A)$.

Let $g \in K_0(A)$ be given. Then there exist $k \in \mathbb{N}$ and projections $p, q \in M_k(A)$ such that $g = [p] - [q]$. For sufficiently large n, there are elements $a, b \in M_n(A_n)$, which we may assume to be positive, such that

$$\|\varphi^{(n)}(a) - p\| < 1/12, \quad \|\varphi^{(n)}(b) - q\| < 1/12.$$

This gives $\|\varphi^{(n)}(a^2 - a)\| = \|\varphi^{(n)}(a)^2 - \varphi^{(n)}(a)\| < 1/4$. Increasing n if necessary, we get $a \in M_k(A_n)$ such that $\|a^2 - a\| < 1/4$ and $\|\varphi^{(n)}(a)^2 - p\| < 1/12$. Similarly, we find $b \in M_k(A_n)$ such that $\|b^2 - b\| < 1/4$ and $\|\varphi^{(n)}(b)^2 - q\| < 1/12$.

It follows from Lemma 8.3.7 that there are projections $p', q' \in M_k(A_n)$ such that $\|a - p'\| < 1/4$ and $\|b - q'\| < 1/4$. Furthermore, $\|\varphi^{(n)}(p') - p\| < 1/4 + 1/12 < 1/2$ so $\varphi^{(n)}(p') \sim p$ and similarly, $\varphi^{(n)}(q') \sim q$. It follows that

$$g = [p] - [q] = [\varphi^{(n)}(p')] - [\varphi^{(n)}(q')] = [\varphi^{(n)}]_0([p'] - [q']).$$

This shows (i).

Since each $[\varphi^{(n)}]_0$ is positive, we have the inclusion $[\varphi^{(n)}]_0(K_0(A_n)_+) \subset K_0(A)_+$. If $g \in K_0(A)_+$, then there is some $k \in \mathbb{N}$ such that $g = [p]$ for some projection $p \in M_k(A)$. Arguing as above, we find some sufficiently large n and projection $p' \in M_k(A_n)$ such that $p \sim \varphi^{(n)}(p')$, which shows the reverse inclusion, proving (ii).

Now, let us show (iii). We have $\ker([\varphi_{m,n}]_0) \subset \ker([\varphi^{(n)}]_0)$ since it holds at the level of the inductive limit of C*-algebras and so passes to the maps at the level of K_0. If $g \in K_0(A)$ satisfies $[\varphi^{(n)}]_0(g) = 0$, then there are $k \in \mathbb{N}$ and projections $p, q \in M_k(A)$ such that $[\varphi^{(n)}]_0([p] - [q]) = 0$. So $[\varphi^{(n)}(p)] + r = [\varphi^{(n)}(q)] + r$ for some $r \in K_0(A_n)$. Using similar approximation arguments as in (i) and (ii), for large enough m, we can find projections $p', q' \in M_k(A_n)$ with $\varphi_{m,n}(p') \sim p$ and $\varphi_{m,n}(q') \sim q$. Then

$$[\varphi_{m,n}(p')] + r = [\varphi^{(n)}(p)] + r = [\varphi^{(n)}(q)] + r = [\varphi_{m,n}(q')] + r.$$

Thus $[\varphi_{m,n}(p')] - [\varphi_{m,n}(q')] = 0$ in $K_0(A_n)$, which is to say $[p'] - [q'] \in \ker([\varphi_{m,n}]_0)$. Hence $\ker([\varphi^{(n)}]_0) = \bigcup_{m=n+1}^{\infty} \ker([\varphi_{m,n}]_0)$. $\square$

12.2.5. Remark. We saw in Theorem 8.3.14 that if A is an AF algebra, then A is stably finite. If A is moreover unital, this in turn implies, by Theorem 12.1.10 and Proposition 12.1.15 that $(K_0(A), K_0(A)_+, [1_A])$ is an ordered abelian group with distinguished order unit. The previous theorem now gives us an alternative proof of this fact, without appealing to stable finiteness.

12.2.6. Recall from Theorem 8.1.2 that any finite-dimensional C*-algebra F is of the form $F \cong M_{n_1} \oplus M_{n_2} \oplus \cdots \oplus M_{n_k}$, for some $k, n_1, \ldots, n_k \in \mathbb{N}$.

Proposition. *Let* $F = M_{n_1} \oplus \cdots \oplus M_{n_m}$ *be a finite-dimensional C*-algebra and let* $e_{ij}^{(k)}$, $1 \leq i, j \leq n_k$, $1 \leq k \leq m$, *be matrix units for* F. *The map*

$$\varphi : (\mathbb{Z}^m, \mathbb{Z}_+^m, (n_1, \ldots, n_m)) \to (K_0(A), K_0(A)_+, [1_A])$$

given by

$$(r_1, \ldots, r_m) \mapsto \sum_{k=1}^{m} r_k [e_{11}^{(k)}]$$

is a unital order isomorphism.

Proof. If $p \in M_\infty(F)$ is a projection, then there are $p_k \in M_\infty(M_{n_k})$, $1 \le k \le m$ such that $p = \sum_{k=1}^{m} p_k$ (this follows from Exercise 8.7.3). Each p_k is a projection in a matrix algebra over M_{n_k}, hence itself is a projection in some larger matrix algebra, where the equivalence classes of projections are determined by their ranks (Exercise 8.7.15 (i)). Thus $p_k \sim r_k [e_{11}^{(k)}]$ for some $r_k \in \mathbb{Z}_+$ and so $p \sim \sum_{k=1}^{m} r_k [e_{11}^{(k)}]$. It follows that the map φ is surjective and that φ is positive. It is also clear that $\varphi((n_1, \ldots, n_m)) = \sum_{k=1}^{m} n_k [e_{11}^{(k)}] = \sum_{k=1}^{m} [1_k] = [1_F]$.

Let $\pi_k : F \to M_{n_k}$ be the surjection onto M_{n_k}. If $\varphi((r_1, \ldots, r_m)) = 0$, then $\sum_{k=1}^{m} r_k [e_{11}^{(k)}] = 0$. It follows that for each $l \in \{1, \ldots, m\}$ we have $0 = [\pi_l]_0(\sum_{k=1}^{m} r_k [e_{11}^{(k)}]) = r_l [e_{11}^{(l)}]$. Thus the direct sum of r_k copies of $e_{11}^{(l)}$ is equivalent to zero. This is only possible if $r_l = 0$. So φ is injective. $\square$

Notice that M_n has the same ordered K-theory for every n, but once we also include the class of the unit, we can tell them apart.

12.3 K_0 for nonunital C*-algebras and half-exactness

What about a nonunital C*-algebra A? We could go ahead and define equivalence of projections in $M_\infty(A)$ and proceed as in the unital case. However, it turns out that this does not have the correct properties. For example, we would not get the isomorphism of Theorem 12.4.7.

12.3.1. Let A be nonunital and let $\tilde{A}$ denote its minimal unitisation (2.2.2). Then there is a split exact sequence

$$0 \longrightarrow A \overset{\iota}{\longrightarrow} \tilde{A} \overset{\pi}{\longrightarrow} \mathbb{C} \longrightarrow 0.$$

We define $K_0(A)$ to be the kernel of the induced map $\pi_0 : K_0(\tilde{A}) \to K_0(\mathbb{C})$.

If we have a unital C*-algebra A to which we attach a new unit, then the sequence above induces a short exact sequence on K-theory and so $\ker(\pi_0) \cong K_0(A)$ (exercise). Thus this definition can also be used for unital C*-algebras.

12.3.2. Let A be a C*-algebra and let $\ell : \tilde{A} \to \tilde{A}$ denote the section of $\pi : \tilde{A} \to \mathbb{C}$ in the split exact sequence

$$0 \longrightarrow A \overset{\iota}{\longrightarrow} \tilde{A} \overset{\overset{\ell}{\frown}}{\underset{\pi}{\longrightarrow}} \mathbb{C} \longrightarrow 0.$$

Let $s : M_\infty(\tilde{A}) \to M_\infty(\tilde{A})$ denote the inflation of the composition map $\ell \circ \pi$.

Theorem. *Let A be a C*-algebra. Then*

$$K_0(A) = \{[p] - [s(p)] \mid p \in M_\infty(\tilde{A}) \text{ a projection}\}.$$

Proof. Let $p \in M_\infty(A)$ be a projection. The $\pi_0([p] - [s(p)]) = [\pi(p)] - [\pi \circ s(p)] = 0$ since $\pi = \pi \circ s$. Thus $\{[p] - [s(p)] \mid p \in M_\infty(\tilde{A}) \text{ a projection}\} \subset K_0(A)$.

Conversely, suppose that $x \in K_0(A)$. Then there are $p, q \in M_\infty(\tilde{A})$ projections such that $[p] - [q] = x$. Without loss of generality, assume that $p, q \in M_n(\tilde{A})$. Then

$$r := \begin{pmatrix} p & 0 \\ 0 & 1_n - q \end{pmatrix}, \qquad t := \begin{pmatrix} 0 & 0 \\ 0 & 1_n \end{pmatrix}$$

are projections in $M_{2n}(\tilde{A})$. Moreover,

$$[r] - [t] = [p] + [1_n - q] - [1_n] = [p] - [q] = x.$$

Then $t = s(t)$ and by assumption, $\pi_0(x) = 0$, so

$$[s(r)] - [t] = [s(r)] - [s(t)] = s_0(x) = \ell_0 \circ \pi_0(x) = 0.$$

It follows that $[s(r)] = [t]$ and so $x = [r] - [s(r)]$, showing the reverse inclusion. $\quad\square$

12.3.3. Let **C*-alg** denote the category whose objects are C*-algebras and whose morphisms are *-homomorphisms. Let **Ab** denote the category whose objects are abelian groups and whose morphisms are group homomorphisms. Then

$$K_0 : \textbf{C*-alg} \to \textbf{Ab}, \quad A \mapsto K_0(A)$$

is a covariant functor which preserves the zero object. Moreover, it is *continuous*, by which we mean that it preserves countable inductive limits, as we saw in Theorem 12.2.4.

12.3.4. The K_0 functor behaves relatively well with respect to exact sequences. Given a short exact sequence of C*-algebras, we do not get a short exact sequence induced on K_0, but we do preserve exactness in the middle. A functor with this property is called *half exact*. However, *split* exact sequences of C*-algebras are completely preserved. This property is called *split exactness*. We will delay the proof of split exactness (Corollary 12.5.8) until after we have learned about the K_1-group and its interaction with K_0.

12.3.5. Theorem. *An exact sequence of C*-algebras*

$$0 \longrightarrow J \overset{\varphi}{\longrightarrow} A \overset{\pi}{\longrightarrow} B \longrightarrow 0$$

induces a sequence of abelian groups

$$K_0(J) \overset{\varphi_0}{\longrightarrow} K_0(A) \overset{\pi_0}{\longrightarrow} K_0(B) ,$$

which is exact at $K_0(A)$, that is, $\operatorname{im}(\varphi_0) = \ker(\pi_0)$.

Proof. We have $\pi_0 \circ \varphi_0 = (\pi \circ \varphi)_0 = 0$, so that $\mathrm{im}(\varphi_0) \subset \ker(\pi_0)$. Suppose that $x \in \ker(\pi_0)$. Then by Theorem 12.3.2, there exists $n \in \mathbb{N} \setminus \{0\}$ and a projection $p_1 \in M_n(A)$ such that $x = [p_1] - [s(p_1)]$. Let $\tilde{\pi}$ denote the inflation of π to $M_\infty(A)$. Since $[\tilde{\pi}(p_1)] - [\tilde{\pi}(s(p_1))] = 0$, there exists a projection $r \in M_m(A)$, for some m, such that $\tilde{\pi}(p_1) \oplus r = \tilde{\pi}(s(p_1)) \oplus r$. Now

$$\tilde{\pi}(p_1) \oplus 1_m \sim \tilde{\pi}(p_1) \oplus r \oplus (1_m - r) \sim \tilde{\pi}(s(p_1)) \oplus r \oplus (1_m - r) \sim \tilde{\pi}(s(p_1)) \oplus 1_m,$$

so we can assume that $r = 1_m$. Let $p_2 := p_1 \oplus 1_m \in M_{m+n}(A)$. Then $x = [p_2] - [s(p_2)]$ and we have

$$\tilde{\pi}(p_2) = \tilde{\pi}(p_1) \oplus 1_m \sim \tilde{\pi}(s(p_1)) \oplus 1_m = s(\tilde{\pi}(p_2)).$$

Let $p := p_2 \oplus 0_{m+n} \subset M_{2(m+n)}(A)$. Then $x = [p] - [s(p)]$ and since $\tilde{\pi}(p_2) \sim (\tilde{\pi}(p_2))$, (i) implies (ii) of Theorem 12.1.5 provides us with a unitary $u \in \mathcal{U}_{2(m+n)}(\tilde{B})$ such that $u\tilde{\pi}(p)u^* = s(\tilde{\pi}(p))$.

Consider the unitary $u \oplus u^* \in M_{4(m+n)}(\tilde{B})$. Since $\tilde{\pi}$ is surjective, there exists a unitary $v \in M_{4(m+n)}(\tilde{A})$ such that $\tilde{\pi}(v) = u \oplus u^*$. Let $q := v(p \oplus 0_{m+n})v^*$, which is a projection in $M_{4(m+n)}(\tilde{B})$. We have

$$\tilde{\pi}(q) = \begin{pmatrix} u & 0 \\ 0 & u^* \end{pmatrix} \begin{pmatrix} \tilde{\pi}(p) & 0 \\ 0 & 0 \end{pmatrix} \begin{pmatrix} u^* & 0 \\ 0 & u \end{pmatrix} = \begin{pmatrix} s(\tilde{\pi}(p)) & 0 \\ 0 & 0 \end{pmatrix}.$$

Thus $s(\tilde{\pi}(q)) = \tilde{\pi}(q)$. Since $q := v(p \oplus 0_{m+n})v^*$, we furthermore see that $x = [q] - [s(q)]$.

By Exercise 12.6.11 (ii), $s(\tilde{\pi}(q)) = \tilde{\pi}(q)$ implies that $q \in \mathrm{im}(\tilde{\varphi})$. So $\tilde{\varphi}(e) = q$ for some $e \in M_{4(m+n)}(\tilde{J})$, and, again by Exercise 12.6.11 (i), since $\tilde{\varphi}$ is injective, e must be a projection. Thus

$$x = [\tilde{\varphi}(e)] - [s(\tilde{\varphi}(e))] = \varphi_0([e] - [s(e)]) \in \mathrm{im}(\varphi).$$

So $\ker(\pi) \subset \mathrm{im}(\varphi_0)$, which proves the theorem. $\qquad\square$

12.4   The abelian group $K_1(A)$

We also associate to a unital C*-algebra A a second abelian group, $K_1(A)$. Set $\mathcal{U}_n(A) := \mathcal{U}(M_n(A))$, the unitary group of the matrix algebra $M_n(A)$, and

$$\mathcal{U}_\infty(A) = \bigcup_{n \in \mathbb{N}} \mathcal{U}_n(A).$$

As in the previous section, we define the orthogonal sum of two unitaries $u \in \mathcal{U}_n(A)$ and $v \in \mathcal{U}_m(A)$ to be

$$u \oplus v = \begin{pmatrix} u & 0 \\ 0 & v \end{pmatrix} \in \mathcal{U}_{n+m}(A) \subset \mathcal{U}_\infty(A).$$

Let $u, v \in \mathcal{U}_\infty(A)$. Then $u \in \mathcal{U}_n(A)$ and $v \in \mathcal{U}_m(A)$ for some $n, m \in \mathbb{N}$. Write $u \sim_1 v$ if there is $k \geq \max\{n, m\}$ such that $u \oplus 1_{k-n}$ homotopic to $v \oplus 1_{k-m}$ in $\mathcal{U}_k(A)$. (That is to say, there is a continuous function $f : [0, 1] \to \mathcal{U}_k(A)$ such that $f(0) = u \oplus 1_{k-n}$ and $f(1) = v \oplus 1_{k-m}$.) Here 1_r is the unit in $M_r(A)$ and, by convention, $w \oplus 1_0 = w$ for any $w \in \mathcal{U}_\infty(A)$.

12.4.1. Let $[u]_1$ denote the $\sim_1$ equivalence class of $u \in \mathcal{U}_\infty(A)$. We will require the following lemma for the proof of Proposition 12.4.3. What follows is called the Whitehead Lemma because it is a C*-algebraic reformulation of the work of J.H.C. Whitehead [129].

12.4.2. Lemma (Whitehead). *Let A be a unital C*-algebra and let $u, v \in \mathcal{U}(A)$. Then*

$$\begin{pmatrix} u & 0 \\ 0 & v \end{pmatrix} \sim_1 \begin{pmatrix} uv & 0 \\ 0 & 1 \end{pmatrix} \sim_1 \begin{pmatrix} vu & 0 \\ 0 & 1 \end{pmatrix} \sim_1 \begin{pmatrix} v & 0 \\ 0 & u \end{pmatrix},$$

as elements in $\mathcal{U}_2(A)$.

Proof. First note that if $x_1 \sim_1 y_1$ and $x_2 \sim_1 y_2$ then $x_1 x_2 \sim_1 y_1 y_2$ by taking $h = fg$ where the functions $f, g : [0, 1] \to \mathcal{U}_n(A)$ are continuous paths from x_1 to y_1 and x_2 to y_2, respectively. It is also clear that $\sim_1$ is transitive. Next we observe that

$$\begin{pmatrix} u & 0 \\ 0 & v \end{pmatrix} = \begin{pmatrix} u & 0 \\ 0 & 1 \end{pmatrix} \begin{pmatrix} 0 & 1 \\ 1 & 0 \end{pmatrix} \begin{pmatrix} v & 0 \\ 0 & 1 \end{pmatrix} \begin{pmatrix} 0 & 1 \\ 1 & 0 \end{pmatrix}$$

and

$$\begin{pmatrix} uv & 0 \\ 0 & 1 \end{pmatrix} = \begin{pmatrix} u & 0 \\ 0 & 1 \end{pmatrix} \begin{pmatrix} 1 & 0 \\ 0 & 1 \end{pmatrix} \begin{pmatrix} v & 0 \\ 0 & 1 \end{pmatrix} \begin{pmatrix} 1 & 0 \\ 0 & 1 \end{pmatrix}.$$

Also,

$$\begin{pmatrix} v & 0 \\ 0 & u \end{pmatrix} = \begin{pmatrix} 0 & 1 \\ 1 & 0 \end{pmatrix} \begin{pmatrix} u & 0 \\ 0 & v \end{pmatrix} \begin{pmatrix} 0 & 1 \\ 1 & 0 \end{pmatrix}.$$

Thus it is sufficient to show that $\begin{pmatrix} 1 & 0 \\ 0 & 1 \end{pmatrix} \sim_1 \begin{pmatrix} 0 & 1 \\ 1 & 0 \end{pmatrix}$, and this follows from Lemma 12.1.4. $\qquad\square$

12.4.3. Proposition. *Let A be a unital C*-algebra. Then $\mathcal{U}_\infty(A)/\sim_1$, with addition given by $[u]_1 + [v]_1 = [u \oplus v]_1$, is an abelian group.*

Proof. For any $u \in \mathcal{U}_\infty(A)$ and any natural number k it is clear that $u \sim_1 u \oplus 1_k$. Since $\oplus$ is associative, we have $(u \oplus v) \oplus w \sim_1 u \oplus (v \oplus w)$. By Lemma 12.4.2, if u and v are both elements of $\mathcal{U}_n(A)$ for some natural number n, we have $u \oplus v \sim_1 uv \sim_1 vu \sim_1 v \oplus u$. For $u \in \mathcal{U}_m(A)$ and $v \in \mathcal{U}_n(A)$, with m not necessarily the same as m, let $z = \begin{pmatrix} 0 & 1_n \\ 1_m & 0 \end{pmatrix}$. Then $u \oplus v$ and z are both elements of $\mathcal{U}_{m+n}(A)$, so

$$\begin{pmatrix} v & 0 \\ 0 & u \end{pmatrix} = \begin{pmatrix} 0 & 1_n \\ 1_m & 0 \end{pmatrix} \begin{pmatrix} u & 0 \\ 0 & v \end{pmatrix} \begin{pmatrix} 0 & 1_n \\ 1_m & 0 \end{pmatrix}^*$$

$$\sim_1 \begin{pmatrix} 0 & 1_n \\ 1_m & 0 \end{pmatrix}^* \begin{pmatrix} 0 & 1_n \\ 1_m & 0 \end{pmatrix} \begin{pmatrix} u & 0 \\ 0 & v \end{pmatrix} = \begin{pmatrix} u & 0 \\ 0 & v \end{pmatrix}.$$

Thus $u \oplus v \sim_1 v \oplus u$.

We show that the addition is well defined. In that case, the above shows that $\mathcal{U}_\infty / \sim_1$ is an abelian group with identity $[1_A]_1$, where the inverse of $[u]_1$ is $[u^*]_1$ since, for $u \in \mathcal{U}_n(A)$, we have $u \oplus u^* \sim_1 uu^* \sim_1 1_n$.

Suppose for u and u' in $\mathcal{U}_m(A)$ there is a continuous path of unitaries $t \mapsto u_t$ with $u_0 = u$ and $u_1 = u'$ and also for v and v' in $\mathcal{U}_m(A)$ there is a continuous path of unitaries $t \mapsto v_t$ with $v_0 = v$ and $v_1 = v'$. Then $t \mapsto u_t \oplus v_t$ is a continuous path of unitaries in $\mathcal{U}_{m+n}(A)$ where $u_0 \oplus v_0 = u \oplus v$ and $u_1 \oplus v_1 = u' \oplus v'$. Thus $u \oplus v \sim_1 u' \oplus v$.

Suppose $u \sim_1 u'$ and $v \sim_1 v'$ for arbitrary $u, u', v, v' \in \mathcal{U}_\infty(A)$. Then there are natural numbers k and l and continuous paths of unitaries from $u \oplus 1_{k-n}$ to $u' \oplus 1_{k-n'}$ and from $v \oplus 1_{l-m}$ to $v' \oplus 1_{l-m'}$. By the above,

$$(u \oplus 1_{k-n}) \oplus (v \oplus 1_{l-m}) \sim_1 (u' \oplus 1_{k-n'}) \oplus (v' \oplus 1_{l-m'}).$$

By the first part of the proof, we have

$$(u \oplus 1_{k-n}) \oplus (v \oplus 1_{l-m}) = ((u \oplus 1_{k-n}) \oplus v) \oplus 1_{l-m} \sim_1 (u \oplus 1_{k-n} \oplus v)$$
$$\sim_1 v \oplus (u \oplus 1_{k-n})$$
$$\sim_1 (v \oplus u) \oplus 1_{k-n}$$
$$\sim_1 v \oplus u \sim_1 u \oplus v.$$

Similarly $(u' \oplus 1_{k-n'}) \oplus (v' \oplus 1_{l-m'}) \sim_1 u' \oplus v'$. Thus $u \oplus v \sim_1 u' \oplus v'$, so the addition is well defined. $\qquad\qquad\square$

12.4.4. Definition. Let A be a unital C^*-algebra. Then the K_1-group of A, $K_1(A)$, is defined to be

$$K_1(A) := \mathcal{U}_\infty(A)/ \sim_1 .$$

12.4.5. As was the case for K_0, any unital $*$-homomorphism $\varphi : A \to B$ between unital C^*-algebras A and B induces a unique group homomorphism $\varphi_1 : K_1(A) \to K_1(B)$. If φ is a $*$-isomorphism then the induced group homomorphism is also an isomorphism. (Exercise.)

12.4.6. Let A be a C^*-algebra. The *suspension* of A is the (nonunital) C^*-algebra defined by

$$SA := \{f \in C([0, 1], A) \mid f(0) = f(1) = 0\}.$$

Let A be a unital C^*-algebra. A projection-valued function $p : [0, 1] \to M_n(A)$ such that $p(0) = p(1) \in M_n$ is called a *normalised loop of projections* over A.

Lemma. *For a unital* C*-*algebra* A *the following hold:*

(i) $K_0(\widetilde{SA})$ *is generated by normalised loops of projections over* A.

(ii) $K_0(SA)$ *is generated by formal differences* $[p] - [q]$ *where* p *and* q *are normalised loops of projections over* A *satisfying* $p(1) = q(1) \in M_n$.

Proof. The key observation is that the unitisation $\widetilde{SA} = \{f \in C_0([0,1], A) \mid f(0) = f(1) \in \mathbb{C}\}$. Then (i) follows, and to see (ii), we just apply the definition of the K_0-group of a nonunital C*-algebra (12.3.1). □

12.4.7. The next theorem gives us an alternative description of $K_1(A)$ for a unital C*-algebra A. While in practice our original description of $K_1(A)$ might give a more concrete picture of the group, reframing K_1 in terms of K_0 means that much of what we proved and will prove for $K_0(A)$ is also true for $K_1(A)$.

Theorem. *If* A *is a unital* C*-*algebra, then* $K_1(A) \cong K_0(SA)$.

Proof. Let $[p] \in K_0(SA)$ where p is a normalised loop of projections over A such that $p(0) = p(1) \in M_n$ for some $n \in \mathbb{N}$. Since conjugation by a unitary does not change the K_0-class, we may assume without loss of generality that $p(0) = p(1) = 1_k \oplus 0_{k'}$ for some $k + k' = n$. The proof of (i) implies (ii) of Proposition 12.1.5 shows that there is a continuous path of unitaries $u : [0,1] \to M_n(A)$ such that $p(t) = u(t)p(1)u(t)^*$. Since $p(1) = p(0) = u(0)p(1)u(0)$, we see that the unitary $u(0)$ commutes with the projection $p(1) = p(0)$. Therefore, since $p(0) = p(1)$ is by assumption diagonal, we deduce that $u(0)$ is of the form

$$u(0) = \begin{pmatrix} v & 0 \\ 0 & w \end{pmatrix}, \quad \text{for } v \in \mathcal{U}_k(A), w \in \mathcal{U}_{k'}(A).$$

In particular, $[v]_1 \in K_1(A)$. Observe that the class $[v]_1$ depends only on the class of $[p] \in K_0(SA)$. Thus the map

$$\varphi : K_0(SA) \to K_1(A), \quad [p] \mapsto [v]_1$$

is a well-defined group homomorphism. To show it is in fact an isomorphism, we will construct an inverse. Let $v \in M_n(A)$ be a unitary. Then

$$u_0 := \begin{pmatrix} v & 0 \\ 0 & v^* \end{pmatrix} \sim_1 \begin{pmatrix} 1_n & 0 \\ 0 & 1_n \end{pmatrix} =: u_1 \in M_n(A),$$

so there is a path of unitaries $u : [0,1] \to M_{2n}(A)$ with $u(0) = u_0$ and $u(1) = u_1$. Define

$$p(t) := u(t)(1_n \oplus 0_n)u(t)^*, \quad q(t) := (1_n \oplus 0_n).$$

Then p and q are normalised loops of projections satisfying $p(1) = q(1)$. It follows that $[p] - [q] \in K_0(SA)$, and so this defines a map

$$\psi : K_1(A) \to K_0(SA) : [u] \mapsto [p] - [q].$$

It is straightforward to check that this map is indeed a well-defined group homomorphism, and that φ and ψ are mutual inverses. Thus $K_0(SA) \cong K_1(A)$. $\square$

12.4.8. We make note of the following theorem, the proof of which we leave as an exercise.

Theorem. *Let A be a unital C*-algebra. Suppose that $A \cong \varinjlim(A_n, \varphi^{(n)})$ is an inductive limit of unital C*-algebras. Then*

$$K_1(A) \cong \varinjlim(K_1(A_n), \varphi_1^{(n)}),$$

is an isomorphism of abelian groups.

12.4.9. To close off this section, we point out the following. By 12.3.3, as was the case for K_0, we have that

$$K_1 : \mathbf{C}^*\text{-alg} \to \mathbf{Ab}, \quad A \mapsto K_1(A)$$

is a continuous covariant functor which preserves the zero object.

12.5   The 6-term exact sequence

In this section we will show that from any short exact sequence of C*-algebras we get a cyclic 6-term exact sequence relating the K_0- and K_1-groups of all three C*-algebras. This will be very useful for calculating the K-theory of a given C*-algebra from simpler, known examples. To keep things relatively brief, some of the details will be omitted.

12.5.1. Definition. Let A be a C*-algebra and $J \subset A$ an ideal. Given a surjective *-homomorphism $\pi : A \to A/J$, the *mapping cone* $M(A, A/J)$ of π is defined to be

$$M(A, A/J) := \{(a, f) \in A \times C([0, 1], A/J) \mid f(0) = 0, f(1) = \pi(a)\}.$$

Observe that the mapping cone is related to the suspension of A/J by the short exact sequence

$$0 \longrightarrow S(A/J) \longrightarrow M(A, A/J) \longrightarrow A \longrightarrow 0 ,$$

where $S(A/J) \to M(A, A/J)$ is the *-homomorphism $f \mapsto (0, f)$ (exercise). With the description of K_1 given by Theorem 12.4.7, this induces a map on K-theory

$$K_1(A/J) = K_0(S(A/J)) \longrightarrow K_0(M(A, A/J)) .$$

The next proposition says that $K_0(M(A, A/J))$ is isomorphic to $K_0(J)$. We omit the proof. For the details, see for example [61, Proposition 4.5.3].

12.5.2. Proposition. *Let A be a C*-algebra, $J \subset A$ an ideal and $\pi : A \to A/J$ a surjective *-homomorphism. Then*

$$J \to M(A, A/J), \quad a \mapsto (a, 0)$$

*is a *-homomorphism and the induced map*

$$K_0(J) \to K_0(M(A, A/J)),$$

is an isomorphism.

12.5.3. Applying the isomorphism in the proposition allows us to define the *index map*, or *boundary map*,

$$\delta_1 : K_1(A/J) \to K_0(J).$$

Proposition. *Suppose that*

$$0 \longrightarrow J \longrightarrow A \longrightarrow A/J \longrightarrow 0$$

is a short exact sequence of C-algebras. Then the induced sequence*

$$K_1(A/J) \xrightarrow{\delta_1} K_0(J) \longrightarrow K_0(A) \longrightarrow K_0(A/J)$$

is exact at $K_0(J)$ and $K_0(A)$.

Proof. Half-exactness of K_0 (Theorem 12.3.5) implies that the sequence is exact at $K_0(A)$. The short exact sequence

$$0 \longrightarrow S(A/J) \longrightarrow M(A, A/J) \longrightarrow A \longrightarrow 0,$$

induces the sequence

$$K_0(S(A/J)) \longrightarrow K_0(M(A, A/J)) \longrightarrow K_0(A),$$

which is exact at $K_0(M(A, A/J))$. By Proposition 12.5.2, this sequence is the same as

$$K_1(A/J) \longrightarrow K_0(J) \longrightarrow K_0(A),$$

which establishes the proposition. $\square$

12.5.4. For calculation purposes, it will be useful to have a description of the index map that does not refer to suspensions or mapping cones. Thus we include, without proof, the following proposition.

Proposition. *Let*

$$0 \longrightarrow J \overset{\varphi}{\longrightarrow} A \overset{\psi}{\longrightarrow} B \longrightarrow 0$$

be an exact sequence of C*-algebras, and, for $n \in \mathbb{N} \setminus \{0\}$, let $u \in \mathcal{U}_n(\tilde{B})$ be a unitary. Then $\delta_1([u]_1) = [p] - [s(p)]$ where $p \in M_{2n}(\tilde{J})$ is a projection such that

$$\tilde{\varphi}(p) = v \begin{pmatrix} 1_n & 0 \\ 0 & 1_n \end{pmatrix} v^*, \qquad \tilde{\psi}(v) = \begin{pmatrix} u & 0 \\ 0 & u^* \end{pmatrix},$$

for a unitary $v \in \mathcal{U}_{2n}(\tilde{A})$.

12.5.5. For $n > 1$ we inductively define the nth suspension of A to be the C*-algebra $S^n A := S(S^{n-1}A)$. Now, generalising the characterisation of $K_1(A)$ as K_0 of the suspension of A, we define higher K-groups K_n for any n by

$$K_n(A) := K_{n-1}(SA) = K_0(S^n(A)).$$

12.5.6. Lemma. *Let A be a C*-algebra, $J \subset A$ an ideal. Then*

$$\cdots \longrightarrow K_1(A) \longrightarrow K_1(A/J) \longrightarrow K_0(J) \longrightarrow \cdots$$

is exact at $K_1(A/J)$.

Proof. Let $M := M(A, A/J)$ be the mapping cone of the map $\pi : A \to A/J$. There is a surjective *-homomorphism

$$\varphi : M \to A, \quad (a, f) \mapsto a,$$

whence we define the mapping cone $M(M, A)$. Let $Z := M(M, A)$. Then

$$0 \longrightarrow SA \longrightarrow Z \longrightarrow M \longrightarrow 0$$

is a short exact sequence. Let $\varphi : M \to A$ denote the map $(a, f) \mapsto a$ given above. By half-exactness of K-theory, the short exact sequence yields the sequence

$$K_0(SA) \longrightarrow K_0(Z) \longrightarrow K_0(M)$$

which is exact at $K_0(Z)$. By Theorem 12.4.7 and Proposition 12.5.2, we know that $K_0(SA) = K_1(A)$ and $K_0(M) = K_0(J)$. Thus, together with Proposition 12.5.3, the result will follow if we can show that $K_0(Z) \cong K_1(A/J)$. To do so, observe that

$$\begin{aligned} \ker(\varphi) &= \{(a, f) \in M \mid a = 0, \} \\ &\cong \{f \in C([0, 1], A/J) \mid f(0) = 0, f(1) = \pi(0) = 0\} \\ &\cong S(A/J). \end{aligned}$$

Thus $Z = M(M, M/(S(A/J)))$, so by Proposition 12.5.2, the K_0 group of the mapping cone Z is isomorphic to the ideal $S(A/J)$, which by Theorem 12.4.7 implies $K_0(Z) = K_0(S(A/J)) = K_1(A/J)$. $\square$

12.5.7. By definition of the higher K-groups, it is now straightforward to deduce the long exact sequence in K-theory. When combined with Bott periodicity (Theorem 12.5.9 below), this will yield a valuable tool for calculating K-groups.

Theorem. *For every short exact sequence of* C*-*algebras*

$$0 \longrightarrow J \longrightarrow A \longrightarrow B \longrightarrow 0$$

there is an exact sequence of abelian groups

$$\cdots \longrightarrow K_n(A) \longrightarrow K_n(A/J) \longrightarrow K_{n-1}(J) \longrightarrow K_{n-1}(A) \longrightarrow \cdots.$$

Proof. Exercise. $\qquad\qquad\qquad\qquad\qquad\qquad\qquad\qquad\qquad\qquad\qquad\qquad\square$

12.5.8. Corollary. *The functor K_0 is split exact.*

It is a remarkable and nontrivial result, due to Atiyah, that although we are able to define an infinite sequence of K-groups associated to any C*-algebra A, in fact all K-theoretic information is contained in K_0 and K_1, the groups for which we have particularly nice concrete realisations in terms of projections and unitaries. This result is called *Bott periodicity* and it tells us that for any C*-algebra A, there is an isomorphism $K_2(A) = K_0(S^2(A)) \cong K_0(A)$. The proof of Bott periodicity is fairly technical and lengthy, so we will not prove it here. However, given the background we have developed, the interested reader should find most proofs in the literature to be accessible. For example, a fairly economical proof, relying on a few more K-theoretical concepts, is given in [61]. A slightly longer but more elementary proof can be found in [103].

12.5.9. Let A be a C*-algebra and let $z : S^1 \to \mathbb{C}$ denote the identity function on $S^1 \subset \mathbb{C}$. Recall that the unitisation of the suspension of A is given by $\widetilde{SA} = \{f \in C_0([0,1], A) \mid f(0) = f(1) \in \mathbb{C}\}$, and thus

$$\widetilde{SA} \cong \{f \in C_0(\mathbb{T}, A) \mid f(1) \in \mathbb{C}\}.$$

For any a projection $p \in M_\infty(A)$ define $u_p \in M_\infty(\widetilde{SA})$ to be the unitary loop $u_p(z) := zp + 1_{\tilde{A}} - p$. The *Bott map* is defined to be

$$\beta_A : K_0(A) \to K_2(A) = K_1(SA), \quad [p] - [s(p)] \mapsto [u_p u_{s(p)}^*]_1.$$

Theorem (Bott Periodicity)**.** *For any* C*-*algebra A the Bott map*

$$\beta_A : K_0(A) \to K_2(A) = K_1(SA)$$

is an isomorphism of abelian groups.

12.5.10. Suppose that

$$0 \longrightarrow J \longrightarrow A \longrightarrow B \longrightarrow 0$$

is a short exact sequence of C*-algebras. We use Bott periodicity to identify $K_0(A/J)$ with $K_2(A/J)$ via the Bott map $\beta_{A/J}$. Combining this with the maps $K_0(A) \to K_0(A/J)$ and $K_2(A/J) \to K_1(J)$ coming from the long exact sequence of Theorem 12.5.7, gives us the *exponential map*

$$\delta_0 : K_0(A/J) \to K_1(J).$$

This in turn yields the 6-term cyclic exact sequence of the next theorem.

12.5.11. Theorem. *For every short exact sequence of* C*-*algebras*

$$0 \longrightarrow J \longrightarrow A \longrightarrow B \longrightarrow 0$$

there is an induced 6-term exact sequence of abelian groups

$$
\begin{array}{ccc}
K_0(J) \longrightarrow & K_0(A) \longrightarrow & K_0(B) \\
\delta_1 \uparrow & & \downarrow \delta_0 \\
K_1(B) \longleftarrow & K_1(A) \longleftarrow & K_1(J).
\end{array}
$$

12.5.12. Suppose that

$$0 \longrightarrow J \xrightarrow{\varphi} A \xrightarrow{\psi} B \longrightarrow 0$$

is a short exact sequence of C*-algebras with A and B unital. Denote by $\tilde{\varphi} : \tilde{J} \to A$ the unital *-homomorphisms extending φ. Let $p \in M_n(B)$ be a projection. Since π is surjective, there is $a \in M_n(A)$ such that $\psi(a) = p$. Replacing a with $a/2 + a^*/2$ if necessary, we may moreover find $a \in M_n(A)_{sa}$ such that $\psi(a) = p$.

As was the case for the index map, we give a more explicit description of the exponential map, which will be useful for the purpose of calculation.

12.5.13. Proposition. *Suppose that*

$$0 \longrightarrow J \longrightarrow A \longrightarrow B \longrightarrow 0$$

is a short exact sequence of C*-*algebras. Let* $\delta_0 : K_0(B) \to K_1(J)$ *be the associated exponential map. Let* $g \in K_0(B)$ *so that* $g = [p] - [s(p)]$ *for some projection* $p \in M_\infty(A)$ *and let* $a \in M_n(\tilde{A})_{sa}$ *satisfy* $\psi(a) = p$. *Then there exists a unique unitary* $u \in M_n(\tilde{I})$ *such that* $\tilde{\varphi}(u) = e^{2\pi i a}$ *and this* u *satisfies* $\delta_0(g) = -[u]_1$.

12.5.14. We have seen above that K-theory is a covariant functor. One also can define the K-homology of a C*-algebra A – abelian groups $K^0(A)$ and $K^1(A)$ – which is a contravariant functor that is dual to K-theory. K-theory and K-homology are related to one another via Kasparov's KK-theory, a bivariant functor whose input is two C*-algebras. KK-theory generalises both K-theory and K-homology in the following way. Given a C*-algebra A, the KK-group $KK_i(A, \mathbb{C}) \cong K^i(A)$, while $KK_i(\mathbb{C}, A) \cong K_i(A)$.

For certain separable nuclear C*-algebras, we can further relate KK-theory to K-theory via the *Universal Coefficient Theorem*, or UCT, for short. The UCT allows one to determine when an element in $KK(A, B)$ determines a homomorphism $K_*(A) \to K_*(B)$. We state what it means for a given separable nuclear C*-algebra to satisfy the UCT here, as it plays an important – if often "behind the scenes" – role in the classification program. We have not (and will not) defined all the terminology.

Definition. Let A be separable nuclear C*-algebra. We say that A *satisfies the UCT* if the following holds: For every separable C*-algebra B, the sequence

$$0 \longrightarrow \mathrm{Ext}^1_{\mathbb{Z}}(K_*(A), K_*(B)) \longrightarrow KK_*(A, B) \longrightarrow \mathrm{Hom}(K_*(A), K_*(B)) \longrightarrow 0$$

is exact.

The UCT is known to hold for all of the nuclear C*-algebras we have covered so far. In fact, Rosenberg and Schochet proved that it holds for all C*-algebras in the so-called *bootstrap class* [104, Theorem 1.17]. The bootstrap class includes commutative C*-algebras, and is closed under most constructions such as countable inductive limits, extensions, tensoring with compacts, tensor products, crossed products by $\mathbb{Z}$, to name a few. It is not known whether or not *every* nuclear C*-algebra satisfies the UCT, and at present, settling the question appears to be a very difficult problem. For a discussion, see for example [138, 6].

12.5.15. We also have a useful way of determining the K-theory of the tensor product of two nuclear C*-algebras, provided one of them has torsion-free K-theory and they both satisfy the UCT.

Theorem. *Suppose that A and B are nuclear C*-algebras and that A satisfies the UCT. If $K_0(A)$ and $K_1(A)$ are both torsion-free or $K_0(B)$ and $K_1(B)$ are both torsion-free, then*

$$K_0(A \otimes B) \cong (K_0(A) \otimes K_0(B)) \oplus (K_1(A) \otimes K_1(B)),$$

and

$$K_1(A \otimes B) \cong (K_0(A) \otimes K_1(B)) \oplus (K_1(A) \otimes K_0(B)),$$

where the tensor product above is the tensor product of abelian groups under their identification as $\mathbb{Z}$-modules.

The full statement of the previous theorem does not require the torsion-free assumption and rather than isomorphisms as above, we get a short exact sequence involving *graded* K-theory (by the graded K-theory of A we mean $K_*(A) := K_0(A) \oplus K_1(A)$ seen as a $\mathbb{Z}/2\mathbb{Z}$-graded abelian group, with $K_0(A)$ in degree zero and $K_1(A)$ in degree one), and can be found, for example, in [8, Theorem 23.1.3]. This more general theorem is called the Künneth Theorem for Tensor Products and was proved by Schochet [110].

12.6 Exercises

12.6.1. Let $p_1, p_2, q_1, q_2, r \in M_\infty(A)$ be projections. Show that

(i) if $p_1 \sim p_2$ and $q_1 \sim q_2$ then $p_1 \oplus q_1 \sim p_2 \oplus q_2$;

(ii) $p_1 \oplus q_1 \sim q_1 \oplus p_1$;

(iii) if $p_1, q_1 \in M_n(A)$ and $p_1 q_1 = 0$, then $p_1 + q_1 \sim p_1 \oplus q_1$;

(iv) $(p_1 \oplus q_1) \oplus r \sim p_1 \oplus (q_1 \oplus r)$.

12.6.2. Show that Murray–von Neumann equivalent projections in a unital C*-algebras need not be unitarily equivalent. (Hint: Suppose A is a C*-algebra containing a nonunitary isometry s and consider $1_A - s^*s$, $1_A - ss^*$.)

12.6.3. Let $n \in \mathbb{N}$. Show that the unitary group of M_n is connected. (Hint: look at the proof of Lemma 12.4.2.)

12.6.4. Let F be a finite-dimensional C*-algebra and A a unital AF algebra.

(i) Show that $K_1(F) = 0$. (Hint: use the previous exercise.)

(ii) Show that $K_1(A) = 0$.

12.6.5. Let A be a unital stably finite C*-algebra. Show that the K_0-class of the unit $[1_A]$ is an order unit for $K_0(A)$.

12.6.6. Let A be a simple unital stably finite C*-algebra. Show that the ordered abelian group $(K_0(A), K_0(A)_+)$ is simple. (Hint: Use Lemma 10.2.2 to show that for any pair of projections $p, q, \in A$ we can find $n \in \mathbb{N}$ and $v \in M_{1,n}(A)$ such that $v^*v \sim p \oplus p \oplus \cdots \oplus p$ and $vv^* \sim q$.)

12.6.7. Let A and B be C*-algebras. Show that any *-homomorphism $\varphi : A \to B$ induces a group homomorphism $\varphi_0 : K_0(A) \to K_0(B)$.

12.6.8. Let X be a compact connected Hausdorff space. Let tr_n denote the standard trace on M_n.

(i) Let $p \in M_n(C(X))$ be a projection. Show that the map $x \to \mathrm{tr}_n(p(x))$ is a constant function in $C(X, \mathbb{Z})$.

(ii) Fix $x \in X$ and show that the map $\text{ev}_x : C(X) \to \mathbb{C}$, defined by $\text{ev}_x(f) = f(x)$, induces a group homomorphism $K_0(C(X)) \to \mathbb{Z}$ such that $(\text{ev}_x)_0([p]) = \text{tr}_n(p)$, and hence the map

$$\dim : K_0(C(X)) \to \mathbb{Z}, \qquad [p]_0 - [q]_0 \mapsto \text{tr}(p(x)) - \text{tr}(q(x)),$$

where tr is the trace in the appropriate matrix algebra and x is any element in X, is well defined and independent of x.

12.6.9. Let X be the Cantor set. Show that

$$K_0(C(X)) \cong C(X, \mathbb{Z}) \quad \text{and} \quad K_1(C(X)) = 0.$$

12.6.10. Let A be a unital C*-algebra and $\tilde{A}$ its unitisation. Show that the split exact sequence

$$0 \longrightarrow A \overset{\iota}{\longrightarrow} \tilde{A} \overset{\pi}{\longrightarrow} \mathbb{C}.$$

induces a split exact sequence in K_0,

$$0 \longrightarrow K_0(A) \overset{\iota_0}{\longrightarrow} K_0(\tilde{A}) \overset{\pi_0}{\longrightarrow} \mathbb{C}.$$

so that $\ker(\pi_0) \cong K_0(A)$, that is, the two definitions for $K_0(A)$ agree.

12.6.11. Let

$$0 \longrightarrow J \overset{\varphi}{\longrightarrow} A \overset{\psi}{\longrightarrow} B \longrightarrow 0$$

be a short exact sequence of C*-algebras.

(i) Show that $\tilde{\varphi} : M_n(\tilde{J}) \to M_n(\tilde{A})$ is injective.
(ii) Show that if $a \in M_n(\tilde{A})$ then $a \in \text{im}(\tilde{\varphi})$ if and only if $\tilde{\psi}(a) = s(\tilde{\psi}_n(a))$.

12.6.12. Let A and B be C*-algebras. Show that $K_i(A \oplus B) \cong K_i(A) \oplus K_i(B)$, $i = 0, 1$.

12.6.13. Let A be a unital C*-algebra. Show that the equivalence relation defined in 12.4 is indeed an equivalence relation on the $\mathcal{U}_\infty(A)$.

12.6.14. Show that any unital $*$-homomorphism $\varphi : A \to B$ between unital C*-algebras A and B induces a unique group homomorphism $K_1(A) \to K_1(B)$. Show that if ϕ is a *-isomorphism then the induced group homomorphism is also an isomorphism.

12.6.15. Let A and B be nonunital C*-algebras and $\varphi : A \to B$ a *-homomorphism.

(i) Show that φ induces a group homomorphism $K_0(A) \to K_0(B)$.
(ii) Suppose that $A \cong \varinjlim(A_n, \varphi_n)$ is an inductive limit. Show that $K_0(A) \cong \varinjlim K_0(A_n, \varphi_0^{(n)})$ as abelian groups.

12.6.16. Let A and B be C*-algebras and $\varphi : A \to B$ a *-homomorphism.

(i) Show that φ induces a *-homomorphism $SA \to SB$.

(ii) Let A be a unital C*-algebra. Suppose that $A \cong \varinjlim(A_n, \varphi^{(n)})$ is an inductive limit of unital C*-algebras. Show that

$$K_1(A) \cong \varinjlim(K_1(A_n), \varphi_1^{(n)}),$$

is an isomorphism of abelian groups.

12.6.17. Let A be a C*-algebra, $J \subset A$ and ideal with surjective *-homomorphism $\pi : A/J$. Let $M(A, A/J)$ denote the mapping cone of π. Show that

$$0 \longrightarrow S(A/J) \longrightarrow M(A, A/J) \longrightarrow A \longrightarrow 0,$$

is a short exact sequence.

12.6.18. Let A be properly infinite (defined in Exercise 10.4.10) and let $s_1, s_2 \in A$ be orthogonal isometries, consider let $t_n := s_2^n s_1$, $1 \leq n \leq 3$.

(i) Let $p, q \in A$ be projections. Show that

$$r = t_1 p t_1^* + t_2(1_A - q)t_2^* + t_3(1 - t_1 t_1^* - t_2 t_2^*)t_3^*$$

is a projection.

(ii) Show that $K_0(A) = \{[p] \mid p \in M_\infty(A) \text{ a projection}\}$.

12.6.19. Calculate the K-theory of the Calkin algebra $\mathcal{B}(H)/\mathcal{K}(H)$.

13 Classification of AF algebras

Approximately finite (AF) C*-algebras were introduced in Chapter 8. The aim of this chapter is to show that separable unital AF algebras can be classified by their pointed ordered K_0-groups. This result is due to Elliott, although a similar classification for AF algebras (not using K-theory) was first obtained by Bratteli [13]. Classification of UHF algebras by supernatural numbers, which we have already seen, can be viewed as a special case. The classification of UHF algebras, due to Glimm [51], predates the results of both Elliott and Bratelli. Elliott's AF classification does not require the algebras be unital, but in what follows we will restrict ourselves to this easier case. The classification of AF algebras was really the launching point for the classification program as we know it today, and we try to indicate how and why in this chapter. After proving the classification theorem in the first section, we look at approximate unitary equivalence of maps and how one might generalise AF classification to other inductive limits. In the final section, we give an overview of Elliott's classification program for C*-algebras by introducing the Elliott invariant and Elliott's conjecture.

13.1 K-theory and classification of AF algebras

Before getting to the classification of AF algebras via Elliott's intertwining argument at the end of the section, we will need a few facts about the K-theory of finite-dimensional C*-algebras and stably finite C*-algebras.

13.1.1. Recall the definition of cancellation from 12.1.12. We have the following immediate implication from Theorem 12.1.14 and Exercise 8.7.19.

Proposition. *Every AF algebra has cancellation.*

13.1.2. Lemma. *Let A be C*-algebra with cancellation. Let $q \in A$ be a projection. Suppose that $p_1, \ldots, p_n \in M_\infty(A)$ are projections satisfying $[q] = [p_1 \oplus \cdots \oplus p_n]$. Then there are are pairwise orthogonal projections $p'_1, \ldots, p'_n \in A$ such that q is Murray–von Neumann equivalent to $\sum_{i=1}^n p'_i$ and each p'_i is Murray–von Neumann equivalent to p_i.*

Proof. It suffices to prove that if $r \in M_n(A)$, $p \in A$ and $q \in A$ satisfy $[r] + [p] = [q]$ then there is a $p' \in A$ such that $p'p = pp' = 0$ and $[r] = [p']$. Let $s \in M_m(A)$ satisfy $[s] = [q] - [p] - [r]$. Since A has the cancellation property, $s \oplus r \sim q - p$ so there is $v \in M_{1,m+n}$ such that $v^*v = r \oplus s$ and $vv^* = q - p$. Let $p' = v(r \oplus 0_m)v^*$. Then $p' \in A$ and with $w = (r \oplus 0_m)v^*$ we have $w^*w = p'$ and $w^*w = r \oplus 0_m$, so $[p'] = [r]$. Since $p' = v(r \oplus 0_m)v^* \leq v(1_{n+m})v = vv^* = q - p$, hence also $p' \leq q$ thus $p'p = pp' = 0$, as required. $\qquad\square$

13.1.3. Let A be a unital C*-algebra. For $u \in A$ a unitary, we define the map

$$\mathrm{ad}\,(u) : A \to A, \quad a \mapsto uau^*.$$

K. R. Strung, *An Introduction to C*-Algebras and the Classification Program*, Advanced Courses in Mathematics - CRM Barcelona, https://doi.org/10.1007/978-3-030-47465-2_13

It is easy to check that this defines a *-automorphism of A. We call a *-automorphism $\varphi : A \to A$ *inner* if there is a unitary $u \in A$ such that $\varphi = \mathrm{ad}\,(u)$.

13.1.4. Proposition. *Let A and B be finite-dimensional C^*-algebras. Suppose that $\varphi : (K_0(A), K_0(A)_+, [1_A]) \to (K_0(B), K_0(B)_+, [1_B])$ is a unital positive homomorphism. Then there exists a *-homomorphism $\Phi : A \to B$ such that $\Phi_0 = \varphi$. Moreover, Φ is unique up to conjugation by a unitary.*

Proof. A is finite-dimensional, so by Theorem 8.1.2 can be written as the direct sum of finitely many matrix algebras. Let

$$A = M_{n_1} \oplus \cdots \oplus M_{n_m}$$

be this decomposition. Let $e_{ij}^{(l)}$ be a set of matrix units for $M_{n_l} \subset M_{n_1} \oplus \cdots \oplus M_{n_m}$, $0 \leq l \leq m$. Denote by 1_l the unit of M_{n_l}. Since φ is positive, $\varphi([1_l]) = [p_l]$ for some projection $p_l \in M_\infty(A)$. Thus

$$[p_1 \oplus \cdots \oplus p_m] = \varphi \left(\sum_{l=1}^m [1_l] \right) = \varphi([1_A]) = [1_B].$$

Since B has cancellation, there are mutually orthogonal projections $q_1, \ldots, q_m \in B$ such that $\sum_{l=1}^m q_l = [1_B]$ and $[\varphi(1_l)] = [p_l] = [q_l]$ for every $1 \leq l \leq m$.

Similarly, we have that $\varphi([e_{11}^{(l)}]) = [p_{11}^{(l)}]$ for some projection $p_{11}^{(l)} \in M_\infty(A)$. Thus

$$[q_l] = \varphi([1_l]) = \varphi(n_l[e_{11}^{(l)}]) = [p_{11}^{(l)} \oplus \cdots \oplus p_{11}^{(l)}].$$

Since $q_l \in B$, there are mutually orthogonal projections $q_{11}^{(l)}, \ldots, q_{n_l, n_l}^{(l)} \in B$ with $[q_{ii}^{(l)}] = [p_{11}^{(l)}] = \varphi([e_{11}^{(l)}])$. Since $q_{ii}^{(l)} \sim q_{11}^{(l)}$ for each $1 \leq i \leq n_l$, there are $v_i \in B$ with $v_i^{(l)}(v_i^{(l)})^* = q_{ii}^{(l)}$ and $(v_i^{(l)})^* v_i^{(l)} = q_{11}^{(l)}$. Put

$$q_{ij}^{(l)} := v_i^{(l)}(v_j^{(l)})^*.$$

One can verify that, for each l, $q_{ij}^{(l)}$ satisfy the matrix relations for M_{n_l}. Since the q_l are pairwise orthogonal, this gives a map from the generators of A to B and hence a map $\Phi : A \to B$. Moreover, we obtain $\Phi_0([e_{11}^{(l)}]) = [\Phi(e_{11}^{(l)})] = [q_{11}^{(l)}] = \varphi([e_{11}^{(l)}])$. Since $[e_{11}^{(l)}]$, $1 \leq l \leq m$, generate $K_0(A)$, this implies $\Phi_0 = \varphi$.

Suppose now that $\Phi, \Psi : A \to B$ are both unital *-homomorphisms satisfying $\Phi_0 = \Psi_0$. Let

$$p_{ij}^{(l)} = \Phi(e_{ij}^{(l)}), \quad q_{ij}^{(l)} = \Psi(e_{ij}^{(l)}), \quad 1 \leq i, j, \leq n_l, \; 1 \leq l \leq m.$$

Then $[p_{ij}^{(l)}] = \Phi_0([e_{ij}^{(l)}]) = \Psi_0([e_{ij}^{(l)}]) = [q_{ij}^{(l)}]$, so $p_{ij}^{(l)} \sim q_{ij}^{(l)}$. Thus there are $v_l \in B$, $1 \leq l \leq m$ satisfying $v_l^* v_l = p_{11}^{(l)}$ and $v_l v_l^* = q_{11}^{(l)}$. Set

$$w := \sum_{l=1}^m \sum_{i=1}^{n_l} q_{il}^{(l)} v_l p_{li}^{(l)}.$$

Then a straightforward calculation shows that w is a unitary and $w p_{ij}^{(l)} w^* = q_{ij}^{(l)}$ for every $1 \leq i, j \leq n_l$ and $1 \leq l \leq m$. It follows that $\Psi(e_{ij}^l) = w \Phi(e_{ij}^{(l)}) w^*$ for every $1 \leq i, j \leq n_l$ and $1 \leq l \leq m$. Since these elements generate A, we have $\Psi = \mathrm{ad}\,(w) \circ \Phi$. $\qquad \square$

13.1.5. Lemma. *Let A, B and C be unital stably finite C*-algebras, with A finite-dimensional. If $\varphi : K_0(A) \to K_0(C)$ and $\psi : K_0(B) \to K_0(C)$ are positive homomorphisms satisfying $\varphi(K_0(A)_+) \subset \psi(K_0(B)_+)$, then there is a positive homomorphism $\rho : K_0(A) \to K_0(B)$ such that $\psi \circ \rho = \varphi$.*

Proof. By Theorem 12.2.6, $(K_0(A), K_0(A)_+)$ is isomorphic to $(\mathbb{Z}^k, \mathbb{N}^k)$ for some $k \in \mathbb{N}$. Let $x_1, \ldots, x_k$ be a basis for $K_0(A)$ as a $\mathbb{Z}$-module. Then $\varphi(K_0(A)_+) \subset \psi(K_0(B)_+)$ implies there are $y_1, \ldots, y_k \in K_0(B)_+$ (again considered as a $\mathbb{Z}$-module) such that

$$\varphi(x_j) = \psi(y_j), \quad 1 \leq j \leq k.$$

Define $\rho : K_0(A) \to K_0(B)$ on generators by $\rho(x_j) := y_j$. Then $\psi \circ \rho = \varphi$ and since $\rho(\mathbb{N}x_1 + \cdots + \mathbb{N}x_k) = \mathbb{N}y_1 + \cdots + \mathbb{N}y_k$, we see that ρ is positive. $\qquad \square$

13.1.6. Proposition (Intertwining). *Let A and B be C*-algebras that can both be written as inductive limits of the form $A = \varinjlim(A_n, \varphi_n)$ and $B = \varinjlim(B_n, \psi_n)$ with each φ_n, ψ_n injective. Suppose that there are *-homomorphisms $\alpha_n : A_n \to B_n$ and $\beta_n : B_n \to A_{n+1}$ making the following diagram commute:*

$$
\begin{array}{ccccccccc}
A_1 & \xrightarrow{\varphi_1} & A_2 & \xrightarrow{\varphi_2} & A_3 & \longrightarrow & \cdots & \longrightarrow & A \\
\alpha_1 \downarrow & \beta_1 \nearrow & \alpha_2 \downarrow & \beta_2 \nearrow & \alpha_3 \downarrow & \beta_2 \nearrow & & & \\
B_1 & \xrightarrow{\psi_1} & B_2 & \xrightarrow{\psi_2} & B_3 & \longrightarrow & \cdots & \longrightarrow & B.
\end{array}
$$

*Then there are *-isomorphisms $\alpha : A \to B$ and $\beta : B \to A$ making*

$$
\begin{array}{ccccccccccc}
A_1 & \xrightarrow{\varphi_1} & A_2 & \xrightarrow{\varphi_2} & A_3 & \longrightarrow & \cdots & \longrightarrow & A \\
\alpha_1 \downarrow & \beta_1 \nearrow & \alpha_2 \downarrow & \beta_2 \nearrow & \alpha_3 \downarrow & \beta_2 \nearrow & & & \beta \downarrow \uparrow \alpha \\
B_1 & \xrightarrow{\psi_1} & B_2 & \xrightarrow{\psi_2} & B_3 & \longrightarrow & \cdots & \longrightarrow & B.
\end{array}
$$

commute.

Proof. Exercise. $\qquad \square$

13.1.7. Lemma. *Let $A = \varinjlim(A_n, \varphi_n)$ be a unital AF algebra and let F be a finite-dimensional C*-algebra. Suppose that there are positive homomorphisms*

$$\alpha : K_0(A_1) \to K_0(F), \quad \gamma : K_0(F) \to K_0(A),$$

such that $\gamma \circ \alpha = [\varphi^{(1)}]_0$. *Then there are* $n \in \mathbb{N}$ *and a positive group homomorphism* $\beta : K_0(F) \to K_0(A_n)$ *such that*

$$K_0(A_1) \xrightarrow{[(\varphi_{1,n})]_0} K_0(A_n) \xrightarrow{[\varphi^{(n)}]_0} K_0(A)$$

$$\begin{array}{ccc} & \alpha \searrow & \beta \uparrow & \nearrow \gamma \\ & & K_0(F) & \end{array}$$

commutes. Moreover, if the maps φ_n *are unital and* $\alpha([1_A]) = [1_F]$, *then also* $\beta([1_F]) = [1_{A_n}]$.

Proof. Let $e_{ij}^{(k)}$, $1 \le k \le r$, $i \le i,j \le m_k$ denote a system of matrix units generating the finite-dimensional C*-algebra F. Then $x_k := \gamma(e_{11}^{(k)}) \in K_0(A)$, so by the continuity of K_0 (Theorem 12.2.4), we can find some $m \in \mathbb{N}$ such that $x_k \in [\varphi^{(m)}]_0(K_0(A_m))$ for each k. Let $\beta' : K_0(F) \to K_0(A_m)$ be the unique group homomorphism extending

$$\beta'([e_{11}^{(k)}]_0) = y_k, \quad 1 \le k \le r.$$

If $g \in K_0(F)_+$ then $g = \sum_{k=1}^r m_k [e_{11}^{(k)}]_0$ for some $m_k \in \mathbb{N}$. Thus β' is positive. Also, β' satisfies $[\varphi^{(m)}]_0 \circ \beta' = \gamma$ and

$$[\varphi^{(m)}]_0 \circ (\beta' \circ \alpha - [\varphi_{1,m}]_0) = \gamma \circ \alpha - [\varphi^{(1)}]_0.$$

From Theorem 12.2.4 (iii), we have

$$\ker([\varphi^{(m)}]_0) = \bigcup_{n=m+1}^{\infty} \ker([\varphi_{m,n}]_0).$$

Thus, for any $g \in K_0(A_1)$ we have $(\beta' \circ \alpha - [\varphi_{1,m}]_0)(g) \in \ker([\varphi^{(m)}]_0)$ so that $(\beta' \circ \alpha - [\varphi_{m,1}]_0)(g) \in \ker([\varphi_{m,n}]_0)$ for some $n \ge m + 1$.

Let $\beta := [\varphi_{m,n}]_0 \circ \beta'$. Then $(\beta \circ \alpha - [\varphi_{1,m}]_0)(g) = 0$ for every $g \in K_0(A_1)$, which is to say, $\beta \circ \alpha = [\varphi_{1,m}]_0$. Moreover,

$$\gamma = [\varphi^{(m)}]_0 \circ \beta' = [\varphi^{(n)} \circ \varphi_{m,n}]_0 \circ \beta' = [\varphi^{(n)}]_0 \circ \beta.$$

From the commutativity of the diagram, we get that if the maps φ_n are unital and $\alpha([1_A]) = [1_F]$, then also $\beta([1_F]) = [1_{A_n}]$. $\square$

13.1.8. Theorem (Elliott [36]). *Suppose that* A *and* B *are unital approximately finite C*-algebras. Any *-isomorphism* $\Phi : A \to B$ *induces an order isomorphism of* K_0-*groups,*

$$\Phi_0 : (K_0(A), K_0(A)_+, [1_A]) \to (K_0(B), K_0(B)_+, [1_B]).$$

*Conversely, if $\varphi : (K_0(A), K_0(A)_+, [1_A]) \to (K_0(B), K_0(B)_+, [1_B])$ is an order isomorphism, then there is a *-isomorphism*

$$\Phi : A \to B$$

satisfying $\Phi_0 = \varphi$.

Proof. Let $(A_n, \psi_n)_{n\in\mathbb{N}}$ and $(B_n, \rho_n)_{n\in\mathbb{N}}$ be inductive limit sequences of finite-dimensional C*-algebras with limits A and B respectively. We may assume that the maps ψ_n and ρ_n are unital and injective (Exercise 8.7.11).

Consider the finite-dimensional C*-algebra A_1. Since $\psi^{(1)}$ is a unital homomorphism, it induces a unital positive map $[\psi^{(1)}]_0 : K_0(A_1) \to K_0(A)$ and thus, by composition with φ we have $\varphi \circ [\psi^{(1)}]_0 : K_0(A_1) \to K_0(B)$. Furthermore, $\varphi \circ [\psi^{(1)}]_0(K_0(A)_+) \subset K_0(B)_+$ so for large enough n_1,

$$\varphi \circ [\psi^{(1)}]_0(K_0(A)_+) \subset [\rho^{(n_1)}]_0(K_0(B_{n_1})_+).$$

Thus, by Lemma 13.1.5 there is

$$\alpha_1 : K_0(A_1) \to K_0(B_{n_1})$$

satisfying $[\rho^{(n_1)}]_0 \circ \alpha_1 = \varphi \circ [\psi^{(1)}]_0$, hence $\varphi^{-1} \circ [\rho^{(n_1)}]_0 \circ \alpha_1 = [\psi^{(1)}]_0$ and we may apply Lemma 13.1.7 to find $m_1 \in \mathbb{N}$ and a map $\beta_1 : K_0(B_{n_1}) \to K_0(A_{m_1})$ with $[\psi^{(m_1)}]_0 \circ \beta_1 = \varphi^{-1} \circ [\rho^{(n_1)}]_0$. Thus $\varphi \circ [\psi^{(m_1)}]_0 \circ \beta_1 = [\rho^{(n_1)}]_0$ so by applying Lemma 13.1.7 again, we have $n_2 > n_1$ and a map $\alpha_2 : A_{m_1} \to B_{n_2}$ such that $\varphi \circ [\psi^{(m_1)}]_0 = [\rho^{(n_1)}]_0 \circ \alpha_2$.

Continuing the same way, we find $n_1, n_2, n_3, \ldots$ and $m_1, m_2, m_3, \ldots$ giving the commutative diagram

$$
\begin{array}{ccccccccc}
K_0(A_1) & \longrightarrow & K_0(A_{m_1}) & \longrightarrow & K_0(A_{m_2}) & \longrightarrow & \cdots & \longrightarrow & K_0(A) \\
\downarrow{\alpha_1} & \nearrow{\beta_1} & \downarrow{\alpha_2} & \nearrow{\beta_2} & \downarrow{\alpha_3} & \nearrow{\beta_3} & & & \varphi^{-1}\uparrow\downarrow\varphi \\
K_0(B_{n_1}) & \longrightarrow & K_0(B_{n_2}) & \longrightarrow & K_0(B_{n_3}) & \longrightarrow & \cdots & \longrightarrow & K_0(B).
\end{array}
$$

By Exercise 8.7.6, the subsequences $(A_{m_1}, \psi_{m_1, m_2})$ and $(B_{n_1}, \rho_{n_1, n_2})$ have inductive limit A and B, respectively. Thus we will relabel A_{m_k} by A_k and B_{n_k} by B_k and relabel the connecting maps accordingly.

By Proposition 13.1.4 there is a *-homomorphism $\sigma_1 : A_1 \to B_1$ such that $[\sigma_1]_0 = \alpha_1$ and $\tilde{\tau}_1 : B_1 \to A_2$ with $[\tilde{\tau}_1]_0 = \beta_1$. By commutativity of the diagram above, we have $[\tilde{\tau}_1]_0 \circ [\sigma_1]_0 = [\psi_1]_0$, so, since $\tilde{\tau}$ are unique up to unitary equivalence, we can find a unitary $v_1 \in B_1$ such that $(\text{ad}\,(v_1) \circ \tilde{\tau}_1) \circ \sigma_1 = \psi_1$. Let $\tau_1 := \text{ad}\,(v_1) \circ \tilde{\tau}_1$. Then the diagram

commutes.

Applying Proposition 13.1.4 again, there is a *-homomorphism $\tilde{\sigma}_2 \colon A_2 \to B_2$ such that $[\tilde{\sigma}_2]_0 = \alpha_2$. Since $[\tilde{\sigma}_2]_0 \circ [\tau_1]_0 = [\rho_1]_0$, there is a unitary $u_2 \in A_2$ such that $\mathrm{ad}\,(u_2) \circ \tilde{\sigma}_2 \circ \tau_1 = \rho_1$. Let $\sigma_2 := \mathrm{ad}\,(u_2) \circ \tilde{\sigma}_2$. Then the diagram

$$
\begin{array}{ccc}
A_1 & \xrightarrow{\;\psi_1\;} & A_2 \\
{\scriptstyle \sigma_1}\downarrow & {\scriptstyle \tau_1}\nearrow & \downarrow{\scriptstyle \sigma_2} \\
B_1 & \xrightarrow[\;\rho_1\;]{} & B_2
\end{array}
$$

commutes. Proceeding in this way, we obtain a commutative diagram

$$
\begin{array}{ccccccccc}
A_1 & \xrightarrow{\;\psi_1\;} & A_2 & \xrightarrow{\;\psi_2\;} & A_3 & \longrightarrow & \cdots & \longrightarrow & A \\
{\scriptstyle \sigma_1}\downarrow & {\scriptstyle \tau_1}\nearrow & \downarrow{\scriptstyle \sigma_2} & {\scriptstyle \tau_2}\nearrow & \downarrow{\scriptstyle \sigma_3} & {\scriptstyle \tau_3}\nearrow & & \nearrow & \\
B_1 & \xrightarrow[\;\rho_1\;]{} & B_2 & \xrightarrow[\;\rho_2\;]{} & B_3 & \longrightarrow & \cdots & \longrightarrow & B
\end{array}
$$

which, by Proposition 13.1.6 gives us *-isomorphisms $\Phi : A \to B$ and $\Psi : B \to A$, and $\Phi_0 = \varphi$. $\square$

13.2 Approximate unitary equivalence

The classification of AF algebras that we have just seen relies on the intertwining of maps of building blocks along the inductive limits. In short, we produce a commutative diagram by lifting maps from K_0 and then apply Proposition 13.1.6. In practice, for more general C*-algebras, it is difficult to produce an exact intertwining. However, as we have often seen, in the theory of C*-algebras it can be enough to have approximate results.

13.2.1. Definition. Let A and B be unital C*-algebras and $\varphi, \psi : A \to B$ unital *-homomorphisms. We say that φ and ψ are *approximately unitarily equivalent*, written $\varphi \approx_{a.u.} \psi$, if there is a sequence of unitaries $(u_n)_{n \in \mathbb{N}}$ in B such that

$$
\lim_{n \to \infty} \|u_n \varphi(a) u_n^* - \psi(a)\| = 0, \quad \text{for every } a \in A.
$$

We can also make sense of the above for nonunital C*-algebras, by taking the unitaries $(u_n)_{n \in \mathbb{N}}$ to be in the multiplier algebra of B. However, for the purposes of this section, we will for the most part only consider unital C*-algebras.

We leave the proof of following facts about approximately unitarily equivalent *-homomorphisms as an exercise.

13.2.2. Proposition. *Let A, B, C and D be unital C*-algebras and suppose that $\varphi : A \to B$, $\psi, \rho, \sigma : B \to C$ and $\chi : C \to D$ are unital *-homomorphisms.*

(i) *If $\psi \approx_{a.u.} \rho$ and $\rho \approx_{a.u.} \sigma$, then $\psi \approx_{a.u.} \sigma$.*

(ii) *If $\psi \approx_{a.u.} \rho$ then $\psi \circ \varphi \approx_{a.u.} \rho \circ \varphi$ and $\chi \circ \psi \approx_{a.u.} \chi \circ \rho$.*

(iii) *Suppose there exist sequences $(\psi_n)_{n \in \mathbb{N}}$ and $(\rho_n)_{n \in \mathbb{N}}$ such that $\psi_n(b) \to \psi(b)$ and $\rho_n(b) \to \rho(b)$ as $n \to \infty$ for every $b \in B$. Then $\psi_n \approx_{a.u.} \rho_n$ for every $n \in \mathbb{N}$ implies $\psi \approx_{a.u.} \rho$.*

13.2.3. Proposition. *Let A and B be unital separable C^*-algebras. Suppose there are an injective *-homomorphism $\varphi : A \to B$ and a sequence $(u_k)_{k \in \mathbb{N}}$ in B such that*

(i) $\lim_{k \to \infty} \|u_k \varphi(a) - \varphi(a) u_k\| = 0$ *for every $a \in A$,*

(ii) $\lim_{k \to \infty} \operatorname{dist}(u_k^* b u_k, \varphi(A)) = 0$ *for every $b \in B$.*

Then there exists a *-homomorphism $\psi : A \to B$ which is approximately unitarily equivalent to φ.

Proof. Since A and B are separable, we can find sequences $(a_k)_{k \in \mathbb{N}} \subset A$ and $(b_k)_{k \in \mathbb{N}} \subset B$ which are dense in A and B respectively. By applying an induction argument, (i) and (ii) imply that, for every $n \in \mathbb{N}$, there exists unitaries $v_1, \ldots, v_n$ contained in $\{u_k \mid k \in \mathbb{N}\}$ and, for $j = 1, \ldots, n$, elements $a_{j,1}, \ldots, a_{j,n} \in A$ such that

$$\|v_n^* \cdots v_1^* b_j v_1 \cdots v_n - \varphi(a_{j,n})\| < 1/n,$$

and

$$\|v_n \varphi(a_j) - \varphi(a_j) v_n\| < 1/2^n, \quad \|v_n \varphi(a_{j,m}) - \varphi(a_{j,m}) v_n\| < 1/2^n,$$

for every $j = 1, \ldots, n$ and $m = 1, \ldots, n - 1$. Now, for $a \in \{a_k \mid k \in \mathbb{N}\}$, we have

$$\|(v_1 \cdots v_n \varphi(a) v_n^* \cdots v_1^*) - (v_1 \cdots v_n v_{n+1} \varphi(a) v_{n+1}^* v_n^* \cdots v_1^*)\|$$
$$< 1/2^n + \|(v_1 \cdots v_n \varphi(a) v_n^* \cdots v_1^*) - (v_1 \cdots v_n \varphi(a) v_{n+1} v_{n+1}^* v_n^* \cdots v_1^*)\|$$
$$= 1/2^n,$$

so $(v_1 \cdots v_n \varphi(a) v_n^* \cdots v_1^*)_{n \in \mathbb{N}}$ defines a Cauchy sequence. Thus we can define

$$\psi(a) := \lim_{n \to \infty} v_1 \cdots v_n \varphi(a) v_n^* \cdots v_1^*,$$

for all $a \in \{a_k \mid k \in \mathbb{N}\}$ and therefore for all $a \in A$. It is easy to see that $\psi : A \to B$ is a *-homomorphism which, by construction, is approximately unitarily equivalent to φ. Moreover, $\|\psi(a_k)\| = \|a_k\|$ for every $k \in \mathbb{N}$ so ψ is injective. Finally,

$$\|\psi(a_{n,j}) - v_1 \cdots v_n \varphi(a_{n,j}) v_n^* \cdots v_1^*\| < \sum_{m=n+1}^{\infty} 1/2^m = 1/2^n,$$

so

$$\|b_j - \psi(a_{n,j})\| \leq 1/2^n + \|v_n^* \cdots v_1^* b_j v_1 \cdots v_n - \varphi(a_{j,n})\|$$
$$< 1/2^n + 1/n.$$

Since the sequence $(b_k)_{k \in \mathbb{N}}$ is dense in B and $\psi(A)$ is closed, we have $\psi(A) = B$. Thus ψ is an isomorphism. $\qquad\square$

The diagram in the next definition should look familiar. Here, we are asking only that our diagram commutes approximately, which is made precise below. For inductive limits that are more general than AF algebras, one can only expect to get such an approximate intertwining.

13.2.4. Definition (Approximate intertwining, cf. [37]). Let A and B be C^*-algebras. Suppose there are inductive limit decompositions $A = \varinjlim(A_n, \varphi_n)$ and $B = \varinjlim(B_n, \psi_n)$. We say that the diagram

$$
\begin{array}{ccccccccc}
A_1 & \xrightarrow{\varphi_1} & A_2 & \xrightarrow{\varphi_2} & A_3 & \longrightarrow & \cdots & \longrightarrow & A \\
{\scriptstyle\alpha_1}\downarrow & {\scriptstyle\beta_1}\nearrow & {\scriptstyle\alpha_2}\downarrow & {\scriptstyle\beta_2}\nearrow & {\scriptstyle\alpha_3}\downarrow & {\scriptstyle\beta_3}\nearrow & & & \\
B_1 & \xrightarrow{\psi_1} & B_2 & \xrightarrow{\psi_2} & B_3 & \longrightarrow & \cdots & \longrightarrow & B
\end{array}
$$

is an *approximate intertwining* if, for every $n \in \mathbb{N}$, there are finite subsets $F_n \subset A_n$, $G_n \subset B_n$ and there is $\delta_n > 0$ such that

(i) $\|\beta_{n+1} \circ \alpha_n(a) - \varphi_n(a)\| < \delta_n$ for every $a \in F_n$,

(ii) $\|\alpha_n \circ \beta_n(b) - \psi_n(b)\| < \delta_n$ for every $b \in G_n$,

(iii) $\varphi_n(F_n) \subset F_{n+1}$, $\alpha_n(F_n) \subset G_n$, and $\beta_n(G_n) \subset F_{n+1}$ for every $n \in \mathbb{N}$,

(iv) $\bigcup_{m=n}^{\infty} \varphi_{n,m}^{-1}(F_m)$ is dense in A_n and $\bigcup_{m=n}^{\infty} \psi_{n,m}^{-1}(G_m)$ is dense in B_n for every $n \in \mathbb{N}$,

(v) $\sum_{n=1}^{\infty} \delta_n < \infty$.

13.2.5. Theorem. *Let A and B be C^*-algebras. Suppose there are inductive limit decompositions $A = \varinjlim(A_n, \varphi_n)$ and $B = \varinjlim(B_n, \psi_n)$ such that the diagram*

$$
\begin{array}{ccccccccc}
A_1 & \xrightarrow{\varphi_1} & A_2 & \xrightarrow{\varphi_2} & A_3 & \longrightarrow & \cdots & \longrightarrow & A \\
{\scriptstyle\alpha_1}\downarrow & {\scriptstyle\beta_1}\nearrow & {\scriptstyle\alpha_2}\downarrow & {\scriptstyle\beta_2}\nearrow & {\scriptstyle\alpha_3}\downarrow & {\scriptstyle\beta_3}\nearrow & & & \\
B_1 & \xrightarrow{\psi_1} & B_2 & \xrightarrow{\psi_2} & B_3 & \longrightarrow & \cdots & \longrightarrow & B
\end{array}
$$

is an approximate intertwining. Then there are $*$-isomorphisms $\alpha : A \to B$ and $\beta : B \to A$ with $\beta = \alpha^{-1}$ and for every $n \in \mathbb{N}$,

(i) $\alpha(\varphi^{(n)}(a)) = \lim_{m \to \infty} \psi^{(m)} \circ \alpha_m \circ \varphi_{n,m}(a)$ *for every $a \in A_n$,*

(ii) $\beta(\psi^{(n)}(b)) = \lim_{m \to \infty} \varphi^{(m+1)} \circ \beta_m \circ \psi_{n,m}(b)$ *for every $b \in B_n$.*

Proof. We must show that for every $n \in \mathbb{N}$ the limits in (i) and (ii) above exist. We show this only for the limit in (i), as case (ii) is entirely analogous. Since the diagram is an approximate intertwining, for every $n \in \mathbb{N} \setminus \{0\}$ there are finite subsets $F_n \subset A_n$, $G_n \subset B_n$ and real numbers $\delta_n > 0$ such that (i)–(v) of Definition 13.2.4 hold. Let $a \in A_n$. Then by (iv) in Definition 13.2.4 we may assume that $a \in \bigcup_{m=n}^{\infty} \varphi_{n,m}^{-1}(F_m)$. Thus we may assume that there is some $m_0 \geq n$ such that $\varphi_{n,m}(a) \in F_m$ for every $m \geq m_0$.

By (i), (ii), (iii) of Definition 13.2.4 we estimate

$$\|\alpha_{m+1} \circ \varphi_m(x) - \psi_m \circ \alpha_m(x)\| < \delta_m + \delta_{m+1}$$

for every $x \in F_m \subset A_m$. From this we have

$$\|\psi^{(m+1)} \circ \alpha_{m+1} \circ \varphi_{n,m+1}(a) - \psi^{(m)} \circ \alpha_m \circ \varphi_{n,m}(a)\|$$
$$= \|\psi^{(m+1)}(\alpha_{m+1} \circ \varphi_m(\varphi_{n,m}(a))) - \psi^{(m+1)}(\psi_m \circ \alpha_m(\varphi_{n,m}(a)))\|$$
$$\leq \|\alpha_{m+1} \circ \varphi_m(\varphi_{n,m}(a)) - \psi_m \circ \alpha_m(\varphi_{n,m}(a))\|$$
$$< \delta_m + \delta_{m+1}.$$

Thus the sequence is Cauchy and hence converges, showing that the limit exists. Then, by continuity, this defines a map $\alpha : A \to B$. Similarly, we use the convergence of the sequence in (ii) to define $\beta : B \to A$, and it is clear that these maps satisfy the requirements; the details are left as an exercise. $\qquad\square$

13.3 The Elliott invariant

What about arbitrary unital C*-algebras? Can they be classified by K_0? The answer is no. As soon as one moves to more complicated C*-algebras, K_0 (even as a unital ordered group) is not enough to distinguish two C*-algebras. For example, suppose that $A_n = C(\mathbb{T}, F_n)$ where F_n is a finite-dimensional C*-algebra and suppose we have *-homomorphisms $\varphi_n : A_n \to A_{n+1}$. The inductive limit $A = \varinjlim(A_n, \varphi_n)$ is called an A$\mathbb{T}$ algebra ("approximately circle" algebra). In the simple unital case, to distinguish two A$\mathbb{T}$ algebras, we need to include the K_1-group and tracial state space.

13.3.1. We also need to consider how the tracial states and K_0 interact. If A is a C*-algebra with a tracial state τ, then $\tau(p) = \tau(q)$ whenever p and q are Murray–von Neumann equivalent. Let $\tau \in T(A)$. If $a \in M_n(A)$ then we extend τ to $M_n(A)$ by $\tau((a_{ij})_{ij}) := \sum_{i=1}^n \tau(a_{ii})$. This gives us an additive map from $V(A) \to \mathbb{R}_{\geq 0}$. From there we can define $\tau_0 : K_0 \to \mathbb{R}$ by putting $\tau_0([p] - [q]) = \tau(p) - \tau(q)$, for $p, q \in V(A)$. This map τ_0 is a *state* on $(K_0(A), K_0(A)_+, [1_A])$.

13.3.2. Definition. The *state space* of an ordered abelian group with distinguished order unit (G, G_+, u) consists of group homomorphisms $\varphi : G \to \mathbb{R}$ such that $\varphi(G_+) \subset \mathbb{R}_{\geq 0}$ and $\varphi(u) = 1$.

States on G can be very useful in helping determine the order structure. Let us look a bit more closely at the states on an ordered abelian group by collecting some results of Goodearl and Handelman [55].

13.3.3. Lemma. *Let G be a partially ordered positive abelian group and $u \in G$ be an order unit. Suppose $H \subset G$ is a subsemigroup containing u and let $f \in S(H, u)$. For $t > 0$ define*

$$p := \sup\left\{ \frac{f(x)}{m} \,\middle|\, x \in H, m > 0, x \leq mt \right\} \quad and \quad q := \inf\left\{ \frac{f(y)}{n} \,\middle|\, y \in H, n > 0, nt \leq y \right\}.$$

Then

(i) $0 \le p \le q < \infty$,

(ii) *if* $g \in S(H + \mathbb{Z}t, u)$ *and* g *extends* f, *then* $p \le g(t) \le q$,

(iii) *if* $p \le r \le q$ *then there exists* $g \in S(H + \mathbb{Z}t, u)$ *extending* f *and satisfying* $g(t) = r$.

Proof. Clearly $p \ge 0$. Since u is an order unit, we have $t \le ku$ for some $k \in \mathbb{N} \setminus \{0\}$. Since f is order-preserving, we have $f(ku) = kf(u) = k$ and $q \le f(ku)/1 = k < \infty$. Now let $x, y \in H$ and $m, n > 0$ such that $x \le mt$ and $nt \le y$. Then $nx \le nmt \le my$, so since f is order-preserving $nf(x) \le mf(y)$, whence $f(x)/m \le f(y)/n$. It follows that $p \le q$, and we have shown (i).

For (ii), suppose $x \in H$, $m > 0$ and $x \le mt$. Since $g \in S(H + \mathbb{Z}t, u)$ extends f, we have $f(x) = g(x) \le mg(t)$. Thus $f(x)/m \le g(t)$, which in turn implies $p \le g(t)$. A similar argument establishes that $g(t) \le q$, showing (ii).

For (iii), if $x \in H$ and $k \in \mathbb{Z}$, put $g(x+kt) = f(x)+kr$. If we show that g is a state, then it will clearly be unique. We have $g(u) = 1$. To show that g is additive and order-preserving, it suffices to show that if $x + kt \ge 0$ for some $x \in H$, $k \in \mathbb{Z}$, then $g(x + kt) = f(x) + kr \ge 0$. If $k = 0$, then $x \ge 0$ so $g(x) = f(x) \ge 0$. If $k < 0$ then $-kt \le x$ so $0 \le r \le q \le f(x)/(-k)$ which implies $0 \le f(x) + kr$. Finally, if $k > 0$ then $-x \le kt$, so $f(-x)/t \le p \le r$ and thus $g(x + kt) = f(x) + rt \ge 0$.   □

13.3.4. Proposition. *Let G be a partially ordered abelian group with order unit u. Suppose that H is a subgroup of G containing u. Then any state on (H, u) can be extended to a state on (G, u).*

Proof. From the previous lemma, there exists a subgroup $K \subset G$ containing H and $g \in S(K, u)$ which extends f. By the Zorn–Kuratowski Lemma, there is a maximal such group, call it M. If $t \in G_+$ but $t \notin M$, then we can use the same procedure as in the previous lemma to extend f to a functional on $M + t\mathbb{Z} \subset M$, contradicting maximality. So $G_+ \subset M$, and since $G = G_+ - G_+$, we are done.   □

13.3.5. For a partially ordered abelian group G with order unit u and $t \in G_+$, define

$$f_*(t) := \sup \left\{ \frac{h}{m} \,\middle|\, h \ge 0, m > 0, hu \le mt \right\},$$

and

$$f^*(t) := \inf \left\{ \frac{k}{n} \,\middle|\, k, n > 0, nt \le ku \right\}.$$

Lemma. *Let G be a partially ordered abelian group with order unit u. Then, for any $t \in G_+$,*

(i) $0 \le f_*(t) \le f^*(t) < \infty$,

(ii) *if* $f \in S(G, u)$ *then* $f_*(t) \le f(t) \le f^*(t)$,

(iii) *if* $f_*(t) \le r \le f^*(t)$ *then there exists a state* $f \in (G, u)$ *such that* $f(t) = r$.

Proof. Let $H := \mathbb{Z}u$ and define $g(nu) = n$. Then H is a subgroup of G containing u and $g \in S(H, u)$. Let

$$p := \sup \left\{ \frac{g(x)}{m} \mid x \in H, m > 0, x \leq mt \right\}$$

and

$$q := \inf \left\{ \frac{g(y)}{n} \mid y \in H, n > 0, nt \leq y \right\}.$$

Then $f_*(t) \geq 0$ and if $h > 0$ and $m > 0$ with $hu \leq mt$, then $g(hu)/m = h/m \leq t$. It follows that so $f_*(t) \leq p$. If $x \in H$ and $x \leq mu$ for $m > 0$. Since $x \in H$, we have $x = hu$ for some $h \in \mathbb{Z}$. Suppose that $h < 0$. Then $g(x)/m = h/m < 0 \leq f_*(t)$. If $h \geq 0$, then $g(x)/m = h/m \leq f_*(t)$ since $hu \leq mt$. Thus also $p \leq f_*(t)$, so $f_*(t) = p$. The fact that $f^*(t) = q$ is similar and left as an exercise.

Now we are in the situation of Lemma 13.3.3, and (i) and (ii) are immediate consequences. Lemma 13.3.3 further implies that $g(t) = r$. Finally, by Proposition 13.3.4, g extends to a state f on all of G, showing (iii). $\qquad\square$

13.3.6. We say that an ordered abelian group with order unit (G, G_+, u) has the *strict ordering from its states* if

$$G_+ = \{0\} \cup \{x \mid f(x) > 0 \text{ for every } f \in S(G)\}$$

If G is simple and its order structure of G is sufficiently well behaved, then the states completely determine the order. In particular this occurs when K_0 is *weakly unperforated*.

13.3.7. Definition. Let (G, G_+) be an ordered abelian group. We say that (G, G_+) is *weakly unperforated* if the following implication holds: if $x \in G$ and there is a positive integer n such that $nx > 0$, then $x > 0$.

For a simple, separable, unital, stably finite C*-algebra A, $(K_0(A), K_0(A)_+)$ is often weakly unperforated. If we plan on using K-theory for classifying C*-algebras as we did for the AF algebras, a well-behaved order structure gives us a better chance at any classification theorem. In the exercises you will show, for example, that the AF algebras have weakly unperforated K_0-groups. In fact, we shall see in the remaining chapters that if there is any hope of classifying a tractable class of C*-algebras, we must ask for weakly unperforated K-theory.

13.3.8. Theorem. *Let (G, G_+, u) be a simple weakly unperforated ordered abelian group with distinguished order unit. Then G has the strict ordering from its states.*

Proof. Let $x \in G_+$. Then, since (G, G_+) is simple (Definition 12.1.17), x is an order unit. Thus $u \leq mx$ for some $m \in \mathbb{N}$, so $f_*(x) > 1/m$ and by Lemma 13.3.5 we have $f(x) > 0$. This shows that

$$G_+ \subseteq \{0\} \cup \{x \mid f(x) > 0 \text{ for every } f \in S(G)\}.$$

Suppose that $f(x) > 0$. Since $S(G)$ is compact in the topology of pointwise convergence, $\inf_{f \in S(G,u)} f(x) > 0$. Combining Lemma 13.3.5 (ii) and (iii), we have $f_*(x) = \inf_{f \in S(G,u)} f(x)$. Thus there exists $h, m > 0$ such that $hu \leq mx$. In particular, $mx > 0$. Since (G, G_+) is weakly unperforated, we have $x > 0$, showing the reverse containment

$$\{0\} \cup \{x \mid f(x) > 0 \text{ for every } f \in S(G)\} \subseteq G_+,$$

which completes the proof.   □

13.3.9. How far, then, can we get if we throw K_1-groups and tracial states into the mix? In fact, quite far! But we will need a few definitions first.

First of all, it makes sense to consider only simple C*-algebras. We should be able to classify these before we can say anything in greater generality. Let us also stick to the separable case. If our C*-algebras are nonseparable, it is unlikely that any invariant will be in any sense computable. Furthermore, as we have often seen so far, it is usually easier to deal with unital C*-algebras.

13.3.10. Definition. Let A be a simple, separable, unital, nuclear C*-algebra. The *Elliott invariant* of A, denoted $\mathrm{Ell}(A)$, is the 6-tuple

$$\mathrm{Ell}(A) := (K_0(A), K_0(A)_+, [1_A], K_1(A), T(A), \rho_A : T(A) \to S(K_0(A))),$$

where $\rho_A : T(A) \to S(K_0(A))$ is as defined in 13.3.14 below.

13.3.11. The reason we ask that A be nuclear, in addition to simple and unital, has to do with von Neumann algebra classification. It turns out the nuclearity of a C*-algebra A is equivalent to a property called *injectivity* of all of the von Neumann algebras resulting from taking the weak closure of every GNS representation of A [23, 24]. Outside of the setting of injective von Neumann algebras, there are no reasonable classification results, whereas for injective von Neumann algebras with separable pre-dual (the weak closure of a GNS representation of a separable C*-algebra will have separable pre-dual) we have the classification results of Connes [26] and Haagerup [57]. Thus, it would be highly unlikely to establish any meaningful classification results in the relatively more complicated setting of the corresponding C*-algebras. The necessity of nuclearity was effectively proved by Dădărlat who constructed separable unital AF algebras which contain unital nonnuclear subalgebras with the same Elliott invariant (as well as real rank and stable rank) as the AF algebra in which they are contained [33].

13.3.12. Restricting to nuclear C*-algebras also means that we do not need to deal with quasitraces. A quasitrace τ on a C*-algebra A is similar to a tracial state but it only satisfies $\tau(a + b) = \tau(a) + \tau(b)$ when a and b commute. A C*-algebra A is *exact* if tensoring a short exact sequence with A preserves exactness for the minimal tensor product norm. By Theorem 6.4.2, any nuclear C*-algebra is exact. In fact, when A is separable, exactness is equivalent to A being the C*-subalgebra of a nuclear C*-algebra [67]. When a C*-algebra is exact, all quasitraces are traces [58].

13.3.13. We have already described the first four pieces of the invariant, so let us say something about the remaining ingredients. The tracial state space of a separable unital C*-algebra is metrisable with respect to the weak*-topology. It is always a Choquet simplex, that is, a simplex X such that any point $x \in X$ can be represented by a unique probability measure on the extreme points. Any metrisable Choquet simplex is realisable as the tracial state space of a simple separable unital nuclear C*-algebra [126].

13.3.14. As we observed above, any tracial state $\tau \in T(A)$ induces a state on $(K_0(A), K_0(A)_+, [1_A])$ by taking $[p] - [q] \mapsto \tau(p) - \tau(q)$ for $p, q \in M_\infty(A)$, where, by abuse of notation, we also use τ to denote the inflation of the given tracial state to the appropriate matrix algebras over A. The final ingredient in the Elliott invariant is the map which takes tracial states to $S(K_0(A), u)$, given by

$$\rho_A : T(A) \to S(K_0(A)), \quad \rho_A(\tau)([p] - [q]) = \tau(p) - \tau(q).$$

The map ρ is always surjective, but need not be injective [11, 59]. Sometimes this map is given as a pairing $\rho : T(A) \times K_0(A) \to \mathbb{R}$ between tracial states and states on $K_0(A)$, defined in the obvious way.

13.3.15. Since the tracial state space of a separable unital C*-algebra is a simplex, in particular it is a compact convex set. Given compact convex sets X, Y, an affine map $f : X \to Y$ is a map satisfying

$$f(\lambda x + (1 - \lambda)y) = \lambda f(x) + (1 - \lambda)f(y),$$

for $x, y \in X$ and $\lambda \in [0, 1]$.

If A and B are two C*-algebras, then a *-homomorphism $\varphi : A \to B$ induces a continuous affine map

$$\varphi_T : T(B) \to T(A), \qquad \tau \mapsto \tau \circ \varphi.$$

Furthermore, if $\varphi_0 : K_0(A) \to K_0(B)$ denotes the map of K_0-groups induced by φ and $\rho_A : T(A) \to S(K_0(A))$, $\rho_B : T(A) \to S(K_0(B))$ are the maps as defined above, then the diagram

$$
\begin{array}{ccc}
T(B) & \xrightarrow{\varphi_T} & T(A) \\
{\scriptstyle \rho_B}\downarrow & & \downarrow{\scriptstyle \rho_A} \\
S(K_0(B)) & \xrightarrow[\cdot \circ \varphi_0]{} & S(K_0(A))
\end{array}
$$

commutes, where $\cdot \circ \varphi_0$ is the map given by $S(K_0(B)) \ni \tau \mapsto \tau \circ \varphi_0 \in S(K_0(A))$.

13.3.16. If $(G_0, (G_0)_+, u)$ is a countable pointed simple ordered abelian group, G_1 a countable abelian group, Δ a metrisable Choquet simplex and $r : \Delta \to S(G)$ a continuous affine map, we call a 6-tuple

$$(G_0, (G_0)_+, u, G_1, \Delta, r : \Delta \to S(G))$$

an *abstract* Elliott invariant.

A map between two Elliott invariants $(G_0, (G_0)_+, u, G_1, \Delta, r : \Delta \to S(G))$ and $(H_0, (H_0)_+, v, \Omega, s : \Omega \to S(H))$ consists of an order unit-preserving group homomorphism $\varphi_0 : G \to H$, a group homomorphism $\varphi_1 : G_1 \to H_1$, and a continuous map $\varphi_T : \Omega \to \Delta$ such that the diagram

$$
\begin{array}{ccc}
\Omega & \xrightarrow{\varphi_T} & \Delta \\
\rho_B \downarrow & & \downarrow \rho_A \\
S(H_0) & \xrightarrow{\cdot \circ \varphi_0} & S(G_0)
\end{array}
$$

commutes. Thus we can make sense of two Elliott invariants being isomorphic.

13.3.17. Combining results from [38] and [101], if $\mathcal{G} = (G_0, (G_0)_+ u, G_1, \Delta, r : \Delta \to S(G))$ is an abstract Elliott invariant such that $(G_0, (G_0)_+, u)$ is weakly unperforated, there exists a simple separable unital nuclear C*-algebra A with $\mathrm{Ell}(A)$ isomorphic to $\mathcal{G}$.

13.3.18. Conjecture (The Elliott Conjecture, 1990). *Let A and B be simple, separable, unital, nuclear, infinite-dimensional C*-algebras. Then if $\varphi : \mathrm{Ell}(A) \to \mathrm{Ell}(B)$ is an isomorphism, there exists a *-homomorphism $\Phi : A \to B$, unique up to approximate unitary equivalence, such that $\mathrm{Ell}(\Phi) = \varphi$.*

It is now known that the conjecture does not hold, at least not in the full generality as stated. We will see why in Chapter 15, where the Jiang–Su algebra $\mathcal{Z}$ is introduced. Assuming weak unperforation of K_0 (Definition 13.3.7), tensoring by the Jiang–Su algebra is undetected by the Elliott invariant for most simple separable unital nuclear C*-algebras, in the sense that if A is such an algebra then $\mathrm{Ell}(A) \cong \mathrm{Ell}(A \otimes \mathcal{Z})$. This tells us that if Elliott's conjecture were true, we would always have $A \cong A \otimes \mathcal{Z}$. Examples show that this need not hold. Thus we either need to restrict the class of C*-algebras further for such a classification theorem or enlarge the invariant. This will be discussed further in the sequel.

13.4 Exercises

13.4.1. [35] Let A and B be unital C*-algebras and suppose that $\alpha : A \to A$ and $\beta : B \to B$ are approximately inner automorphisms, that is, α and β are approximately unitarily equivalent to the identity maps id_A and id_B, respectively. Show that $\alpha \otimes \beta : A \otimes_{\min} B \to A \otimes_{\min} B$ is approximately inner.

13.4.2. Let A and B be C*-algebras.

(i) Show that the minimal tensor product of the multiplier algebra of A and B, that is, $M(A) \otimes_{\min} M(B)$, is a unital C*-subalgebra of $M(A \otimes_{\min} B)$.

(ii) By taking unitaries in multiplier algebras, extend Exercise 13.4.1 to arbitrary A and B (that is, for A and B not necessarily unital).

13.4.3. Let A be a unital C*-algebra with $T(A) \neq \emptyset$. Show that every $\tau \in T(A)$ induces a state on $(K_0(A), K_0(A)_+, [1_A])$.

13.4.4. An ordered abelian group (G, G_+) is unperforated if, for any $x \in G$, if there is $n > 0$ such that $nx \geq 0$ then $x \geq 0$.

(i) Show that unperforation implies weak unperforation.

(ii) Let A be a unital AF algebra. Show that $(K_0(A), K_0(A), [1_A])$ is unperforated. (Hint: Show that the K_0-group of a finite-dimensional C*-algebra is unperforated and use Theorem 12.2.4.)

(iii) Show that an unperforated ordered abelian group must be torsion-free.

13.4.5. Let A and B be unital C*-algebras. Show that a unital *-homomorphism $\varphi : A \to B$ induces a morphism $\mathrm{Ell}(A) \to \mathrm{Ell}(B)$. Show that if $\varphi : A \to B$ is an isomorphism, then so is the induced map.

13.4.6. Given a unital C*-algebra A, we say that *projections separate tracial states* on A if, for every $\tau, \tau' \in T(A)$ with $\tau \neq \tau'$, there exists $n \in \mathbb{N}$ and a projection $p \in M_n(A) \cong M_n \otimes A$ with $\tau(p) \neq \tau'(p)$.

(i) Suppose A is a unital C*-algebra where projections separate tracial states. Show that $\rho : T(A) \to SK_0(A)$ is injective.

(ii) Show that if A is unital and either A has real rank zero or A has a unique tracial state, then projections separate tracial states.

13.4.7. Show that if A is unital, simple and purely infinite, then the Elliott invariant reduces to $(K_0(A), [1_A]_0, K_1(A))$.

14 The Cuntz semigroup and strict comparison

We saw in Chapter 12 that via the K_0-group of a C*-algebra we are able to determine important information by studying the structure of its projections. The more projections a C*-algebra has, the more information we can gain from its K-theory. For example, when a simple, stably finite, nuclear C*-algebra has real rank zero, then all tracial information can be read off from the ordered K_0-group since the pairing map will be an affine homeomorphism, and in that case the Elliott invariant reduces to K_0 and K_1 [9]. However, unlike von Neumann algebras, C*-algebras need not have any nontrivial projections at all! The Cuntz semigroup is the analogue for positive elements of the Murray–von Neumann semigroup of projections. As the name suggests, the idea for considering equivalence classes of positive elements rather than just projections was introduced by Cuntz [30]. The equivalence relation he defined there is now know as Cuntz equivalence. This was briefly introduced in the Exercises of Chapter 3. In Cuntz' original work, he actually looked at the Grothendieck group of the semigroup we define in Definition 14.1.2 below, however usually too much information is lost upon passing from the Cuntz semigroup to the Grothendieck group, so the group is not often used.

The chapter begins with the definition of the Cuntz semigroup and a number of technical results about Cuntz subequivalence. In Section 14.2, we compare Cuntz subequivalence with Murray–von Neumann equivalence and describe the subsemigroup of purely positive elements. In Section 14.3 we consider the order structure on the Cuntz semigroup and show that in many cases, the Murray–von Neumann semigroup and tracial state space can be recovered from the Cuntz semigroup. In the final section, we gather some remarks about the category **Cu**, which contains the stabilised Cuntz semigroup of any C*-algebra.

14.1 Cuntz equivalence and the Cuntz semigroup

Let A be a separable C*-algebra. For two positive elements a, b in A we say that a is *Cuntz subequivalent* b and write $a \precsim b$ if there are $(r_n)_{n \in \mathbb{N}} \subset A$ such that $\lim_{n \to \infty} \|r_n b r_n^* - a\| = 0$. We write $a \sim b$ and say a and b are *Cuntz equivalent* if $a \precsim b$ and $b \precsim a$. We saw in Exercise 3.4.4 that Cuntz equivalence is indeed an equivalence relation; symmetry is of course automatic so one only needs to check transitivity and reflexivity. Many of the results in this section come from [100, 30]; see also [4] for a nice survey paper on the Cuntz semigroup.

14.1.1. For a C*-algebra A, define $M_\infty(A) = \bigcup_{n \in \mathbb{N}} M_n(A)$ as in Section 12.1, and for $a \in M_m(A)_+$ and $b \in M_n(A)_+$, define

$$a \oplus b := \begin{pmatrix} a & 0 \\ 0 & b \end{pmatrix} \in M_{m+n}(A)_+.$$

For $n \in \mathbb{N}$, let 0_n denote the direct sum of n copies of 0 (equivalently, 0_n is the zero element in $M_n(A)$). Suppose $a, b \in M_\infty(A)$. Then $a \in M_m(A)$ and $b \in M_n(A)$

for some $m, n \in \mathbb{N}$. We say that a is Cuntz subequivalent to b, again written $a \precsim b$ if $a \oplus 0_{\max\{n-m,0\}} \precsim b \oplus 0_{\max\{m-n,0\}}$ as elements in $M_{\max\{m,n\}}(A)$. It is straightforward to verify that if $a, a', b, b' \in M_\infty(A)$ satisfy $a \precsim a'$ and $b \precsim b'$ then $a \oplus a' \precsim b \oplus b'$.

14.1.2. Definition. Let A be a separable C*-algebra. The *Cuntz semigroup* of A is defined by
$$W(A) := M_\infty(A)_+/\sim,$$
with addition given by $[a] + [b] = [a \oplus b]$ for $a, b \in M_\infty(A)$.

14.1.3. Observe that for a commutative C*-algebra $A = C_0(X)$ and $f, g \in C_0(X)_+$, we have $f \precsim g$ if and only if $\mathrm{supp}(f) \subset \mathrm{supp}(g)$. Indeed, one direction is trivial since for any $r \in C_0(X)$ the function $\mathrm{supp}(rgr^*) \subset \mathrm{supp}(g)$.

For the other direction, let $K_n = \{x \in X \mid f(x) \geq 1/n\}$, which is a compact subset of X. Clearly $K_n \subset \mathrm{supp}(g)$. By continuity of g, there is some $\delta > 0$ such that $g(x) > \delta$ for every $x \in K_n$. The set $U_n = \{x \in X \mid g(x) > \delta\}$ is open and, by construction, contains K_n. Let $h_n \in C_0(X)$ be a continuous function satisfying $(h_n)|_{K_n} = 1$ and $(h_n)|_{X \setminus U_n} = 0$. Then
$$r_n(x) \begin{cases} h_n(x)/g(x) & x \in U_n, \\ 0 & x \in X \setminus U_n, \end{cases}$$
is a well-defined positive continuous function, and
$$\|f - (r_n f)^{1/2} g (r_n f)^{1/2}\| = \|f - r_n f g\| \leq 1/n \to 0 \text{ as } n \to \infty,$$
so $f \precsim g$.

This useful observation will allow us to compare commuting elements via the functional calculus.

14.1.4. For a positive element a in a C*-algebra A and $\epsilon > 0$, we denote by $(a - \epsilon)_+$ the positive element obtained from applying the functional calculus to the function $(t - \epsilon)_+ : [0, \|a\|] \to [0, \|a\|]$ defined by
$$(t - \epsilon)_+ := \begin{cases} 0, & t \in [0, \epsilon) \\ t - \epsilon & t \in [\epsilon, \|a\|]. \end{cases}$$

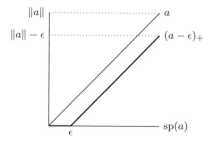

We leave as an exercise to the reader to check that whenever $\epsilon_1, \epsilon_2 > 0$ then for any $a \in A_+$ we have

$$(a - (\epsilon_1 + \epsilon_2))_+ = ((a - \epsilon_1)_+) - \epsilon_2)_+ \, .$$

14.1.5. Lemma. *Let A be a separable C*-algebra and $a, b \in A$. The following hold:*

(i) *If $a \geq 0$, then for any $\epsilon > 0$ we have $(a - \epsilon)_+ \precsim a$.*

(ii) *If $a \geq 0$ and $n \in \mathbb{N} \setminus \{0\}$, we have $a^n \sim a$.*

(iii) *For arbitrary $a \in A_+$ we have $a^*a \sim aa^*$.*

(iv) *If $0 \leq a \leq b$ then $a \precsim b$.*

Proof. For (i) and (ii), one simply applies the observation above in the commutative C*-algebra $C^*(a)$, since any positive function $f \in C_0(\mathrm{sp}(a))$ has the same support as f^n.

For (iii), we have $(a^*a)^2 = a^*(aa^*)a$ so $(a^*a)^2 \precsim aa^*$ and similarly $(aa^*)^2 \precsim a^*a$. Using (ii),

$$a^*a \sim (a^*a)^2 \precsim aa^* \sim (aa^*)^2 \precsim a^*a,$$

hence $a^*a \sim aa^*$.

To show (iv), since $0 \leq a \leq b$, we have $0 \leq a^{1/2} \leq b^{1/2}$ (Theorem 3.1.12). Thus $a^{1/2} \in \overline{b^{1/2}Ab^{1/2}}$.

Let $(u_n)_{n \in \mathbb{N}}$ be an approximate unit for $\overline{b^{1/2}Ab^{1/2}}$. Approximate each u_n up to $1/n$ by an element of the form $b^{1/2}v_nb^{1/2}$. Then $(b^{1/2}v_nb^{1/2})_{n \in \mathbb{N}}$ is also an approximate unit for $\overline{b^{1/2}Ab^{1/2}}$. Let $z_n := a^{1/2}b^{1/2}v_n$. Then $z_nb^{1/2} \to a^{1/2}$ as $n \to \infty$, and we compute

$$\|a - z_nbz_n^*\| = \|a - z_nb^{1/2}b^{1/2}z_n^*\| \to \|a - a^{1/2}a^{1/2}\| = 0, \text{ as } n \to \infty,$$

which is to say, $a \precsim b$. $\qquad\qquad\qquad\qquad\qquad\qquad\qquad\qquad\qquad\qquad\square$

14.1.6. The next lemma tells us about the behaviour of Cuntz subequivalence under addition. As one might expect, addition of orthogonal elements in a C*-algebra A is equivalent to their orthogonal direct sum as elements in $M_\infty(A)$.

Lemma. *Let A be a C*-algebra and let $a, b \in A_+$. Then $a + b \precsim a \oplus b$ in $M_2(A)_+$ and if $ab = ba = 0$ then $a + b \sim a \oplus b$.*

Proof. First of all, we have

$$\begin{pmatrix} 1 & 1 \\ 0 & 0 \end{pmatrix} \begin{pmatrix} a & 0 \\ 0 & b \end{pmatrix} \begin{pmatrix} 1 & 0 \\ 1 & 0 \end{pmatrix} = \begin{pmatrix} a+b & 0 \\ 0 & 0 \end{pmatrix},$$

so $a + b \precsim a \oplus b$. Let

$$x = \begin{pmatrix} a^{1/2} & b^{1/2} \\ 0 & 0 \end{pmatrix}.$$

Then

$$a + b \sim \begin{pmatrix} a+b & 0 \\ 0 & 0 \end{pmatrix} = xx^* \sim x^*x = \begin{pmatrix} a & a^{1/2}b^{1/2} \\ b^{1/2}a^{1/2} & b \end{pmatrix},$$

so if $ab = ba = 0$, we have $a + b \sim a \oplus b$. $\qquad\square$

14.1.7. Proposition. *Let A be a separable C*-algebra, $a, b \in A_+$ and let $\epsilon > 0$. Suppose that $\|a-b\| < \epsilon$. Then there exists $z \in A$ with $\|z\| \leq 1$ such that $(a-\epsilon)_+ = zbz^*$.*

Proof. For $r > 1$, define $g_r : [0, \|b\|] \to \mathbb{R}_+$ by $g_r(t) := \min\{t, t^r\}$. Then

$$\lim_{r \to 1} g_r(b) = b$$

and since

$$\|a - g_r(b)\| \leq \|a - b\| + \|b - g_r(b)\| < \epsilon + \|b - g_r(b)\|,$$

we can choose $r_0 > 0$ to be sufficiently close to 1 so that, for some $0 < \delta < \epsilon$, we have $\|a - g_{r_0}(b)\| < \delta + \epsilon$ and $\|b - g_{r_0}(b)\| < \delta$. Let $c := g_{r_0}(b)$. Then

$$a = a - c + c \leq \|a - c\| \cdot 1_{\tilde{A}} + c < \delta \cdot 1_{\tilde{A}} + c,$$

so $(a - \delta)_+ \precsim (c + \delta - \delta)_+ = c$. Let $f := f(a) \in C^*(a)$ be defined by applying the functional calculus to

$$f(t) = \begin{cases} ((t - \epsilon)/(t - \delta))^{1/2} & t \in [\epsilon, \|a\|], \\ 0 & t \in [0, \epsilon). \end{cases}$$

Then $\|f\| < 1$ and $f(a - \epsilon)_+ f = (a - \epsilon)_+$, so $(a - \epsilon)_+ \leq fcf$.

Set $d := c^{1/2}f$. By Exercise 5.4.11, there is $u \in A''$ such that $d = u|d|$. Let

$$x = u(a - \epsilon)_+^{1/2}.$$

Now, $d^*d = fcf \geq (a - \epsilon)_+$ so applying Proposition 3.3.4 to $(a - \epsilon)_+^{1/2}$, d^*d and $t = 1/4$, we can find $v \in A$ such that $(a - \epsilon)_+^{1/2} = d^*d^{1/4}v$. As in the proof of Lemma 14.1.5 (iv), we can find $(y_n)_{n \in \mathbb{N}}$ such that $\lim_{n \to \infty}(d^*d)^{1/2}y_n = (d^*d)^{1/4}$. Then

$$x = u(a - \epsilon)_+^{1/2} = u(d^*d)^{1/4}v = \lim_{n \to \infty} u(d^*d)^{1/2}y_n v = \lim_{n \to \infty} dy_n v,$$

so we see that $x \in A$. Define

$$z_n := x^*(1/n + b^r)^{-1/2}b^{(r-1)/2}.$$

We will show that the sequence $(z_n)_{n \in \mathbb{N}}$ converges in A to some element z and that $zbz^* = (a - \epsilon)_+$.

Observe that

$$xx^* = u(a - \epsilon)_+ u^* \leq ud^*du^* = u|d||d|u^* = dd^* = c^{1/2}f^2c^{1/2} \leq c \leq b^{r_0},$$

and

$$b^{(r_0-1)/2}b^{(r_0-1)/2} = b^{r_0-1},$$

so we may apply Lemma 3.3.3 with respect to x^*, $b^{(r_0-1)/2}$, and b^{r_0} with $t_1 = 1$ and $t_2 = (r_0 - 1)/r_0$ to see that the sequence $(z_n)_{n \in \mathbb{N}}$ converges to some $z \in A$. We have

$$\|z_nb^{1/2} - x^*\|^2 = \|x^*((1/n + b^{r_0})^{-1/2}b^{r_0} - 1)\|^2$$
$$\leq \|b^{r_0/2}(1/n + b^{r_0})^{-1/2}b^{r_0/2} - 1)\| \to 0$$

as $n \to \infty$, giving $zb^{1/2} = x^*$. Thus

$$zbz^* = x^*x = (a - \epsilon)_+^{1/2}u^*u(a - \epsilon)_+^{1/2} = (a - \epsilon)_+,$$

proving the claim.

It now only remains to show that z is a contraction. We have

$$z_n^*z_n = b^{(r_0-1)/2}(1/n + b^{r_0})^{-1/2}xx^*(1/n + b^{r_0})^{-1/2}b^{(r_0-1)/2} \leq 1$$

so indeed $\|z_n\| = \|z_n^*z_n\|^{1/2} \leq 1$. $\square$

14.1.8. Lemma. *For any positive element a in a C*-algebra A, the set $\{b \mid b \precsim a\}$ is norm-closed in A.*

Proof. Exercise. $\square$

14.1.9. Theorem. *Let A be a C*-algebra and suppose $a, b \in A_+$ are two positive elements. Then the following are equivalent:*

(i) $a \precsim b$,

(ii) *for every $\epsilon > 0$, we have $(a - \epsilon)_+ \precsim b$,*

(iii) *for every $\epsilon > 0$ there is a $\delta > 0$ such that $(a - \epsilon)_+ \precsim (b - \delta)_+$.*

Proof. Suppose that $a \precsim b$. Then there exists a sequence $(r_n)_{n \in \mathbb{N}}$ such that $\|r_nbr_n^* - a\| \to 0$ as $n \to \infty$. In particular, for any $\epsilon > 0$ we can find $N \in \mathbb{N}$ with $\|r_Nbr_N^* - a\| < \epsilon$. Thus by Proposition 14.1.7, there exists a contraction $z \in A$ such that $(a - \epsilon)_+ = zr_Nbr_N^*z^* \precsim b$, showing that (i) implies (ii).

Let $\epsilon > 0$ and suppose that (ii) holds. Then, since $(a-\epsilon/2)_+ \precsim b$, there exists some $r \in A$ such that $\|rbr^* - (a-\epsilon/2)_+\| < \epsilon' < \epsilon/2$. The sequence $((b-1/n)_+)_{n \in \mathbb{N}}$

is monotone increasing and $\|(b - 1/n)_+ - b\| \to 0$ as $n \to \infty$. Let $N \in \mathbb{N}$ satisfy $1/N < \|r\|^{-2}(\epsilon/2 - \epsilon')$, and choose $\delta > 0$ with $\delta \leq 1/N$. Then

$$\|(a - \epsilon/2)_+ - r(b - \delta)_+ r^*\| \leq \|(a - \epsilon/2)_+ - b\| + \|r(b - \delta)_+ r^*\|$$
$$< \epsilon' + \|r\|^2 \|(b - 1/N)_+\|$$
$$= \epsilon/2.$$

Thus, by Proposition 14.1.7 there exists a contraction $z \in A$ with $(a - \epsilon)_+ = z(b - \delta)_+ z^* \precsim (b - \delta)_+$, showing that (ii) implies (iii).

Finally, suppose that (iii) holds. Then $(a - \epsilon)_+ \precsim b$ for every $\epsilon > 0$. Since $(a - \epsilon)_+ \to a$ as $\epsilon \to 0$, it follows from Lemma 14.1.8 that $a \precsim b$. □

14.1.10. Recall that a unital C*-algebra A has stable rank one if the invertible elements of A are dense in A (Definition 8.6.6). As we saw in Section 3.3, in general, C*-algebras do not admit polar decompositions. If $a \in A$, where we consider A to be a concrete C*-algebra on some Hilbert space H, and its polar decomposition in $\mathcal{B}(H)$ is $a = v|a|$, then in general all we get is that $v \in A''$, as was shown in Exercise 5.4.11. In [90], Pedersen showed that in a unital C*-algebra of stable rank one, one can get close to a polar decomposition: there exists a unitary $u \in A$ such that, for any $f \in C(\mathrm{sp}(|a|))$, $f \geq 0$ which vanishes on a neighbourhood of zero then $vf(|a|) = uf(|a|)$. We will use this fact in the proof of the following proposition.

Proposition. *If A is a unital C*-algebra with stable rank one, then $a \precsim b$ if and only if for every $\epsilon > 0$ there is a unitary $u \in A$ such that $u^*(a - \epsilon)_+ u \in \overline{bAb}$.*

Proof. Let $\epsilon > 0$. Suppose there is a unitary $u \in A$ such that $u^*(a - \epsilon)_+ u \in \overline{bAb}$. Then $(u^*(a - \epsilon)_+ u)^{1/2} \in \overline{b^{1/2}Ab^{1/2}}$. Let $(b^{1/2}x_n b^{1/2})_{n \in \mathbb{N}}$ be an approximate unit for $\overline{b^{1/2}Ab^{1/2}}$. Set $z_n := (u^*(a - \epsilon)_+ u)^{1/2} b^{1/2} x_n$. Then

$$\|u^*(a - \epsilon)_+ u - z_n b z_n^*\|$$
$$= \|u^*(a - \epsilon)_+ u - (u^*(a - \epsilon)_+ u)^{1/2} b^{1/2} x_n b x_n b^{1/2} (u^*(a - \epsilon)_+ u)^{1/2})\| \to 0$$

as $n \to \infty$. Thus $(uz_n)b(uz_n)^* \to (a - \epsilon)_+$ as $n \to \infty$, which is to say that $(a - \epsilon)_+ \precsim b$. Since ϵ was arbitrary, by Theorem 14.1.9 we have $a \precsim b$.

Conversely, suppose that $a \precsim b$ and let $\epsilon > 0$. Let $r \in A$ satisfy $rbr^* = (a - \epsilon/2)_+$ and let $z := rb^{1/2}$. Let $v \in A''$ such that z^* has polar decomposition $z^* = v(zz^*)^{1/2}$. Observe that $v(zz^*)v^* = z^*z$, and, approximating by polynomials, we see that for any $f \in C(\mathrm{sp}(zz^*)) = C(\mathrm{sp}(z^*z))$ vanishing at zero, we have $vf(zz^*)v^* = f(z^*z)$. Since $(zz^* - \epsilon/2)_+$ vanishes on a neighbourhood of zero, there exists a unitary $u \in A$ satisfying

$$u(zz^* - \epsilon/2)_+ = v(zz^* - \epsilon/2)_+.$$

Thus

$$u(zz^* - \epsilon/2)_+ u^* = v(zz^* - \epsilon/2)_+ v^* = (z^*z - \epsilon/2)_+,$$

and so

$$u(a - \epsilon)_+ u^* = (b^{1/2} r^* r b^{1/2} - \epsilon/2)_+ \le b^{1/2} r^* r b^{1/2} \in \overline{bAb}.$$

Thus $u(a - \epsilon)_+ u^* \in \overline{bAb}$, as required. $\square$

14.1.11. Remark. Notice that the "if" direction in the proof of the previous theorem does not require that A has stable rank one.

14.2   Projections and purely positive elements

Let p and q be projections in a separable C*-algebra A. Then $p, q \in A_+$ and it is easy to see that if p and q are Murray–von Neumann (sub)equivalent (Definition 3.3.1) then they are Cuntz (sub)equivalent. In the case of subequivalence, the converse turns out to be true as well. If p and q are projections which are Cuntz subequivalent then they must be Murray–von Neumann subequivalent.

14.2.1. Proposition. *Let p and q be projections in a separable C*-algebra A. Then p is Murray–von Neumann subequivalent to q if and only if p is Cuntz subequivalent to q.*

Proof. Suppose that p is Cuntz subequivalent to q. Let $0 < \epsilon < 1$. Then there exists $r \in A$ such that $\|rqr^* - p\| < \epsilon$ and we may assume that $\|r\| \le 1$. By Proposition 14.1.7, there is $z \in A$ satisfying $(p - \epsilon)_+ = zrq(zr)^*$. Observe that since $\mathrm{sp}(p) \in \{0, 1\}$ there is some $\lambda > 0$ such that $(p - \epsilon)_+ = \lambda p$. Let $w = \lambda^{-1/2} zr$. Then $p = wqw^* = (wq)(wq)^*$ and $(wq)^*(wq) = qw^*wq$ is a projection. Now $qw^*wq \le \|w\|^2 q$ but since qw^*wq is projection, we must have $\|w\|^2 = 1$. Thus qw^*wq is a subprojection of q which is Murray–von Neumann equivalent to p. We leave the other direction as an exercise. $\square$

We would also like to see what happens when we compare projections to arbitrary positive elements. The next lemma is relatively elementary, but will often be useful.

14.2.2. Lemma. *Let A be a unital C*-algebra and let $a, p \ge 0$ with p a projection and $\|a\| \le 1$. Then $0 \le a \le p$ if and only if $pa = ap = a$.*

Proof. Since $0 \le 1 - p \le 1$ we get that $0 \le (1 - p)a(1 - p) \le (1 - p)p(1 - p) = 0$ by Proposition 3.1.11 (a). Thus $(1 - p)a(1 - p) = 0$. Since $a \ge 0$, we can rewrite $0 = (1 - p)a(1 - p) = (a^{1/2}(1 - p))^*(a^{1/2}(1 - p))$ which then implies $a^{1/2}(1 - p) = 0$ and hence $a^{1/2}a^{1/2}(1 - p) = a - ap = 0$. The fact that $pa - a = 0$ is similar.

Now if $pa = ap = a$, then $p - a = p - pa = p - pap = p(1 - a)p$. Since $\|a\| \le 1$, we have $(1 - a) \ge 0$ so $p - a = ((1 - a)^{1/2}p)^*((1 - a)^{1/2}p) \ge 0$. $\square$

The next lemma generalises Proposition 14.2.1 to the case that a projection is Cuntz subequivalent to an arbitrary positive element.

14.2.3. For $0 < \epsilon < 1$ define $f_\epsilon : [0,1] \to [0,1]$ by

$$
f_\epsilon(t) := \begin{cases} 0 & t \in [0, \epsilon/2], \\ \text{linear} & t \in (\epsilon/2, \epsilon), \\ 1 & t \in [\epsilon, 1]. \end{cases}
$$

Lemma. *Let A be a separable C^*-algebra and $a \in A_+$. If $p \in A$ is a projection that is Cuntz subequivalent to a, then there exist a $\delta > 0$, a positive scalar $\lambda > 0$ and a projection $q \leq \lambda a$ such that p is Murray–von Neumann equivalent to q and $f_\delta(a)q = q$.*

Proof. Given $0 < \epsilon_0 < 1$, there is $\lambda_0 > 0$ such that $(p - \epsilon_0)_+ = \lambda_0 p$. Since $p \precsim a$, by Theorem 14.1.9 there exists $\epsilon_1 > 0$ such that $\lambda_0 p = (p - \epsilon_0)_+ \precsim (a - \epsilon_1)_+$. We use Proposition 14.1.7 to find $z \in A$ such that $z(a - \epsilon_1)_+ z^* = (p - \epsilon_0)_+ = \lambda_0 p$. Let $w = \lambda_0^{-1/2} z$. Then $p = w(a - \epsilon_1)w^*$ and $q := (w(a - \epsilon_1)_+^{1/2})^*(w(a - \epsilon_1)_+^{1/2})$ is a projection satisfying $q \sim p$. Setting $\lambda := \|w\|^2$, we see that $q \leq \lambda a$. Finally, let $\delta > 0$ satisfy $\delta < \epsilon_1$. Then $f_\delta(a)(a - \epsilon_1)_+ = (a - \epsilon_1)_+$, which in turn implies $f_\delta(a)q = q$.

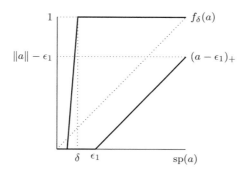

14.2.4. We saw in Exercise 3.4.4 that Cuntz subequivalence in A is transitive: for $a, b, c \in A_+$, if $a \precsim b$ and $b \precsim c$ then $a \precsim c$. Since positive elements $a, b \in A$ are Cuntz equivalent if, by definition $a \precsim b$ and $b \precsim a$, whenever $a \sim a'$, $b \sim b'$ and $a \precsim b$ we have $a' \precsim b'$. Thus for $x, y \in W(A)$, we write $x \leq y$ if $a, b \in M_\infty(A)_+$ satisfy $x = [a]$, $y = [b]$ and $a \precsim b$. It is easy to see that $\leq$ is reflexive, transitive and symmetric. Thus $\leq$ defines a partial order (Definition 3.1.2) on $W(A)$.

14.2.5. Recall that $V(A)$ denotes the Murray–von Neumann semigroup of A (Definition 12.1.2). Given $p \in M_\infty(A)$, we will denote its equivalence class in $V(A)$ by $\langle p \rangle$ to avoid confusion with Cuntz equivalence classes. Let $\sim_{MvN}$ denote Murray–von Neumann equivalence. Suppose that p, p', q, q' are projections $M_\infty(A)$ such that $p \sim_{MvN} p'$, $q \sim_{MvN} q'$ and p is Murray–von Neumann subequivalent to q. Then p' is also Murray–von Neumann subequivalent to q and there are $v, w \in M_\infty(A)$ such that

$$
p' = v^*v, \quad q = w^*w, \quad q' = ww^*, \quad vv^* \leq q.
$$

Let $y = wv$. Then

$$yy^* = wvv^*w^* \leq wqw^* = q',$$

and

$$y^*y = v^*w^*wv = v^*qv = v^*(vv^*)qv = v^*vv^*v = p',$$

so p' is Murray–von Neumann subequivalent to q'. Thus we may define a relation $\leq$ on $V(A)$ by putting $\langle p \rangle \leq \langle q \rangle$ if p is Murray–von Neumann subequivalent to q. Unlike for $W(A)$, the relation $\leq$ on $V(A)$ is not necessarily a partial order. However, it is a *preorder*, which is to say, $\leq$ is both reflexive and transitive, but not necessarily symmetric. Indeed, in the Cuntz algebra $\mathcal{O}_n$, $n \geq 3$, for any two nonzero projections $p, q \in \mathcal{O}_n$ one has $p \precsim_{MvN} q \precsim_{MvN} p$, but not all projections are equivalent. See [31] for more on the K-theory of the Cuntz algebras. On the other hand, we will see below that symmetry will hold in a stably finite C*-algebra.

14.2.6. For $a \in M_\infty(A)$, we will denote by $[a]$ the Cuntz equivalence class of a in $W(A)$. Define a map

$$V(A) \to W(A), \quad \langle p \rangle \mapsto [p].$$

Since Murray–von Neumann equivalence implies Cuntz equivalence, this map is a well-defined semigroup morphism. Whenever this map is injective, then Murray–von Neumann equivalence and Cuntz equivalence agree on projections. This is always the case when A is stably finite, because in this case, $(V(A), \leq)$ is also partially ordered.

14.2.7. Proposition. *Let A be stably finite. Then $\leq$ is a partial order on $V(A)$.*

Proof. Since $\leq$ is a preorder, we only need to show that $\leq$ is symmetric, that is, if $x, y \in V(A)$ satisfy $x \leq y$ and $y \leq x$ then $x = y$. Let $p, q \in M_\infty(A)$ with $p \precsim_{MvN} q$ and $q \precsim_{MvN} p$. Then $p \sim_{MvN} r$ for some projection r with $r \leq q$. Let $p' = q - r$. Then p' is clearly self-adjoint and $(p')^2 = (q-r)^2 = q^2 - qr - rq + r^2 = q - r - r + r = q - r = p'$, so p' is a projection, and $p \oplus p' \sim_{MvN} r \oplus p' \sim_{MvN} q$. Similarly, there is a projection $q' \in M_\infty(A)$ such that $q \oplus q' \sim_{MvN} p$. Then

$$p \oplus p' \oplus q' \sim_{MvN} q \oplus q' \sim_{MvN} p.$$

Since A is stably finite, $M_n(A)$ is finite for every $n \in \mathbb{N}$ and by Proposition 12.1.7 this means in particular that p cannot be equivalent to a proper subprojection. Thus $p' \oplus q' = 0$, so $p' = q' = 0$ and therefore $p \sim_{MvN} q$. Thus $\leq$ is symmetric and hence a partial order. □

For the map defined in 14.2.6, we obtain the following.

14.2.8. Corollary. *Let A be a separable, stably finite C*-algebra. Then the map $V(A) \to W(A)$ is injective.*

Let A be stably finite. Then we identify $V(A)$ as a subsemigroup of $W(A)$. We would like to determine when a positive element which is *not* a projection is in the $W(A)$ equivalence class of a projection.

14.2.9. In a partially ordered abelian semigroup $(P, \leq)$, we say that $x \in P$ is *positive* if $x + y \geq y$ for every $y \in P$. The semigroup P is said to be positive if every $x \in P$ is positive. If P has a zero element (that is, P is a monoid), then P is positive if and only if $x \geq 0$ for every $x \in P$. For example, if P is an abelian semigroup equipped with the *algebraic order*, which is defined by $x \leq y$ if and only if there exists $z \in P$ with $x + z = y$, then P is positive. It is easy to see that the Cuntz semigroup is positive. The subequivalence relation on the Cuntz semigroup extends the algebraic order in the sense that if $x + z = y$ then $x \leq y$, however we may have that $x \leq y$ and there is no such z with $x + z = y$ (see exercises).

On the other hand, the Murray–von Neumann semigroup *does* agree with the algebraic order. Indeed, we have used this fact many times already. If $p \precsim_{MvN} q$ then $p \sim_{MvN} r \leq q$ for some projection r, and then we have $p \oplus (q - r) \sim_{MvN} q$. For a separable stably finite C*-algebra A, we saw that $V(A) \subset W(A)$, so in that case, when restricted to projections in $M_\infty(A)$, the semigroup $W(A)$ will of course be algebraic. This turns out to be true for any separable C*-algebra, regardless of whether or not it is stably finite. In fact, we can do even better, as the next proposition shows [92, Proposition 2.2].

14.2.10. Proposition. *Let A be a separable C*-algebra, $a \in M_\infty(A)_+$ and $p \in M_\infty(A)$ a projection. Suppose that $p \precsim a$. Then there exists $b \in M_\infty(A)_+$ such that $p \oplus b \sim a$.*

Proof. Let $n \in \mathbb{N}$ such that $p \in M_n(A)$. By Lemma 14.2.3, there exist a projection $q \in M_n(A)_+$ and $\lambda > 0$ such that $p \sim_{MvN} q$ and $q \leq \lambda a$. Let $r = q - (1_{\tilde{A}} - q)$ and observe that $rar \geq 0$. Then

$$\lambda a \sim a \leq a + rar = 2(qaq + (1_{\tilde{A}} - q)a(1_{\tilde{A}} - q)) \sim qaq \oplus (1_{\tilde{A}} - q)a(1_{\tilde{A}} - q).$$

It follows from Lemma 14.1.5 (iv) that $a \precsim qaq \oplus (1_{\tilde{A}} - q)a(1_{\tilde{A}} - q)$. Let $b := (1_{\tilde{A}} - q)a(1_{\tilde{A}} - q)$. Then $qaq \leq \|a\|q \sim q \sim p$, so $a \precsim p \oplus b$. To conclude the proof, we must show $p \oplus b \precsim a$. Now $q \leq \lambda a$ which means that $q \in \overline{aAa}$, the hereditary C*-subalgebra generated by a. Thus $(1_{\tilde{A}} - q)a^{1/2} \in \overline{aAa}$ whence also $b \in \overline{aAa}$. Now it follows from Proposition 14.1.10 and Remark 14.1.11 that $p \oplus b \sim a$. □

14.2.11. For a C*-algebra A, let

$$W(A)_+ := \{x \in W(A) \mid x \neq [p] \text{ for any projection } p \in M_\infty(A)\},$$

that is, $W(A)_+$ consists of equivalence classes of elements which are *never* equivalent to a projection. When A is stably finite, we obtain

$$W(A) = V(A) \sqcup W(A)_+.$$

An element $x \in W(A)_+$ is called *purely positive*.

Let A be a separable unital C*-algebra which is simple or has stable rank one. We will show that in this case, $W(A)_+$ is a subsemigroup of $W(A)$ with a certain absorbing property. For the case of stable rank one, we first require the following lemma.

14.2.12. Lemma. *Let A be a separable unital C*-algebra with stable rank one. Let $a \in M_\infty(A)$. Then $[a] \in W(A)$ is purely positive if and only if $0 \in \mathrm{sp}(a)$ and 0 is not an isolated point of $\mathrm{sp}(a)$.*

Proof. Suppose that $a \in M_n(A)_+$ for some $n \in \mathbb{N} \setminus \{0\}$ and that $[a] \in W(A)_+$. If $0 \in \mathrm{sp}(a)$ is an isolated point, then there exists $\epsilon > 0$ such that $(0, \epsilon) \cap \mathrm{sp}(a) = \emptyset$. Similarly, if $0 \notin \mathrm{sp}(a)$, then since $\mathrm{sp}(a) \cup \{0\}$ is closed, again there exists $\epsilon > 0$ such that $(0, \epsilon) \cap \mathrm{sp}(a) = \emptyset$. Let $p = f_{\epsilon/2}(a)$, where $f_{\epsilon/2}$ is defined as in 14.2.3. Then p is a projection and since $f_{\epsilon/2}$ and $\mathrm{id}_{\mathrm{sp}(a)}$ have the same support on $\mathrm{sp}(a)$, we have $[p] = [a]$ by 14.1.3. This contradicts the fact that $[a] \in W(A)_+$. So $0 \in \mathrm{sp}(a)$ and 0 is not an isolated point of $\mathrm{sp}(a)$.

Conversely, suppose $0 \in \mathrm{sp}(a)$ and 0 is not an isolated point of $\mathrm{sp}(a)$ but $[a] = [p]$ for some projection p, which we may assume is an element of $M_n(A)_+$. By Lemma 14.2.3 there exists $\delta > 0$ and positive scalar λ and a projection $q \leq \lambda a$ such that $p \sim q$ and $f_\delta(a)q = q$. Since 0 is not an isolated point, $f_\delta(a)$ is not a projection. In particular is not equal to q, so by Lemma 14.2.2 we have $q < f_\delta(a)$. Since A has stable rank one, so too does $M_n(A)$ (Proposition 12.1.13). Let $0 < \epsilon < \delta/2$. Then $f_\delta(a) \leq (a - \epsilon)_+$. Since $a \precsim q$, Proposition 14.1.10 provides a unitary $u \in M_n(A)$ such that $u^*(a - \epsilon)_+ u \in qAq$, and, since qAq is hereditary, this in turn implies that $u^* f_\delta(a) u \in qAq$. Now

$$qu^* f_\delta(a) uq \leq qu^* 1_A uq = q,$$

which implies

$$uqu^* + u(f_\delta(a) - q)u^* \leq q.$$

Since $f_\delta(a)$ is not a projection, $u^* f_\delta(a) u - q > 0$. Then $q \sim uqu^*$ is a proper subprojection of q. But $M_n(A)$ has stable rank one and so is stably finite (Exercise 8.7.21). This contradicts Proposition 12.1.7, so $[a]$ is purely positive. $\qquad \square$

14.2.13. Proposition. *Let A be a separable unital C*-algebra which is simple or has stable rank one. Then the following hold:*

(i) *$W(A)_+$ is a subsemigroup of $W(A)$ and if $x \in W(A)_+$ then we also have $x + y \in W(A)_+$ for any $y \in W(A)$.*

(ii) *$V(A) = \{x \in W(A) \mid \text{whenever } x \leq y, \text{ there exists } z \text{ such that } x + z = y\}$.*

Proof. Let $x \in W(A)_+$, $y \in W(A)$ and choose representative $a, b \in M_n(A)$. Then $\mathrm{sp}(a)$ contains zero and zero is not an isolated point. Thus $0 \in \mathrm{sp}(a \oplus b)$ and is not an isolated point (Exercise 4.4.14). Thus $x + y \in W(A)_+$, showing (i).

For (ii), let $X := \{x \in W(A) \mid \text{if } x \leq y, \text{ there exists } z \text{ such that } x + z = y\}$. We know that $V(A) \subseteq X$ by Proposition 14.2.10, so we only need the reverse inclusion. Suppose $x \in W(A)_+$. Let $x = [a]$ for $a \in M_n(A)$. Then $a \sim \|a\|^{-1} a \leq 1_{M_n(A)}$. But $1_{M_n(A)}$ is a projection, and by (i), $x + y \in W(A)_+$ for every $y \in W(A)$. Thus there is no y satisfying $x + y = [1_{M_n(A)}]$. Thus $x \notin V(A)$. Since $W(A)$ is the disjoint union of $V(A)$ and $W(A)_+$, the result follows. $\qquad \square$

14.3   Comparing the Cuntz semigroup to the Elliott invariant

We begin this section by looking at the property of *almost unperforation*. This property is equivalent to a property which allows one to use traces to measure the size of positive elements, called *strict comparison of positive elements*. The presence of these two equivalent properties will be essential to establishing an important representation theorem which relates the Cuntz semigroup to the Elliott invariant. It also has deep, and perhaps surprising, connections to the Jiang–Su algebra of Chapter 15 and the nuclear dimension of Chapter 17. This section is mostly based on the papers [92, 20].

14.3.1. Let X be a metric space. A function $f : X \to \mathbb{R} \cup \{\infty\}$ is *lower semi-continuous* if $f^{-1}((t, \infty))$ is open, for every $t \in \mathbb{R}$. Rephrasing this property in terms of convergent sequences, f is lower semicontinuous if and only if, whenever $(x_n)_{n \in \mathbb{N}} \subset X$ is a sequence converging to $x \in X$, we have $f(\lim_{n \to \infty} x_n) \le \liminf_{n \to \infty} f(x_n)$. Indeed, if this is the case let $t \in \mathbb{R}$ then consider the set $f^{-1}((-\infty, t))$. Let $x \in \overline{f^{-1}((-\infty, t))}$ and let $(x_n)_{n \in \mathbb{N}} \subset f^{-1}((-\infty, t))$ be a sequence converging to x. Then $f(x) \le \liminf_{n \to \infty} f(x_n) \le t$, so $x \in f^{-1}((-\infty, t])$, which is to say, $f^{-1}((-\infty, t])$ is closed. Thus $f^{-1}((t, \infty))$ is open, so f is lower semicontinuous. On the other hand, if f is lower semicontinuous and $x_n \to x$, then for large enough x we have $x \in f^{-1}((t, \infty))$ for every $t < f(x)$, so $\liminf_{n \to \infty} f(x_n) \ge f(x)$.

14.3.2. Proposition. *Let X be a metric space.*

(i) *The pointwise supremum of lower semicontinuous functions $(f_n : X \to \mathbb{R} \cup \{\infty\})_{n \in \mathbb{N}}$ is again lower semicontinuous.*

(ii) *Every lower semicontinuous $f : X \to [0, \infty]$ is the pointwise supremum of continuous real functions.*

Proof. Let $(f_n : X \to \mathbb{R} \cup \{\infty\})$ be a sequence of lower semicontinuous functions and suppose that $f(x) := \sup_{n \in \mathbb{N}} f_n(x)$ defines a function $f : X \to \mathbb{R} \cup \{\infty\}$. Then, for every $t \in \mathbb{R}$ the set $f^{-1}((t, \infty)) = \bigcup_{n \in \mathbb{N}} f_n^{-1}((t, \infty))$ is open in X, thus f is lower semicontinuous. This shows (i).

Now suppose that $f : X \to [0, \infty]$ is lower semicontinuous. Let

$$C(f) := \{g : X \to \mathbb{R} \mid g \text{ continuous}, g \le f\},$$

and let $\tilde{f}$ denote the pointwise supremum of $C(f)$. Let $x \in X$ and $\epsilon > 0$. The set $F := f^{-1}((-\infty, f(x) - \epsilon]) \subset X$ is closed and does not contain x. Since continuous functions on a metric space separate points, we can find a continuous function $g : X \to [0, 1]$ such that $g|_F = 0$ and $g(x) = 1$. Let $h := (f(x) - \epsilon)g$. Then $h \in C(F)$, so $\tilde{f}(x) \ge f(x) - \epsilon$. Since x and ϵ were arbitrary, we have $\tilde{f} = f$, showing (ii).   □

14.3.3. Definition. A *subadditive rank function* on a C*-algebra A is a function

$$D : A \to [0, 1]$$

satisfying

   (i) $\sup\{D(a) \mid a \in A\} = 1$,
  (ii) if $ab = ba = a^*b = ba^* = 0$ then $D(a + b) = D(a) + D(b)$,
 (iii) for every $a, b \in A$, we have $D(a) = D(a^*a) = D(aa^*) = D(a^*)$,
 (iv) if $a, b \in A_+$ and $a \precsim b$ then $D(a) \leq D(b)$,
  (v) for every $a, b \in A$ we have $D(a + b) \leq D(a) + D(b)$.

14.3.4. Let A be a unital C*-algebra with $T(A) \neq \emptyset$. For $\tau \in T(A)$, the map

$$d_\tau : A \to [0, 1], \quad a \mapsto \lim_{n \to \infty} \tau((a^*a)^{1/n}),$$

defines a subadditive rank function on A.

14.3.5. We saw in 11.3.11 that any tracial state on a commutative C*-algebra $C(X)$ induces a Borel probability measure on the space X. In particular, if A is a unital C*-algebra with a tracial state τ, then, for any normal element $a \in A$, τ induces a measure on the spectrum of a. Thus we might expect a similar relationship between d_τ and measures on the spectrum of normal elements. Indeed, we have the following, established in [9, Proposition I.2.1].

Proposition. *Let A be a C*-algebra and suppose that $D : A \to [0, 1]$ is a subadditive rank function. Let $a \in A_+$. Suppose that $U \subset \mathrm{sp}(a)$ is a σ-compact open subset. For any $f \in C_0(\mathrm{sp}(a))$ such that $\mathrm{coz}(f) := \{x \in X \mid f(x) \neq 0\} = U$, set*

$$\mu(U) := D(f(a)).$$

Then μ is a finitely additive, σ-compact probability measure on $\mathrm{sp}(a)$ defined on the σ-algebra generated by σ-compact open subsets.

Proof. We show that $\mu(U)$ does not depend on the choice of the function f. Since $\mathrm{coz}(f^*f) = \mathrm{coz}(f)$ and $D(f) = D(f^*f)$, we may assume that $f \geq 0$. Suppose that $g \in C_0(\mathrm{sp}(a))$ satisfies $\mathrm{coz}(g) := \{x \in X \mid g(x) \neq 0\} = U$. Again, we may assume that $g \geq 0$. Then $\mathrm{supp}(f) = \mathrm{coz}(f) = \mathrm{coz}(g) = \mathrm{supp}(g)$, so $f \precsim g$ by 14.1.3. It follows from Definition 14.3.3 that $D(f) = D(g)$. Thus μ is well defined. We leave the rest of the details to the reader. $\square$

14.3.6. Definition. Let A be a C*-algebra. A *dimension function* on A is a function $d : M_\infty(A)_+ \to [0, \infty]$ satisfying

   (i) if a and b are orthogonal then $d(a + b) = d(a) + d(b)$,
  (ii) if $a \precsim b$ then $d(a) \leq d(b)$.

We say that d is *lower semicontinuous* if

$$d(a) = \lim_{\epsilon \to 0^+} d((a - \epsilon)_+), \quad a \in M_\infty(A)_+.$$

14.3.7. Let A be a unital C*-algebra with nonempty tracial state space $T(A)$. Let $\tau \in T(A)$. If $a \in M_n(A)$ then we extend τ to $M_n(A)$ by $\tau((a_{ij})_{ij}) := \sum_{i=1}^{n} \tau(a_{ii})$. This allows us to extend the rank function d_τ to matrix algebras in the obvious way. We will also denote this extension by $d_\tau : M_\infty(A)_+ \to [1, \infty)$. Thus

$$d_\tau(a) := \lim_{n \to \infty} \tau(a^{1/n}), \quad a \in M_\infty(A)_+.$$

We leave it as an exercise to show that this is a lower semicontinuous dimension function on A. In fact, any subadditive rank function on a C*-algebra A can always be extended to a lower semicontinuous dimension function [9, Theorem II.3.1]. The next theorem is stated without proof. It follows from work of Blackadar–Handelman [9] together with the difficult result of Haagerup [58] that all quasitraces on an exact C*-algebra are traces.

14.3.8. Theorem. *Let A be a simple unital exact C*-algebra. Then every lower semicontinuous dimension function on A is of the form d_τ for some tracial state $\tau \in T(A)$.*

14.3.9. Recall that when A is a separable, unital C*-algebra, its tracial state space is a metrisable Choquet simplex with respect to the weak-* topology hence is a compact convex set (13.3.13). The definition for an affine map was given in 13.3.15. In particular, if X is any compact convex set, an affine function $f : S \to \mathbb{R}$ is a function satisfying

$$f(\lambda x + (1 - \lambda)y) = \lambda f(x) + (1 - \lambda)f(y),$$

for $x, y \in X$ and $\lambda \in [0, 1]$.

14.3.10. Definition. A positive partially ordered semigroup $(P, \leq)$ is *almost unperforated* if the following holds: Let $x, y \in P$ and suppose there exist $n, m \in \mathbb{N}$ with $n > m$, such that $nx \leq my$. Then $x \leq y$.

14.3.11. For a positive partially ordered abelian semigroup P and $x \in P$, define

$$S(P, x) := \{f : P \to [0, \infty] \mid f(x) = 1, f \text{ is additive and order-preserving}\}.$$

Note that any dimension function normalised at 1_A gives an element in $S(W(A), [1_A])$, and conversely, from any $f \in S(W(A), [1_A])$, we can define a normalised dimension function on A.

14.3.12. The purpose of the next few results is to show that almost unperforation of an ordered abelian semigroup can be rephrased in terms of the behaviour of its states. Recall that the definition of a (partially) ordered abelian group (G, G_+)

was given in Definition 12.1.8, and an order unit for (G, G_+) is an element $u \in G$ such that, for every $x \in G(P)$ there exists $n \in \mathbb{N}$ such that $-nu \leq x \leq nu$. The state space of an ordered abelian group was defined in 13.3.2. In a partially ordered abelian semigroup P, we can similarly define an order unit $u \in P$ as an element such that, for any $x \in P$, there exists $n \in \mathbb{N}$ such that $x \leq nu$.

14.3.13. Given a positive partially ordered abelian semigroup P, let $G(P)$ denote the Grothendieck group of P, that is, the group of formal differences $\langle x \rangle - \langle y \rangle$ where $x, y \in P$. We define the positive cone of $G(P)$ to be

$$G(P)_+ := \{\langle x \rangle - \langle y \rangle \in G(P) \mid x + z \geq y + z \text{ for some } z \in P\}.$$

We leave it as an exercise to show that this gives us an ordered abelian group.

14.3.14. Lemma. *Let P be a partially ordered abelian semigroup with order unit u. Suppose that $t \in P$ and $f(t) < f(u)$ for every state $f \in S(P, u)$. Then there exists $m \in \mathbb{N}$ and $z \in P$ such that $mt + z \leq mu + z$.*

Proof. Let $G(P)$ be the partially ordered abelian group as given in 14.3.13. We will prove the lemma by contradiction. Assume that for every $m \in \mathbb{N}$ and every $z \in P$, it is never the case that $mt + z \leq mu + z$. Then $m\langle t \rangle \leq m\langle u \rangle$ never holds either. Let $f_*(t)$ and $f^*(t)$ be defined as in 13.3.5. Suppose that $0 < k < n$ and $n\langle t \rangle \leq k\langle u \rangle$. But $k\langle t \rangle \leq n\langle t \rangle$, so $k\langle t \rangle \leq k\langle u \rangle$, which contradicts the assumption. Thus

$$f^*(\langle t \rangle) := \inf \left\{ \frac{k}{n} \,\middle|\, k, n > 0, n\langle t \rangle \leq k\langle u \rangle \right\} \geq 1.$$

It follows that there exists r such that $0 \leq f_*(\langle t \rangle) \leq 1 \leq r \leq f^*(\langle t \rangle)$ and a state $f \in (G, \langle u \rangle)$ such that $f(\langle t \rangle) = r$. Define $d_f \in S(P, u)$ by setting $d_f(x) = f(\langle x \rangle)$. Then one checks that $d_f \in S(P, u)$. But then d_f is a state satisfying $d_f(t) > 1 = d_f(u)$, a contradiction. $\qquad\square$

14.3.15. Proposition. *A positive partially ordered abelian semigroup P is almost unperforated if and only if the following holds: when $x, y \in P$ satisfy $x \leq ny$ for some $n \in \mathbb{N}$ and $f(x) < f(y)$ for every $f \in S(P, y)$, then $x \leq y$.*

Proof. Let $P_0 := \{x \in P \mid x \leq ny \text{ for some } n \in \mathbb{N}\}$. Then P_0 is a subsemigroup of P with the property that $x \in P_0$ and $z \leq x$ then $z \in P_0$. Observe that y is an order unit for P_0. Suppose $f \in S(P_0, y)$. Then f extends to a state $\bar{f}$ on $S(P, y)$ by setting $\bar{f}(x) = \infty$ for $x \in P \setminus P_0$. Thus, without loss of generality, we may assume that y is an order unit.

First assume that P is almost unperforated. Suppose that $x \leq ny$ and $f(x) < f(y)$ for every $f \in S(P, y)$. Since the set $S(P, y)$ is compact in the topology of norm convergence, we can find $c < 1$ such that $f(x) < c < 1 = f(y)$ for every $f \in S(P, y)$. Let $p < q \in \mathbb{N}$ such that $c < p/q$. Then $f(x) < p/q$ so $f(qx) < p = f(py)$. By the previous lemma, there is $m \in \mathbb{N}$ and $z \in P$ such that $mqx + z \leq mpy + z$. Observe that

$$2mqx + z = mqx + (mqx + z) \leq mqx + (mpy + z)$$
$$= mpy + (mqx + z) \leq 2mpy + z.$$

Iterating this, we see that $kmqx + z \leq kmpy + z$ for every $k \in \mathbb{Z}$. Since y is an order unit, there exists $l \in \mathbb{N}$ such that $z \leq ly$. Thus

$$kmqx \leq kmqx + z \leq (kmp + l)y.$$

For sufficiently large k, we have $kmq \geq kmp + l$, so almost unperforation implies that $x \leq y$, as required.

Now suppose P has the property that if $x, y \in P$ satisfy $x \leq ny$ for some $n \in \mathbb{N}$ and $f(x) < f(y)$ for every $f \in S(P, y)$, then $x \leq y$. Let $x, y \in P$ be elements with $mx \leq ny$, for $m > n$. Then $x \leq mx \leq ny$ so $mf(x) \leq nf(y) = n$. Thus $f(x) \leq n/m < 1 = f(y)$ for every $f \in S(W, y)$ and we conclude that $x \leq y$. Thus P is almost unperforated. $\qquad\square$

14.3.16. Definition. Let A be a simple separable unital C*-algebra with $T(A) \neq \emptyset$. We say that A has *strict comparison of positive elements* if the following implication holds for positive elements $a, b \in M_\infty(A)$:

$$\text{if } d_\tau(a) < d_\tau(b) \text{ for every } \tau \in T(A), \text{ then } a \precsim b.$$

We saw in Section 14.2 that when A is a separable stably finite C*-algebra, the Cuntz semigroup contains the Murray–von Neumann semigroup of projections $V(A)$ as a subsemigroup, and $W(A)$ decomposes into the disjoint union

$$W(A) = V(A) \sqcup W(A)_+$$

where $W(A)_+$ is the subsemigroup of purely positive elements. In this way the Cuntz semigroup contains information about one part of the Elliott invariant (Definition 13.3.10), namely $K_0(A)$. As it turns out, in many cases we can also extract tracial information.

14.3.17. For a compact convex set Δ, let

$$\mathrm{LAff}(\Delta)_+ := \{f : \Delta \to \mathbb{R}_{\geq 0} \mid f \text{ lower semicontinuous, affine}\}.$$

Observe that $\mathrm{LAff}(\Delta)_+$ is a semigroup with respect to pointwise addition of functions. We also define the subsemigroup $\mathrm{Aff}(\Delta)_+ \subset \mathrm{LAff}(\Delta)_+$ to be those functions in $\mathrm{LAff}(\Delta)_+$ which are continuous and the subsemigroup $\mathrm{LAff}_b(\Delta)_+ \subset \mathrm{LAff}_b(\Delta)_+$ to be those functions in $\mathrm{LAff}(\Delta)_+$ which are bounded above. All of these semigroups can be given an order: $f \leq g$ if and only if $f(x) \leq g(x)$ for every $x \in \Delta$. For each of these semigroups, we denote $\mathrm{LAff}(\Delta)_{++}$, $\mathrm{Aff}(\Delta)_{++}$ and $\mathrm{LAff}_b(\Delta)_{++}$ to be the subset of strictly positive functions in the respective semigroups $\mathrm{LAff}(\Delta)_+$, $\mathrm{Aff}(\Delta)_{++}$ and $\mathrm{LAff}_b(\Delta)_{++}$.

14.3.18. Recall that if A is a unital exact C*-algebra, then every quasitrace is a trace (13.3.11), and that a nuclear C*-algebra is always exact. The tracial state space of A, $T(A)$, is compact convex (Exercise 5.4.13).

For any $[p] \in V(A)$ a projection, we define a positive affine function

$$\hat{p} : T(A) \to \mathbb{R}_{\geq 0}, \quad \tau \mapsto \tau(p).$$

Here, by abuse of notation, τ means the inflation of τ to the appropriate matrix algebra over A. Note that $\hat{p}$ is well defined because $\hat{p}(\tau)$ does not depend on the choice of representative for the class $[p]$.

14.3.19. Let A be a unital separable exact C*-algebra. Let

$$\widetilde{W}(A) := V(A) \sqcup \mathrm{LAff}_b(T(A))_{++}.$$

We define a semigroup structure on $\widetilde{W}(A)$ as follows:

  (i) If $[p], [q] \in V(A)$, then addition is the usual addition in $V(A)$.
 (ii) If $f, g \in \mathrm{LAff}_b(T(A))_{++}$, then $(f + g)(\tau) = f(\tau) + g(\tau)$, $\tau \in T(A)$.
(iii) If $[p] \in V(A)$ and $f \in \mathrm{LAff}_b(T(A))_{++}$, then $[p] + f = \hat{p} + f$, where addition on the right-hand side is as elements in $\mathrm{LAff}_b(T(A))_{++}$.

In a similar manner, we define a relation $\leq$ on $\widetilde{W}(A)$:

 (iv) $\leq$ restricts to the usual order on $V(A)$,
  (v) if $f, g \in \mathrm{LAff}_b(T(A))_{++}$, then $f \leq g$ if and only if $f(\tau) \leq g(\tau)$ for every $\tau \in T(A)$,
 (vi) if $[p] \in V(A)$ and $g \in \mathrm{LAff}_b(T(A))$, then $f \leq [p]$ if and only if $f(\tau) \leq \hat{p}(\tau)$ for every $\tau \in T(A)$,
(vii) if $[p] \in V(A)$ and $g \in \mathrm{LAff}_b(T(A))$, then $[p] \leq f$ if and only if $\hat{p}(\tau) \leq f(\tau)$ for every $\tau \in T(A)$.

14.3.20. Let S and T be partially ordered semigroups. An *order-embedding* is a semigroup map $\varphi : S \to T$ satisfying $\varphi(x) \leq \varphi(y)$ if and only if $x \leq y$. If an order-embedding φ is surjective, we call φ an *order-isomorphism*.

We will show that if A is a simple, separable, unital, nuclear C*-algebra whose Cuntz semigroup $W(A)$ is almost unperforated and weakly divisible (Definition 14.3.35 below), then there is an order-isomorphism $\varphi : W(A) \to \widetilde{W}(A)$.

14.3.21. Lemma. *Let A be a C*-algebra and $\tau \in T(A)$ a faithful tracial state. Let $a \in A_+$. Then for any $\epsilon < \delta$ with $\epsilon, \delta \in \mathrm{sp}(a)$, we have*

$$d_\tau((a - \delta)_+) < d_\tau((a - \epsilon)_+).$$

Proof. Of course, since $\epsilon < \delta$, we have $(a-\delta)_+ \precsim (a-\epsilon)_+$, whence $d_\tau((a-\delta)_+) \leq d_\tau((a-\epsilon)_+)$. The point is to show that this inequality is strict. By the previous proposition, d_τ induces a measure μ on $\mathrm{sp}(a)$ satisfying $\mu(U) = d_\tau(f(a))$ for any $f \in C_0(\mathrm{sp}(a))$ with $\mathrm{coz}(f) = U$. Let $V := \{t \in (\delta, \epsilon] \subset \mathrm{sp}(a) \mid (t-\epsilon)_+ > 0\}$. Then $\mathrm{coz}((t-\delta)_+) = \mathrm{coz}((t-\epsilon)_+) \sqcup V$. Since $\epsilon \in \mathrm{sp}(a) \cap V$, there exists $f \in C_0(\mathrm{sp}(a))_+$ with $\mathrm{coz}(f) \subset V$ and $f(\epsilon) \neq 0$. Since $f \neq 0$ and τ is faithful, $\tau(f) > 0$. Now

$$\mu(V) \geq \mu(\mathrm{coz}(f)) = d_\tau(f) = \lim_{n\to\infty} \tau(f^{1/n}) > 0,$$

so we have

$$d_\tau((a-\epsilon)_+) - d_\tau((a-\delta)_+) = \mathrm{coz}((t-\delta)_+) - \mathrm{coz}((t-\delta)_+) = \mu(V) > 0,$$

showing that the inequality is strict. $\qquad\square$

14.3.22. Recall that the subsemigroup of purely positive elements $W(A)_+$ consists of equivalence classes not containing projections. Let us say that a positive element $a \in A_+$ is *purely positive* if a is not Cuntz equivalent to a projection (so that $[a] \in W(A)_+$).

Lemma. *Let A be a simple C*-algebra with $T(A) \neq 0$ and strict comparison. Let $a, b \in A_+$ with a purely positive. Suppose that $d_\tau(a) \leq d_\tau(b)$ for every $\tau \in T(A)$. Then $a \precsim b$.*

Proof. By Lemma 14.2.12, since $[a] \in W(A)_+$, we know that $0 \in \mathrm{sp}(a)$ and 0 is not an isolated point. Thus there is a strictly decreasing sequence of positive real numbers $(\epsilon_n)_{n\in\mathbb{N}} \subset \mathrm{sp}(a)$ which converge to zero. Since A is simple, every tracial state is faithful and by the previous lemma, we have $d_\tau((a-\epsilon_n)_+) < d_\tau(a) \leq d_\tau(b)$ for every $n \in \mathbb{N}$. By strict comparison, this in turn implies that $(a-\epsilon_n)_+ \precsim b$ for every $n \in \mathbb{N}$. Lemma 14.1.8 tells us that $\{x \in A_+ \mid x \precsim b\}$ is norm closed, so, since $\|(a-\epsilon_n)_+ - a\| \to 0$ as $n \to \infty$, we also have $a \precsim b$. $\qquad\square$

14.3.23. Now we will show that when A is a simple C*-algebra with stable rank one and strict comparison, then there is an order-embedding of $W(A)_+$ into $\mathrm{LAff}_b(T(A))_{++}$. This was first shown in [92, Proposition 3.3].

Proposition. *Let A be a simple C*-algebra with stable rank one. Then the map*

$$\iota : W(A)_+ \to \mathrm{LAff}_b(T(A))_{++},$$

defined by $\iota([a])(\tau) = d_\tau(a)$ is a semigroup homomorphism. Moreover, if A has strict comparison, then ι is an order-embedding.

Proof. Since A is simple, every tracial state is faithful. Thus $\iota([a])$ is a strictly positive function for every $a \in A$. Thus $\mathrm{im}(\iota) \subset \mathrm{LAff}_b(T(A))$, as claimed. We saw in Proposition 14.2.13 that $W(A)_+$ is a semigroup. If $[a], [b] \in W(A)_+$ then, for every $\tau \in T(A)$,

$$\iota([a] \oplus [b])(\tau) = \iota([a \oplus b])(\tau) = d_\tau(a \oplus b) = d_\tau([a]) + d_\tau([b]) = \iota([a])(\tau) + \iota([b])(\tau),$$

which shows that ι is a semigroup homomorphism.

If A has strict comparison, then $[a] \leq [b]$ means $a \precsim b$ which, by Lemma 14.3.22, holds if and only $d_\tau(a) \leq d_\tau(b)$. Thus if A has strict comparison, ι is an order-embedding. □

14.3.24. We would like to tie together Proposition 14.3.23 and Corollary 14.2.8 to find an order embedding of the whole Cuntz semigroup $W(A)$ into $\widetilde{W}(A)$. We need one more lemma about the interaction between elements in $V(A)$ and $W(A)_+$ with respect to the order structure on $W(A)$.

Lemma. *Let A be a simple, exact C*-algebra with strict comparison. Suppose that $p \in A$ is a projection and $a \in A_+$ is a purely positive element. Then $[p] \leq [a]$ if and only if $d_\tau(p) < d_\tau(a)$ for every $\tau \in T(A)$.*

Proof. Since A has strict comparison, the fact that $d_\tau(p) < d_\tau(a)$ for every $\tau \in T(A)$ implies $[p] \leq [a]$ is automatic. So we show the other implication. The proof is by contradiction. Let $[p] \leq [a]$ and suppose that there exists $\tau \in T(A)$ such that $d_\tau(a) \leq d_\tau(p)$. Let $0 < \epsilon < 1$ be given. Then, by Theorem 14.1.9 (iii) there is a $\delta > 0$ such that

$$(p - \epsilon)_+ \precsim (a - \delta)_+.$$

Since p is a projection and therefore $\mathrm{sp}(p) \in \{0, 1\}$, there exists $\lambda > 0$ such that $(p - \epsilon)_+ = \lambda p$. In particular, $p \sim (p - \epsilon)_+$. Thus

$$d_\tau(p) = d_\tau((p - \epsilon)_+) \leq d_\tau((a - \delta)_+).$$

By assumption $d_\tau(a) \leq d_\tau(p)$, so, applying Lemma 14.3.21 for the first inequality,

$$d_\tau((a - \delta)_+) < d_\tau(a) \leq d_\tau(p).$$

Hence $d_\tau(p) < d_\tau(p)$, a contradiction. □

14.3.25. Theorem. *Let A be a separable, simple, unital, exact C*-algebra with stable rank one and strict comparison. Then there is an order-embedding*

$$\varphi : W(A) \to \widetilde{W}(A),$$

such that $\varphi|_{V(A)} = \mathrm{id}_{V(A)}$ and $\varphi|_{W(A)_+} = \iota$.

Proof. Since A has stable rank one, there is a decomposition $W(A) = V(A) \sqcup W(A)_+$ into subsemigroups $V(A)$ and $W(A)_+$ (Proposition 14.2.13). Thus the map φ is well defined. We show that φ is an order-embedding. This amounts to verifying that the image of φ satisfies conditions (iv)–(vii) of 14.3.19. We leave this verification as an exercise for the reader. □

14.3.26. Let A be a C*-algebra. An element $a \in A_+$ is called *strictly positive* if $\overline{aA} = A$. If a is strictly positive, then we also have $\overline{aAa} = A$ (exercise). Thus, by Theorem 3.2.7 if A is separable, this just says if a is strictly positive then the hereditary C*-subalgebra generated by a is all of A.

14.3.27. Lemma. *Let A be a separable, unital C*-algebra. Suppose that $(a_n)_{n\in\mathbb{N}}$ is sequence of positive elements of A satisfying*

$$\overline{a_1 A a_1} \subset \overline{a_2 A a_2} \subset \overline{a_3 A a_3} \subset \cdots .$$

Let $B := \overline{\bigcup_{n=1}^{\infty} \overline{a_n A a_n}}$. Then if a_∞ is a strictly positive element of B we have $[a_\infty] = \sup_{n\in\mathbb{N}}[a_n]$. Furthermore, $d_\tau(a_\infty) = \sup_{n\in\mathbb{N}} d_\tau(a_n)$ for every $\tau \in T(A)$.

Proof. First, let us show that $[a_n] \leq [a_\infty]$, that is, that for every $n \in N$ and every $\epsilon > 0$ there exists $r \in A$ such that $\|ra_\infty r^* - a_n\| < \epsilon$. This is satisfied if $a_n \in \overline{a_\infty A a_\infty}$. Let us show that $\overline{a_\infty A a_\infty} = B$. In that case, because $a_n \in \overline{a_n A a_n}$ by Theorem 3.2.7, the result follows. Let $a \in A$ and suppose that $b, c \in A_\infty$. Then, for every $n \in \mathbb{N}$ there are $x_n, y_n \in \overline{a_n A a_n}$ such that $\lim_{n\to\infty} \|x_n - b\| = 0$ and $\lim_{n\to\infty} \|y_n - c\| = 0$. Then $x_n a^{1/2} \to ba^{1/2}$ and $a^{1/2} y_n \to a^{1/2}c$ so $x_n a y_n \to bac$. Since $x_n a y_n \in \overline{a_n A a_n}$ it follows that $bac \in B$. Since a_∞ is a strictly positive element of B, we have $B = \overline{a_\infty B a_\infty}$. Thus $\overline{a_\infty A a_\infty} = BAB \subset B$. Clearly $B \subset BAB$, so indeed $\overline{a_\infty A a_\infty} = B$.

Now suppose that there is $[b] \in W(A)$ such that $[a_n] \leq [b]$ for every n. For every $n \in \mathbb{N}$ choose $\epsilon_n > 0$ such $\epsilon_{n+1} < \epsilon_n$ and find $x_n \in \overline{a_n A a_n}$ such that $\|x_n - a_\infty\| < \epsilon_n$. Then $(a_\infty - \epsilon_n)_+ \precsim x_n$ by Proposition 14.1.7, so

$$[(a_\infty - \epsilon_n)_+] \leq [x_n] \leq [a_n] \leq [b].$$

Theorem 14.1.9 (ii) now implies that $[a_\infty] \leq [b]$, so we see that

$$[a_\infty] = \sup_{n\in\mathbb{N}}[a_n].$$

Finally, since $\lim_{n\to\infty} x_n = a_\infty$, for any $\tau \in T(A)$ the dimension function d_τ is lower semicontinuous so we have

$$\sup_{n\in\mathbb{N}} d_\tau(a_n) \leq d_\tau(a_\infty) \leq \lim\inf_{n\to\infty} d_\tau(x_n) \leq \lim\inf_{n\to\infty} d_\tau(a_n) = \sup_{n\in\mathbb{N}} d_\tau(a_n),$$

which shows that final statement. $\qquad\square$

14.3.28. The next lemma looks at what happens when A also has stable rank one. In this case, we can read off supremum from a Cuntz-increasing sequence of positive elements. In what follows, for any positive element $a \in A_+$ we denote $\mathrm{Her}(a) := \overline{aAa}$.

Lemma. *Let A be a separable unital C*-algebra with stable rank one. Let $(a_n)_{n\in\mathbb{N}} \subset A$ such that $[a_1] \leq [a_2] \leq \cdots$. Then there exists $a_\infty \in A_+$ such that $[a_\infty] = \sup_{n\in\mathbb{N}}[a_n]$. Furthermore, $d_\tau(a_\infty) = \sup_{n\in\mathbb{N}} d_\tau([a_n])$ for every $\tau \in T(A)$.*

Proof. Let $\epsilon_1 = 1/2$. For $n \geq 1$, apply Theorem 14.1.9 (iii) to a_{n-1}, a_n and $n^{-1}\epsilon_{n-1}$ to find $\delta_n > 0$ such that

$$(a_{n-1} - n^{-1}\epsilon_{n-1})_+ \precsim (a_n - \delta_n)_+.$$

Let $\epsilon_n := \min\{1/n, \epsilon_{n-1}, \delta_n\}$. Then

$$(a_j - \epsilon_j/k)_+ \precsim (a_n - \epsilon_n)_+$$

for every $1 \leq j \leq n-1$ and $1 \leq k \leq n$.

Now, by our choice of ϵ_n, we have that $(a_2 - \epsilon_1/2)_+ \precsim (a_2 - \epsilon_2)_+$, so by Proposition 14.1.10, since A has stable rank one, there exists a unitary $u \in A$ such that, for every $\delta > 0$, $u^*((a_2 - \epsilon_1/2)_+ - \delta)_+ u^* \in \mathrm{Her}((a_2 - \epsilon_2)_+)$. In particular, this is true for $\delta = \epsilon_1/2$. Let u_1 denote the associated unitary. Then $((a_2 - \epsilon_1/2)_+ - \epsilon_1/2)_+ \in u_1 \mathrm{Her}((a_2 - \epsilon_2)_+) u_1^*$, and hence, by Exercise 14.5.1, $(a_2 - \epsilon_1)_+ \subset u_1 \mathrm{Her}((a_2 - \epsilon_2)_+) u_1^*$. Thus

$$\mathrm{Her}((a_2 - \epsilon_1)_+) \subset u_1 \mathrm{Her}((a_2 - \epsilon_2)_+) u_1^*.$$

Iterating this, we get

$$(a_2 - \epsilon_1)_+ \subset u_1 \mathrm{Her}((a_2 - \epsilon_2)_+) u_1^* \subset u_1 u_2 \mathrm{Her}((a_3 - \epsilon_3)_+) u_2^* u_1^* \subset \cdots$$
$$\cdots \subset (u_1 u_2 \cdots u_{n-1}) \mathrm{Her}((a_n - \epsilon_n)_+) (u_1 u_2 \cdots u_{n-1})^* \subset \cdots$$

For $n \in \mathbb{N}$, Let $b_n := (u_1 u_2 \cdots u_{n-1})(a_n - \epsilon_n)_+ (u_1 u_2 \cdots u_{n-1})^* \in A_+$. Note that $[b_n] = [(a_n - \epsilon_n)_+]$. Then

$$\overline{b_1 A b_1} \subset \overline{b_2 A b_2} \subset \overline{b_3 A b_3} \subset \cdots,$$

so we may apply Lemma 14.3.27 to find $a_\infty \in A$ such that

$$[a_\infty] = \sup_{n \in \mathbb{N}}[b_n] = \sup_{n \in \mathbb{N}}[(a_n - \epsilon_n)_+],$$

which satisfies $d_\tau(a_\infty) = \sup_{n \in \mathbb{N}} d_\tau((a_n - \epsilon_n)_+) \leq \sup_{n \in \mathbb{N}} d_\tau(a_n)$, for every tracial state $\tau \in T(A)$.

We will show that in fact we also have $[a_\infty] = \sup_{n \in \mathbb{N}}[a_n]$. Let $n \in \mathbb{N}$. For any $m < n$ we have, by construction,

$$[(a_m - \epsilon_m/(n-1))_+] \leq [(a_n - \epsilon_{n+1})_+] \leq [a_\infty].$$

Thus, as $n \to \infty$ we have $[(a_m - \epsilon)_+] \leq [a_\infty]$ for any $\epsilon > 0$. By Theorem 14.1.9 (iii) this means $[a_m] \leq [a_\infty]$. So $[a_\infty] = \sup_{m \in \mathbb{N}}[(a_m - \epsilon_m)_+] \leq \sup_{m \in \mathbb{N}}[a_m] \leq [a_\infty]$, whence $[a_\infty] = \sup_{m \in \mathbb{N}}[a_m]$. It follows that, for any $\tau \in T(A)$ we have $d_\tau(a_\infty) = \sup_{n \in \mathbb{N}} d_\tau(a_n)$, as required. $\qquad\square$

Now we extend this result to the whole Cuntz semigroup.

14.3.29. Theorem. *Let A be a unital separable C*-algebra with stable rank one. Let $\{[a_n]\}_{n \in \mathbb{N}} \subset W(A)$ be a bounded sequence, that is, there exists $k \in \mathbb{N}$ such that $[a_n] \leq k[1_A]$ for every $n \in \mathbb{N}$. Then there exists $a_\infty \subset M_\infty(A)_+$ such that $[a_\infty] = \sup_{n \in \mathbb{N}}[a_n]$ and, for every $\tau \in T(A)$, $d_\tau(a_\infty) = \sup_{n \in \mathbb{N}} d_\tau(a_n)$.*

Proof. Let $\{[a_n]\}_{n\in\mathbb{N}} \subset W(A)$ be sequence such that, for some $k \in \mathbb{N}$, $[a_n] \leq k[1_A]$ for every $n \in \mathbb{N}$. Proceeding as in the proof of the previous lemma, we may find a sequence of strictly positive real numbers $(\epsilon_n)_{n\in\mathbb{N}}$ such that

$$[(a_n - \epsilon_n)_+] \leq [(a_{n+1} - \epsilon_{n+1})_+],$$

for every $n \in \mathbb{N}$ and such that, whenever there exists $[b]$ with $[(a_n - \epsilon_n)_+] \leq [b]$ for every $n \in \mathbb{N}$ then also $[a_n] \leq [b]$, for every $n \in \mathbb{N}$. Now, since $[a_n] \leq k[1_A] = [1_A \otimes 1_{M_k}]$, using Proposition 14.1.7 we find $c_n \in M_\infty(A)$ such that

$$(a_n - \epsilon_n)_+ = c_n(1_A \otimes 1_{M_k})c_n^*.$$

Put

$$x_n := (1_A \otimes 1_{M_k})c_n c_n^*(1_A \otimes 1_{M_k}).$$

Then we have $x_n \in M_k(A)$ and $[x_n] = [(a_n-\epsilon_n)_+] \leq [a_n] \leq [a_{n+1}]$ for every $n \in \mathbb{N}$. Since A has stable rank one, so does $M_k(A)$ (Proposition 12.1.13). Thus we may apply Lemma 14.3.28: there exists $a_\infty \in M_k(A)_+$ such that $[a_\infty] = \sup_{n\in\mathbb{N}}[x_n]$. Thus $[a_\infty]$ is the supremum of the sequence $\{[(a_n - \epsilon_n)_+]\}_{n\in\mathbb{N}}$. By choice of the sequence $(\epsilon_n)_{n\in\mathbb{N}}$, this in turn implies that $[a_\infty] = \sup_{n\in\mathbb{N}}[a_n]$. As in the proof of Lemma 14.3.28, we also have $d_\tau(a_\infty) = \sup_{n\in\mathbb{N}} d_\tau(a_n)$ for every $\tau \in T(A)$. $\square$

14.3.30. Corollary. *Let A be a separable, unital C^*-algebra with stable rank one. Suppose that $x \in W(A)$ satisfies $x \leq [1_A]$. Then there exists $a \in A_+$ such that $x = [a]$.*

Proof. Let $k \in \mathbb{N}$ and $b \in M_k(A)_+$ satisfy $x = [b]$. For every $n \in \mathbb{N}$, find $c_n \in M_\infty(A)$ satisfying

$$(b - 1/n)_+ = c_n 1_a c_n^*.$$

Then $a_n := 1_A c_n^* c_n 1_A \in A$ and satisfies $a_n \sim (b - 1/n)_+$. Furthermore, the sequence $\{[a_n]\}_{n\in\mathbb{N}}$ is increasing, so we apply Lemma 14.3.28 to find $a \in A_+$ satisfying $[a] = \sup_{n\in\mathbb{N}}[a_n]$. Then

$$[a] = \sup_{n\in\mathbb{N}}[a_n] = \sup_{n\in\mathbb{N}}[(b - 1/n)_+] = [b] = x,$$

as required. $\square$

14.3.31. Corollary. *Let A be a separable unital C^*-algebra with stable rank one. Let $\{[a_n]\}_{n\in\mathbb{N}} \in W(A)$ be a bounded increasing sequence with supremum $[a]$. Then there is a projection $p \in M_\infty(A)$ with $[p] = [a]$ if and only if there exists $n_0 \in \mathbb{N}$ such that $[a_n] = [p]$ for every $n \geq n_0$.*

Proof. The "if" direction is clear. Suppose that we have such a sequence with $\sup_{n\in\mathbb{N}}[a_n] = [a] = [p]$. Without loss of generality, assume that $(a_n)_{n\in\mathbb{N}} \subset A_+$, $a \in A_+$ and $p \in A_+$. For any $n \in \mathbb{N}$ we have $a_n \precsim p$. As in the proof of Lemma 14.3.27 we can find a sequence of $\epsilon_n > 0$ decreasing to zero with

$$[(p - \epsilon_n)_+] \leq [a_n] \leq [p].$$

Of course, for sufficiently large n, we have $[(p-\epsilon_n)_+] = p$. Let $n_0 \in \mathbb{N}$ be sufficiently large that $[(p - \epsilon_n)_+] = [p]$ for every $n \geq n_0$. Then for every $n \geq n_0$ we have $[p] \leq [a_n] \leq [p]$, in which case $[p] = [a_n]$. □

The next proposition follows from Proposition 14.3.2.

14.3.32. Proposition. *Let X be a compact convex set and $f \in \mathrm{LAff}(X)_{++}$. Then there is a strictly increasing sequence of continuous functions $(f_n)_{n\in\mathbb{N}} \subset \mathrm{Aff}(X)$ such that $f = \sup_{n\in\mathbb{N}} f_n$.*

To prove the two lemmas before the final theorem of this chapter, we require the next result. We state it here without proof because the proof requires knowledge of things we have not covered, such as von Neumann algebra pre-duals. The result can be found in [19, Corollary 3.10].

14.3.33. Lemma. *Let A be a separable, unital C*-algebra with $T(A) \neq \emptyset$. Then for every continuous $f \in \mathrm{Aff}_b(T(A))$ there exists $a \in A_{sa}$ such that $f(\tau) = \tau(a)$ for every $\tau \in T(A)$. Furthermore, A is simple and $f(\tau) > 0$ for every $\tau \in T(A)$, then we can choose $a \in A_+$.*

14.3.34. Lemma. *Let X be a locally compact metric space and let $A = C_0(X)$. Let $U \subset X$ be an open subset and denote by χ_U the indicator function. Then $\chi_U \subset M$ where M is the enveloping von Neumann algebra of A. Let $g \in C_0(X)$ be a positive function such that $g(x) > 0$ if and only if $x \in U$. Then, for any $\tau \in T(A)$,*

$$d_\tau([g]) = \tau_M(\chi_U),$$

where τ_M denotes the extension of τ to M.

Proof. Exercise. □

14.3.35. Definition. Let A be a separable C*-algebra. We say $W(A)$ is *weakly divisible* if for every $x \in W(A)_+$ and $n \in \mathbb{N}$, there exists $y \in W(A)_+$ such that $ny \leq x \leq (n+1)y$.

14.3.36. Lemma. *Let A be a simple, separable, unital C*-algebra with nonempty tracial state space $T(A)$. Suppose that $W(A)$ is weakly divisible. Let*

$$\iota : W(A)_+ \mapsto \mathrm{LAff}_b(T(A))_{++}$$

be given by $\iota([a]) = d_\tau(a)$. Then for every $f \in \mathrm{Aff}(T(A))_{++}$ and $\epsilon > 0$ there exists $x \in W(A)_+$ such that $|f(\tau) - \iota(x)(\tau)| < \epsilon$ for every $\tau \in T(A)$.

Proof. Let $f \in \mathrm{Aff}(T(A))$. Since A is simple, separable and unital with $T(A) \neq \emptyset$, we may apply Lemma 14.3.33 to find $a \in A_+$ such that $f(\tau) = \tau(a)$ for every $\tau \in A$. For an open set $U \subset \mathrm{sp}(a)$, let χ_U denote the indicator function. We can find $n \in \mathbb{N}$ and, for $1 \leq i \leq n$ rational numbers $r_i = p_i/q_i$ for $p_i, q_i \in \mathbb{N}$ along with open subsets $U_i \subset \mathrm{sp}(a)$ such that

$$\sup_{t \in \mathrm{sp}(a)} \left| t - \sum_{i=1}^{n} r_i \chi_{U_i}(t) \right| < \epsilon/2.$$

For every $1 \leq i \leq n$, choose positive functions $g_i \in C_0(\mathrm{sp}(a))$ such that $g_i(t) > 0$ if and only if $t \in U_i$. From the previous lemma, we have $d_\tau([g_i]) = \tau(\chi_{U_i})$ for any $\tau \in T(A)$. Let $a_i = g_i(a)$. Then

$$\left| \sum_{i=1}^{n} r_i d_\tau([a_i]) - \tau(a) \right| < \epsilon/2.$$

By assumption, $W(A)$ is weakly divisible, so we can find $x_i \in W(A)_+$ such that $q_i x_i \leq [a_i] \leq (q_i + 1)x_i$. Define

$$x := \sum_{i=1}^{n} p_i x_i.$$

We claim that x satisfies the conclusions of the lemma. For every $\tau \in T(A)$ and every $1 \leq i \leq n$ we have

$$\frac{1}{q_i + 1} d_\tau([a_i]) \leq d_\tau(x_i) \leq \frac{1}{q_i} d_\tau([a_i]).$$

Multiplying by p_i and summing over i we see that

$$\sum_{i=1}^{n} \frac{p_i}{q_i + 1} d_\tau([a_i]) \leq d_\tau(x) \leq \sum_{i=1}^{n} r_i d_\tau([a_i]).$$

Replacing p_i and q_i by kp_i and kq_i, respectively, for sufficiently large k, we have, for every $\tau \in T(A)$,

$$\left| d_\tau(x) - \sum_{i=1}^{n} r_i d_\tau([a_i]) \right| < \epsilon/2.$$

It follows that

$$|\iota(x)(\tau) - f(\tau)| = |d_\tau(x) - \tau(a)|$$

$$= \left| d_\tau(x) - \sum_{i=1}^{n} r_i d_\tau([a_i]) + \left| \sum_{i=1}^{n} r_i d_\tau([a_i]) - \tau(a) \right| \right.$$

$$< \left| d_\tau(x) - \sum_{i=1}^{n} r_i d_\tau([a_i]) \right| + \epsilon/2$$

$$< \epsilon/2 + \epsilon/2 = \epsilon,$$

so x satisfies the conclusions, as claimed. $\qquad\square$

14.3.37. Lemma. *Let $f \in \mathrm{LAff}_b(T(A))_{++}$ and $\delta > 0$ such that $f(\tau) \geq \delta$ for every $\tau \in T(A)$. Then there exists a sequence $(f_n)_{n \in \mathbb{N}} \subset \mathrm{Aff}(T(A))$ satisfying the following:*

(i) $\sup_{n \in \mathbb{N}} f_n(\tau) = f(\tau)$ *for every $\tau \in T(A)$,*

(ii) *for every $\tau \in T(A)$ and every $n \in \mathbb{N}$ we have*

$$f_{n+1}(\tau) - f_n(\tau) \geq \frac{\delta}{2}\left(\frac{1}{n} - \frac{1}{n-1}\right).$$

Proof. Since δ satisfies $f(\tau) \geq \delta$ for every $\tau \in T(A)$ and $\in \mathrm{LAff}_b(T(A))_{++}$, also the function $f - \frac{\delta}{2} \in \mathrm{LAff}_b(T(A))_{++}$. Thus, by Proposition 14.3.32, $f - \frac{\delta}{2}$ is the supremum of a strictly increasing sequence $(g_n)_{n \in \mathbb{N}} \subset \mathrm{Aff}(T(A))_{++}$, that is,

$$\sup_{n \in \mathbb{N}} g_n(\tau) = f(\tau) - \frac{\delta}{2},$$

for every $\tau \in T(A)$. Define

$$f_n(\tau) := g_n(\tau) + \frac{\delta}{2} - \frac{\delta}{2n}.$$

Then the sequence $(f_n)_{n \in \mathbb{N}}$ satisfies properties (i) and (ii) (exercise).   □

Finally, we come to the main theorem of this chapter: provided a C*-algebra A has good enough structural properties (made precise in the statement of the theorem), then $W(A) \cong \widetilde{W}(A)$ and the isomorphism preserves the Murray–von Neumann semigroup and sends the purely positive elements to functions in $\mathrm{LAff}_b(T(A))_{++}$. In the next chapter, we will see that the properties of A we demand for the theorem to hold are satisfied by a large class of simple, separable, nuclear C*-algebras, namely those that absorb the Jiang–Su algebra tensorially. This theorem tells us that in many cases, the Cuntz semigroup together with the K_1-group contains the same information as the Elliott invariant.

14.3.38. Theorem. *Let A be a simple, separable, unital, exact C*-algebra with stable rank one. Suppose that A has strict comparison and $W(A)$ is weakly divisible. Then the map*

$$\iota : W(A)_+ \to \mathrm{LAff}_b(T(A))_{++}, \quad \iota([a])(\tau) = d_\tau(a)$$

is surjective. Thus the map $\varphi : W(A) \to \widetilde{W}(A)$ defined in Theorem 14.3.25 is an order-isomorphism.

Proof. Let $f \in \mathrm{LAff}_b(T(A))_{++}$ and let $\delta > 0$ satisfy $f(\tau) \geq \delta$ for every tracial state $\tau \in T(A)$. By Lemma 14.3.37, there is a sequence $(f_n)_{n \in \mathbb{N}} \subset \mathrm{Aff}(T(A))$ such that $\sup_{n \in \mathbb{N}} f_n(\tau) = f(\tau)$ for every $\tau \in T(A)$, and for every $n \in \mathbb{N}$,

$$f_{n+1}(\tau) - f_n(\tau) \geq \frac{\delta}{2}\left(\frac{1}{n} - \frac{1}{n-1}\right), \quad \text{for every } \tau \in T(A).$$

Let $0 < \epsilon_n < \frac{\delta}{4}(\frac{1}{n} - \frac{1}{n-1})$. By Lemma 14.3.36 there exists $x_n \in W(A)_+$ such that $|f_n(\tau) - \iota(x_n)(\tau)| < \epsilon_n$ for every $\tau \in T(A)$. Then

$$\iota(x_n)(\tau) < f_n(\tau) + \epsilon_n \leq f_{n+1}(\tau) - \epsilon_n < \iota(x_{n+1})(\tau),$$

for every $\tau \in T(A)$. Since A has strict comparison, this implies $x_n \leq x_{n+1}$ for every $n \in \mathbb{N}$. By Theorem 14.3.29, there exists $x \in W(A)$ with $x := \sup_{n \in \mathbb{N}} x_n$ and that $\iota(x)(\tau) = f(\tau)$ for every $\tau \in T(A)$. Finally, since $(x_n)_{n \in \mathbb{N}}$ is strictly increasing, x cannot be an element of $V(A)$. Hence $x \in W(A)_+$, which shows that ι is surjective.   □

14.4   Further remarks on the Cuntz semigroup

As it was for K-theory and the Elliott invariant, we have seen that the Cuntz semigroup is an isomorphism invariant. Furthermore, in the previous section, we saw that in certain cases, two of the same ingredients go into building both the Elliott invariant and the Cuntz semigroup, namely the Murray–von Neumann semigroup of projections and the tracial state space. The Cuntz semigroup as we have presented it, however, lacks one of the properties of K-theory that makes it such a useful invariant: the map that takes C*-algebras to abelian semigroups by mapping a C*-algebra A to its Cuntz semigroup $W(A)$ does not respect inductive limits. This can make the computation of the Cuntz semigroup difficult.

In [27], Coward, Elliott and Ivanescu set out to correct this problem. Recall that the first result in the second section of this chapter was that Murray–von Neumann subequivalence and Cuntz subequivalence agree on projections. Extending the ideas of a Murray–von Neumann-type equivalence from projections to positive elements was a main motivation for introducing the Cuntz equivalence relation in the first place. One reason for this is that for a von Neumann algebra M, an understanding of projections yields a great deal of information about the structure of M. A von Neumann algebra M always contains many projections. For example, we saw in Proposition 5.2.11 that if $a \in M$ then so are the support projections of a. When we try to adapt this to the C*-algebra setting, however, we run into the problem that we may not have any nontrivial projections at all! Indeed, we will meet such C*-algebras in the next chapter.

The approach to K-theory of Chapter 12 constructed the Murray–von Neumann semigroup $V(A)$ of a C*-algebra A via equivalence classes of projections over matrix algebras. Then $K_0(A)$ was defined to be the Grothendieck group of $V(A)$. An equivalent construction can be given in terms of finitely generated projective A-modules, or equivalently finitely generated projective Hilbert A-modules (modules endowed with an A-valued "inner product", subject to certain properties). Loosely speaking, if M is a finitely generated projective A-module, then there exists an A-module N such that $M \oplus N \cong A^n \, (= A \oplus A \oplus \cdots \oplus A)$, which we can use to find an idempotent $e_m \in M_n(A)$ (that is, $e_M^2 = e_M$) such that $e(A^n) \cong M$. From e

we construct a projection $p_M \in M_n(A)$ satisfying $p_M e_M = p_M$ and $e_M p_M = e_M$. Then M and M' are isomorphic if and only if $p_M \sim_{M_vN} p_{M'}$.

It is this A-module picture that is generalised by Coward, Elliott and Ivanescu [27]. They adapt Cuntz equivalence to the setting of countably generated Hilbert C*-modules and use this to construct a semigroup we will denote by $\mathrm{Cu}(A)$ and show that $\mathrm{Cu}(A)$ is an object in a category called **Cu**, whose objects are partially ordered semigroups, equipped with some extra structure.

14.4.1. Let $(P, \leq)$ be a partially ordered semigroup. We define the *compact containment* relation $\ll$ as follows. For $x, y \in P$, write $x \ll y$ if, whenever $y_1 \leq y_2 \leq \cdots$ is an increasing sequence with $\sup_n y_n \geq y$, then there exists $n_0 \in \mathbb{N}$ such that $x_n \leq y_n$ for every $n \geq n_0$.

14.4.2. Definition (The category **Cu**). An object in **Cu** is a partially ordered abelian semigroup $(P, \leq)$ with zero element satisfying the following axioms:

(i) If $x_1 \leq x_2$ and $y_1 \leq y_2$ then $x_1 + x_2 \leq y_1 + y_2$.

(ii) Every increasing sequence (or equivalently, countably upward directed subset) in P has a supremum.

(iii) For every $x \in P$ the set $x^{\ll} := \{y \in P \mid y \ll x'\}$ is upward directed with respect to $\leq$ and $\ll$ and contains a sequence $(x_n)_{n \in \mathbb{N}}$ such that $x_1 \ll x_2 \ll \cdots$ and $x = \sup_{n \in \mathbb{N}} x_n$.

(iv) If P_1 and P_2 are countable upward directed sets then $P_1 + P_2$ is upwarded directed and $\sup_{n \in \mathbb{N}}(P_1 + P_2) = \sup_{n \in \mathbb{N}} P_1 + \sup_{n \in \mathbb{N}} P_2$.

(v) If $x_1 \ll x_2$ and $y_1 \ll y_2$ then $x_1 + x_2 \ll y_1 + y_2$.

A morphism in the category **Cu** is a semigroup morphism $\varphi : P \to S$ between objects $P, S \in$ **Cu** satisfying the following

(i) $\varphi(0) = 0$,

(ii) if $x \leq y$ then $\varphi(x) \leq \varphi(y)$,

(iii) if $P_1 \subset P$ is a countable upward directed subset then $\varphi(\sup_{n \in \mathbb{N}} p_n) = \sup_{n \in \mathbb{N}}(\varphi(p_n))$,

(iv) if $x \ll y$ then $\varphi(x) \ll \varphi(y)$.

We call a semigroup $P \in$ **Cu** a **Cu**-semigroup, and a morphism φ between **Cu**-semigroups is called a **Cu**-morphism.

14.4.3. We saw in Section 12.2 that it is possible to define inductive limits in the category of (ordered) abelian groups. In [27], it is shown that this is also possible in the category **Cu**.

Theorem. *Suppose that P_n are* **Cu**-*semigroups and for every $n \in \mathbb{N}$ there exists a* **Cu**-*morphism $\varphi_n : P_n \to P_{n+1}$. Then the inductive limit $P = \varinjlim(P_n, \varphi_n)$ exists and is an element of* **Cu**.

14.4.4. Theorem. *Let A and B be separable C*-algebras. Then*

(i) $\mathrm{Cu}(A) \in \mathbf{Cu}$ *and any* *-*homomorphism* $\varphi : A \to B$ *induces a* **Cu**-*morphism* $\varphi_* : \mathrm{Cu}(A) \to \mathrm{Cu}(B)$,

(ii) *if* $A = \varinjlim(A_n, \varphi_n)$ *then* $\mathrm{Cu}(A) \cong \varinjlim(\mathrm{Cu}(A), \varphi_*)$,

(iii) $\mathrm{Cu}(A) \cong W(A \otimes \mathcal{K})$.

In view of the above, we often call $\mathrm{Cu}(A) = W(A \otimes \mathcal{K})$ the *stabilised Cuntz semigroup* of A.

14.5 Exercises

14.5.1. Let A be a separable C*-algebra and suppose $\epsilon_1, \epsilon_2 > 0$. Show that for any $a \in A_+$ we have

$$(a - (\epsilon_1 + \epsilon_2))_+ = ((a - \epsilon_1)_+ - \epsilon_2)_+$$

14.5.2. Let p and q be projections in a unital C*-algebra A. Show that $p \precsim q$ if and only if p is Murray von–Neumann equivalent to a projection $r \leq q$.

14.5.3. Complete the proof of Theorem 14.3.25 by showing that the order structure on $\varphi(W(A)) \subset \widetilde{W(A)}$ satisfies (iv) to (vii) of 14.3.19.

14.5.4. Let A be a C*-algebra and $a \in A_+$.

(i) Show that a is strictly positive if and only if $\overline{aAa} = A$.

(ii) If $A = C(X)$ show that $f \in A$ is a strictly positive element if and only if it is a strictly positive function, that is, $f(x) > 0$ for every $x \in A$.

(iii) If a is strictly positive, show that $\phi(a) > 0$ for every $\phi \in S(A)$.

14.5.5. Let A be a C*-algebra with $T(A) \neq \emptyset$ and let $p \in A$ be a projection. Show that $d_\tau(p) = \tau(p)$ for every $\tau \in T(A)$.

14.5.6. Show that the sequence $(f_n)_{n \in \mathbb{N}}$ defined in the proof of Lemma 14.3.37 satisfies properties (i) and (ii) of the statement of Lemma 14.3.37.

14.5.7. Fill in the details of the proof of Theorem 14.3.25.

14.5.8. [Kirchberg] Let A and B be separable unital C*-algebras and $D \subset A \otimes_{\min} B$ a nonzero hereditary C*-subalgebra.

(i) Let $h \in D$ be a nonzero positive element. Show that there are pure states $\varphi \in S(A)$ and $\psi \in S(B)$ such that $\varphi \otimes \psi(h) \neq 0$. Show that $b_1 = (\varphi \otimes \mathrm{id}_B)(h) \in \mathbb{C} \otimes B \cong B$ is a nonzero positive element. Show that if we replace h by some multiple of h, we may assume that $\|b_1\| = 1$.

(ii) Use Theorem 11.2.7 and Proposition 14.1.7 to show that there exists a positive element $a_1 \in A$ with $\|a_1\| = 1$ and $z \in A \otimes_{\min} B$ such that

$$z^*(a_1^{1/2} \otimes 1_B)h(a_1^{1/2} \otimes 1_B)z = (a_1 \otimes b_1 - 1/4)_+.$$

(iii) Let $1/2 < \delta < 1$. Identifying $C^*(a_1, 1_A) \cong C(X)$ and $C^*(b_1, 1_B) \cong C(Y)$, show that $\operatorname{supp}((a_1 - \delta)_+ \otimes (b_1 - \delta)_+) \subset \operatorname{supp}((a_1 \otimes b_1 - 1/4)_+$ as positive continuous functions in $C(X \times Y)$. Thus find a positive element s of $C^*(a_1, 1_A) \otimes C^*(b_1, 1_B)$ satisfying

$$s^{1/2}((a_1 \otimes b_1) - 1/4)_+ s^{1/2} = (a_1 - \delta)_+ \otimes (b_1 - \delta)_+.$$

(iv) Let $x := h^{1/2}(a_1^{1/2} \otimes 1_B)zs$. Show that x^*x is a simple tensor and $xx^* \in D$.

14.5.9. Let A be a simple, separable, unital, nuclear C^*-algebra with property (SP) (Definition 11.2.12) and let B be a simple unital purely infinite C^*-algebra. Show that $A \otimes B$ is purely infinite. In particular, this holds whenever A is AF or purely infinite.

14.5.10. Let A be a finite-dimensional C^*-algebra. Show that $W(A) = V(A)$.

14.5.11. Let A be a simple unital purely infinite C^*-algebra.

(i) Suppose that $a, b \in A_+ \setminus \{0\}$. Show that $a \precsim b$.

(ii) Show that $W(A) = \{0, \infty\}$ where $\infty + \infty = \infty$.

15 The Jiang–Su algebra

The C*-algebra of compact operators $\mathcal{K}(H)$ and the C*-algebra $C(X)$ with X contractible both have the same K-theory as the complex numbers, that is, $K_0(\mathbb{C}) = \mathbb{Z}$ and $K_1(\mathbb{C}) = 0$. We would like to know if there is such a C*-algebra that falls within the scope of Elliott classification in the sense of Conjecture 13.3.18, that is, one which is simple, separable, unital, nuclear and infinite-dimensional. Neither $\mathcal{K}$ nor $C(X)$ satisfy all these conditions since $\mathcal{K}$ is nonunital and $C(X)$ is either nonsimple or finite-dimensional. This would entail that the K-theory of any simple, separable, unital, nuclear C*-algebra A is isomorphic (as a pair of groups, but not necessarily ordered groups) to the K-theory of $A \otimes \mathcal{Z}$. If moreover, we could arrange that $\mathcal{Z}$ has a unique tracial state, then, for any simple C*-algebra A, the tensor product $A \otimes \mathcal{Z}$ would have a tracial state space affinely homeomorphic to the tracial state space of A. (Note that $\mathcal{K}(H)$ does not admit a tracial state, and unless X is a single point, $C(X)$ has more than one tracial state.) This is almost enough to conclude that if A is a simple separable unital nuclear C*-algebra, then $\mathrm{Ell}(A) \cong \mathrm{Ell}(A \otimes \mathcal{Z})$ (we will need to say something about the order structure of K_0, namely that it is weakly unperforated, see Definition 13.3.7 below). This is the question that motivated Jiang and Su to construct their celebrated algebra, which we now call the Jiang–Su algebra [62].

In Section 15.1 we look at the building blocks of the Jiang–Su algebra, the *dimension-drop algebras*. Section 15.2 contains the rather technical construction of the Jiang–Su algebra. Section 15.3 shows that the K_0-group of a simple, separable, unital, stably finite C*-algebra is always weakly unperforated. In 15.4 we show that the Cuntz semigroup of a C*-algebra tensored with the Jiang–Su algebra is almost unperforated, and therefore such tensor products have strict comparison of positive elements.

15.1 Dimension-drop algebras

We start this section by defining a dimension-drop algebra.

15.1.1. Definition. Let $p, q, d \in \mathbb{N}\backslash\{0\}$ with both p and q dividing d. The *dimension-drop algebra* $I(p, d, q)$ is defined to be

$$I(p, d, q) := \{f \in C([0,1], M_d) \mid f(0) \in M_p \otimes 1_{d/p}, f(1) \in 1_{d/q} \otimes M_d\}.$$

If p and q are relatively prime, then we call $I(p, pq, q)$ a *prime dimension-drop algebra*. In this case, we sometimes simply write $I(p, q)$, since it is understood that $d = pq$.

15.1.2. Proposition. *The dimension-drop algebra $I(p, d, q)$ has no nontrivial projections if and only if p and q are relatively prime.*

Proof. Exercise. $\qquad\square$

K. R. Strung, *An Introduction to C*-Algebras and the Classification Program*, Advanced Courses in Mathematics - CRM Barcelona, https://doi.org/10.1007/978-3-030-47465-2_15

15.1.3. Proposition. *Let $p, q, d \in \mathbb{N} \setminus \{0\}$ with both p and q dividing d. Let g be the greatest common divisor of p and q, and set $n := dr/(pq)$. The K-theory of the dimension-drop algebra $I(p, d, q)$ is given by*

$$K_0(I(p, d, q)) \cong \mathbb{Z}, \qquad K_1(I(p, d, q)) \cong \mathbb{Z}/n\mathbb{Z}.$$

In particular, if $I(p, q)$ is a prime dimension-drop algebra, $K_1(I(p, q)) = 0$.

Proof. Let $A := I(p, d, q)$. Then $J := C_0((0, 1), M_d)$ is an ideal in A and $J = SM_d$, the suspension of M_d as defined in 12.4.6. For an element $f \in A$ write $f(0) = f_0 \otimes 1_{d/p}$ for $f_0 \in M_p$ and also $f(1) = 1_{d/q} \otimes f_1$ for $f_1 \in M_q$. Then

$$\pi : A \to M_p \oplus M_q, \quad f \mapsto (f_0, f_1),$$

defines a surjective *-homomorphism giving the short exact sequence

$$0 \longrightarrow J \overset{\iota}{\longrightarrow} A \overset{\psi}{\longrightarrow} M_p \oplus M_q \longrightarrow 0.$$

By Theorem 12.4.6 this induces the cyclic 6-term exact sequence

$$
\begin{array}{ccccc}
K_0(J) & \overset{\iota_0}{\longrightarrow} & K_0(A) & \overset{\psi_0}{\longrightarrow} & K_0(M_p \oplus M_q) \\
{\scriptstyle \delta_1}\big\uparrow & & & & \big\downarrow{\scriptstyle \delta_0} \\
K_1(M_p \oplus M_q) & \underset{\psi_1}{\longleftarrow} & K_1(A) & \underset{\iota_1}{\longleftarrow} & K_1(J).
\end{array}
$$

Since $M_p \oplus M_q$ is finite-dimensional, $K_0(M_p \oplus M_q) \cong \mathbb{Z} \oplus \mathbb{Z}$ by Theorem 12.2.6, and by Exercise 12.6.4 (i), $K_1(M_p \oplus M_q) = 0$. Since $J = SM_d$, we have $K_0(J) = K_0(SM_d) = K_1(M_d) = 0$ by Theorem 12.4.7 and, using Bott periodicity, $K_1(J) = K_1(SM_d) = K_2(M_d) = K_0(M_d) = 0$. Thus the 6-term exact sequence above becomes

$$0 \longrightarrow K_0(A) \overset{\psi_0}{\longrightarrow} \mathbb{Z} \oplus \mathbb{Z}$$
$$\big\downarrow{\scriptstyle \delta_0}$$
$$0 \longleftarrow K_1(A) \underset{\iota_1}{\longleftarrow} \mathbb{Z}.$$

Let $r_d \in M_d$ be a rank one projection. Then $[r_d]$ is a generator for $K_0(M_d) \cong \mathbb{Z}$ and $d[r_d] = [1_{M_d}]$. Let $u := u(t) = \exp(2\pi i t) \cdot 1_{M_d}$, $t \in [0, 1]$. Then u is a unitary in $\widetilde{SM_d} = \{f \in C_0([0, 1], A) \mid f(0) = f(1) \in \mathbb{C}\}$ and, by definition of the Bott map (12.5.9) we have $\beta_{M_d}([1_{M_d}]) = -[u]_1$.

Let $a \in I(p, d, q)$ be the function $a(t) = t \cdot 1_{M_d}$. Then a is self-adjoint and satisfies $\psi(a) = (0, 1_{M_q})$.

Moreover, a satisfies $\tilde{\iota}(u) = \exp(2\pi \iota(t 1_{M_d})) = \exp(2\pi i t 1_{M_d}) = \exp(2\pi i a)$. It follows from Proposition 12.5.13 that $\delta_0([(0, 1_{M_q})]) = -[u]_1$.

Now consider the self-adjoint element $1_{I(p,d,q)} - a \in I(p,d,q)$. For $t \in [0,1]$ we have $1_{I(p,d,q)} - a(t) = (1 - t) \cdot 1_{M_d}$. We have $\psi(1_{I(p,d,q)} - a) = (1_{M_p}, 0)$ and $\tilde{\iota}(u^*) = \exp(2\pi i(1_{I(p,d,q)} - a))$, so $\delta_0([(1_{M_p}, 0)]) = -[u^*]_1$.

Let r_p, r_q, r_d denote rank one projections in M_p, M_q and M_d respectively. Note that $[(r_p, 0)]$ and $[(0, r_q)]$ generate $K_0(M_p \oplus M_q) \cong \mathbb{Z} \oplus \mathbb{Z}$. Now

$$r\beta_{M_d}([r_d]) = \beta_{M_d}(r[r_d]) = \beta_{M_d}([1_{M_d}]) = -[u]_1$$

and similarly,

$$p\delta_0([(r_p, 0)]) = -[u^*]_1 \quad \text{and} \quad q\delta_0([(0, r_q)]) = -[u]_1.$$

Since $[r_d]$, and hence $\beta_{M_d}([r_d])$, is a generator of $K_1(SM_d) \cong \mathbb{Z}$, $\delta_0([(0, r_q)])$ generates the ideal $\frac{d}{q}\mathbb{Z}$ and $\delta_0([(r_p, 0)])$ generates $\frac{d}{p}\mathbb{Z}$. Since the greatest common divisor or d/p and d/q is n, the image of δ_0 is $n\mathbb{Z}$. By exactness, $\ker(\iota_1) \cong n\mathbb{Z}$ and ι_1 is surjective. It follows that $K_1(I(p,d,q)) \cong \mathbb{Z}/\ker(\iota_1) \cong \mathbb{Z}/n\mathbb{Z}$.

Finally, the kernel of ψ_0 is zero, so $K_0(I(p,d,q)) \cong \operatorname{im}(\psi) = \ker(\delta_0)$. Since $\operatorname{im}(\delta_0) \cong n\mathbb{Z} \cong \mathbb{Z}$, we have $\ker(\delta_0) \cong \mathbb{Z}$. Thus $K_0(I(p,d,q)) \cong \mathbb{Z}$. □

15.1.4. Proposition. *Let $I(p,d,q)$ be a dimension-drop algebra. Then for any ideal $I \subset I(p,d,q)$ there is a unique closed subset $E_I \subset [0,1]$ such that*

$$I = \{f \in I(p,d,q) \mid f(x) = 0 \text{ for all } x \in E_I\}.$$

Proof. Exercise.   □

15.2   Jiang and Su's construction

The Jiang–Su algebra, denoted $\mathcal{Z}$, is a simple C*-algebra isomorphic to an inductive limit of prime dimension-drop algebras. Since K-theory is continuous with respect to inductive limits and a prime dimension-drop algebra $I(p,q)$ has $K_0(I(p,q)) = \mathbb{Z}$ and $K_1(I(p,q)) = 0$, it follows that if we can arrange that an inductive limit of such C*-algebras is simple and has unique trace, then we will have constructed a C*-algebra that is infinite-dimensional, simple, separable and unital, but has the same K-theory and tracial state space as $\mathbb{C}$. We follow the exposition in Jiang and Su's original paper [62].

15.2.1. Proposition. *There exists an inductive sequence (A_n, φ_n) of prime dimension-drop algebras $A_n := I(p_n, d_n, q_n)$, such that, for every $m, n \in \mathbb{N}$ the maps*

$$\varphi_{m,n} := \varphi_{n-1} \circ \cdots \varphi_{m+1} \circ \varphi_m : A_m \to A_n$$

are injective and of the form

$$\varphi_{m,n} = \mathbf{u}^* \begin{pmatrix} f \circ \xi_1^{(m,n)} & 0 & \cdots & 0 \\ 0 & f \circ \xi_2^{(m,n)} & \cdots & 0 \\ \vdots & \vdots & & \vdots \\ 0 & 0 & \cdots & f \circ \xi_k^{(m,n)} \end{pmatrix} \mathbf{u},$$

for every $f \in A_m$, where $k = d_m/d_n$, $\mathbf{u}$ is a continuous path of unitaries in $M_{p_n q_n}$, and ξ_i is a sequence of continuous paths in $[0, 1]$ satisfying

$$|\xi_i^{(m,n)}(x) - \xi_i^{(m,n)}(y)| \leq 1/2^{n-m},$$

for every $x, y \in [0, 1]$.

Proof. Let $A_m = I(p_m, d_m, q_m)$ with $d_m = p_m q_m$ be a prime dimension-drop algebra. Choose integers $k_0 > 2q_m$ and $k_1 > 2p_m$ such that $(k_0 p_m, k_1 q_m) = 1$. Let $p_{m+1} := k_0 p_m$, $q_{m+1} := k_1 q_m$, and $d_{m+1} := p_{m+1} q_{m+1}$ and set

$$A_{m+1} := I(p_{m+1}, d_{m+1}, q_{m+1}),$$

which is again a prime dimension-drop algebra.

We will construct the map $\varphi_m : A_m \to A_{m+1}$ by first describing the boundary conditions. Let $k = k_0 k_1$ and let r_0 be the positive integer remainder of k divided by q_{m+1} so that r_0 satisfies

$$0 < r_0 \leq q_{m+1}, \quad r_0 \equiv k \bmod q_{m+1}.$$

Define

$$\xi_i^{(m)}(0) = \begin{cases} 0 & \text{if } 1 \leq i \leq r_0, \\ 1/2 & \text{if } r_0 < i \leq k. \end{cases}$$

We have that $k - r_0$ is a multiple of q_{m+1} and, since

$$r_0 q_m \equiv k q_m \equiv k_0 q_{m+1} \equiv 0 \bmod q_{m+1},$$

it follows that $r_0 q_m$ is a multiple of q_{m+1}. Put

$$a := (k - r_0)/q_{m+1}, \quad b_m := r_0 q_m/q_{m+1}.$$

If $f \in A_m$, we have the matrix

$$\begin{pmatrix} f(\xi_1^{(m)}(0)) & 0 & \cdots & 0 \\ 0 & f(\xi_2^{(m)}(0)) & \cdots & 0 \\ \vdots & \vdots & \ddots & \vdots \\ 0 & 0 & \cdots & f(\xi_k^{(m)}(0)) \end{pmatrix}$$

where there are $r_0 q_m$ blocks of $p_m \times p_m$ matrices followed by $(k - r_0)$ blocks of $p_m q_m \times p_m q_m$ matrices. Since

$$a p_m q_m + b p_m = p_{m+1},$$

and

$$a q_{m+1} = k - r_0, \quad b q_{m+1} = r_0 q_m,$$

we can also view this matrix as a block diagonal matrix of q_{m+1} matrices in M_{bp_m} followed by q_{m+1} matrices in $M_{ap_m q_m}$.

Now take a unitary $u_0^{(m+1)} \in M_{d_{m+1}}$ where conjugation swaps the matrices in $M_{ap_m a_m}$ in between the matrices in M_{bp_m}, giving a block diagonal matrix of q_{m+1} blocks of size $p_{m+1} \times p_{m+1}$. Thus

$$
\rho_0(f) = (u_0^{(m+1)})^* \begin{pmatrix} f(\xi_1^{(m)}(0)) & 0 & \cdots & 0 \\ 0 & f(\xi_2^{(m)}(0)) & \cdots & 0 \\ \vdots & \vdots & \ddots & \vdots \\ 0 & 0 & \cdots & f(\xi_k^{(m)}(0)) \end{pmatrix} u_0^{(m+1)}, \quad f \in A_m,
$$

defines a *-homomorphism $\rho_0 : A_m \to M_{p_{m+1}} \otimes 1_{q_{m+1}}$.

Now let us construct the map at the endpoint 1. The construction is similar. Let r_1 be the integer satisfying

$$
0 < r_1 \le p_{m+1}, \quad r_1 \equiv k \bmod p_{m+1}.
$$

Define

$$
\xi_1(1) = \begin{cases} 1/2 & \text{if } 1 \le i \le k - r_1, \\ 1 & \text{if } k - r_1 < i \le k. \end{cases}
$$

As before, $k - r_1$ and $r_1 p_m$ are both multiples of p_{m+1}, so there exists a unitary $u_1^{(m+1)} \in M_{d_{m+1}}$ such that

$$
\rho_1(f) = (u_1^{(m+1)})^* \begin{pmatrix} f(\xi_1^{(m)}(1)) & 0 & \cdots & 0 \\ 0 & f(\xi_2^{(m)}(1)) & \cdots & 0 \\ \vdots & \vdots & \ddots & \vdots \\ 0 & 0 & \cdots & f(\xi_k^{(m)}(1)) \end{pmatrix} u_1^{(m+1)}, \quad f \in A_m,
$$

defines a *-homomorphism $\rho_1 : A_m \to 1_{p_{m+1}} \otimes M_{q_{m+1}}$.

To define φ_m, we need to connect ρ_0 and ρ_1 along the interval. Let ξ_i be given by

$$
\xi_i(t) = \begin{cases} t/2 & \text{if } 1 \le i \le r_0, \\ 1/2 & \text{if } r_0 < i \le k - r_1, \\ (t+1)/2 & \text{if } k - r_1 < i \le k. \end{cases}
$$

Since the unitary group of $M_{d_{m+1}}$ is connected (Exercise 12.6.3), we can find a continuous path of unitaries $\mathbf{u}^{(m+1)}$ connecting $u_0^{(m+1)}$ and $u_1^{(m+1)}$. Set

$$
\varphi_m = (\mathbf{u}^{(m+1)})^* \begin{pmatrix} f \circ \xi_1 & 0 & \cdots & 0 \\ 0 & f \circ \xi_2 & \cdots & 0 \\ \vdots & \vdots & \ddots & \vdots \\ 0 & 0 & \cdots & f \circ \xi_k \end{pmatrix} \mathbf{u}^{(m+1)}, \quad f \in A_m.
$$

Note that φ_m is injective, since $\varphi_m(f) = 0$ if and only if $f = 0$, and moreover φ_m is unital. Repeating this construction gives us an inductive sequence

$$A_1 \xrightarrow{\varphi_1} A_2 \xrightarrow{\varphi_2} \cdots \xrightarrow{\varphi_{m-1}} A_m \xrightarrow{\varphi_m} A_{m+1} \xrightarrow{\varphi_{m+1}} \cdots ,$$

such that each induced map $\varphi_{m,n} : A_m \to A_n$ is unital and injective. It is easy to see that each $\varphi_{m,n}$ has the form

$$\varphi_{m,n} = (\mathbf{u}^{(n)})^* \begin{pmatrix} f \circ \xi_1^{(m,n)} & 0 & \cdots & 0 \\ 0 & f \circ \xi_2^{(m,n)} & \cdots & 0 \\ \vdots & \vdots & \ddots & \vdots \\ 0 & 0 & \cdots & f \circ \xi_k^{(n)} \end{pmatrix} \mathbf{u}^{(n)}, \quad f \in A_m,$$

where $\mathbf{u}^{(n)}$ is the continuous path of unitaries as constructed above with respect to the embedding $A_{n-1} \to A_n$. Moreover, $\xi_i^{(m,n)}$ is a composition of $m - n$ functions $[0,1] \to [0,1]$, each of which is of the form

$$\xi(t) = t/2, \quad \xi(t) = 1/2, \quad \text{or} \quad \xi(t) = (1+t)/2.$$

Thus

$$|\xi_i^{(m,n)}(x) - \xi_i^{(m,n)}(y)| \leq 1/2^{n-m},$$

for every $x, y \in [0,1]$. $\square$

15.2.2. Let $A = I(p,d,q)$ be a dimension-drop algebra. For any *-homomorphism $\varphi : A \to B$ into a C*-algebra B, we write $\mathrm{sp}(\varphi) := E_{\ker(\varphi)}$, where $E_{\ker(\varphi)}$ is the unique closed subset of $[0,1]$ given by Proposition 15.1.4 with respect to the ideal $\ker(\varphi)$.

15.2.3. Given a *-homomorphism $\varphi : I(p,d,q) \to I(p',d',q')$ and $t \in [0,1]$, define the *-homomorphism $\varphi^t : I(p,d,q) \to M_{d'}$ by

$$\varphi^t(f) = \varphi(f)(t), \quad f \in I(p,d,q).$$

Proposition. *Let $(A_n, \varphi_n)_{n \in \mathbb{N}}$ be an inductive sequence of dimension-drop algebras, with each φ_n unital and injective. Suppose that for every integer $m > 0$ and any nonempty open subset $U \subset [0,1]$ there is an integer N such that for every $n \geq N$ and any $t \in [0,1]$ we have*

$$\mathrm{sp}(\varphi_{m,n}^t) \cap U \neq \emptyset.$$

Then $A = \varinjlim A_n$ is simple.

Proof. First, we claim that for any nonzero $f \in A_m$ there is an N such that $\varphi_{m,n}(f)(t) \neq 0$ for every $n \geq N$ and every $t \in [0,1]$. Observe that, for any fixed m, we have $\bigcup_{n>m} \bigcup_{t \in [0,1]} \mathrm{sp}(\varphi_{m,n}^t) = [0,1]$. Note that if $\varphi_{m,n}(f)(t) = 0$, then $f \in \ker(\varphi_{m,n}^t)$ so $f|_{\mathrm{sp}(\varphi_{m,n}^t)} = 0$. Suppose that there is a nonzero $f \in A_m$, for some m, such that for every $n > m$ and $\varphi_{m,n}(f)(t) = 0$ for every $t \in [0,1]$. Then $f|_{\mathrm{sp}(\varphi_{m,n}^t)} = 0$ for every $t \in [0,1]$ and every $n > m$, which is to say that $f|_{\bigcup_{m>n} \bigcup_{t \in [0,1]} \mathrm{sp}(\varphi_{m,n}^t)} = 0$, contradicting the fact that f is nonzero. This proves the claim.

Now let J be a nonzero ideal of A. By Proposition 8.2.9, $J = \overline{\bigcup_{n \in \mathbb{N}} J \cap A_n}$. Let $f \in J \cap A_m$ be a nonzero function. (Here, since the connecting maps are injective we have identified A_n with its image in A to simplify notation.) Then there is an N such that $\varphi_{m,n}(f)(t) \neq 0$ for every $n \geq N$ and every $t \in [0,1]$. But $\varphi_{m,n}(J \cap A_m) \subset J \cap A_n$, so we have $\varphi_{m,n}(f) \in J \cap A_n$ for every $n \geq N$. Since $\varphi_{m,n}(f)(t) \neq 0$ for every $t \in [0,1]$, the functional calculus allows us to define $1_{[0,1]}(f) = 1$. Then $1 \in J \cap A_n$ so $J = A_n$ for all sufficiently large n. Hence $J = A$, and A is simple. $\qquad\square$

15.2.4. Proposition. *Let $(A_n, \varphi_n)_{n \in \mathbb{N}}$ be an inductive limit of the form given in Proposition 15.2.1. Then the limit*

$$A := \varinjlim(A_n, \varphi_n)$$

is a simple C*-algebra.

Proof. By construction, the maps φ_n are injective, and it also easy to see that they are unital. We will show that A satisfies the condition of Proposition 15.2.3. By definition, $x \in \mathrm{sp}(\varphi_{m,n}^t)$ if and only if $f(x) = 0$ for every $f \in \ker(\varphi_{m,n}^t)$. Now $f \in \ker(\varphi_{m,n}^t)$ if and only if

$$\begin{pmatrix} f \circ \xi_1^{(m,n)}(t) & 0 & \cdots & 0 \\ 0 & f \circ \xi_2^{(m,n)}(t) & \cdots & 0 \\ \vdots & \vdots & \ddots & \vdots \\ 0 & 0 & \cdots & f \circ \xi_k^{(m,sn)}(t) \end{pmatrix} = 0,$$

so we must have that $\xi_i^{(m,n)}(t) \in \mathrm{sp}(\varphi_{m,n}^t)$ for every $1 \leq i \leq k$. Now

$$\bigcup_{i=1}^{k} \xi^{(m,n)}([0,1]) = [0,1] \quad \text{and} \quad |\xi_1^{(m,n)}(x) - \xi_i^{(m,n)}(y)| \leq 1/2^{n-m},$$

so we have

$$\mathrm{dist}(z, \mathrm{sp}(\varphi_{m,n}^t)) \leq 1/2^{n-m}$$

for any $z, t \in [0,1]$. Thus, given any open subset $U \subset [0,1]$, we can easily find N such that $\mathrm{sp}(\varphi_{m,n}^t) \cap U \neq \emptyset$ for every $n \geq N$ and $t \in [0,1]$, showing that A is simple. $\qquad\square$

15.2.5. Proposition. *Let $I = (p, d, q)$ be a dimension drop algebra. Then a tracial state $\tau \in T(I(p,d,q))$ is extreme if and only if there exists $x \in [0,1]$ such that $\tau(f) = \mathrm{tr}_d(f(x))$, where tr_d is the normalised trace on M_d.*

Proof. Exercise.   □

15.2.6. Proposition. *Let $(A_n, \varphi_n)_{n \in \mathbb{N}}$ be an inductive limit of the form given in Proposition 15.2.1. Then the limit*

$$A := \varinjlim(A_n, \varphi_n)$$

has a unique tracial state.

Proof. Let $m \in \mathbb{N}$ and $f \in A_m$. We claim that given any $\epsilon > 0$ there is an $N \in \mathbb{N}$ such that, for every $n \geq N$ and any two distinct extreme tracial states $\tau_1^{(n)}, \tau_2^{(n)} \in T(A_n)$, we have

$$|\tau_1^{(n)}(\varphi_{m,n}(f)) - \tau_2^{(n)}(\varphi_{m,n}(f))| < \epsilon.$$

To prove the claim, choose $\delta > 0$ sufficiently small so that $\|f(x) - f(y)\| < \epsilon/2$ whenever $|x - y| < \delta$. Let $n \in \mathbb{N}$ be large enough that $2^{n-m} \leq 1/\delta$. Then, for each $\xi_i^{(m,n)}$ we have

$$\|f(\xi_i^{(m,n)}(x)) - f(\xi_i^{(m,n)}(y))\| < \epsilon,$$

for every $x, y \in [0,1]$. Since $\tau_1^{(n)}$ and $\tau_2^{(n)}$ are extreme tracial states, there exists $x_1, x_2 \in [0,1]$ such that $\tau_j^{(n)}(f) = \mathrm{tr}_{d_n}(f(x_j))$ for every $f \in A_n$, $j \in \{1,2\}$. Thus

$$|\tau_1^{(n)}(\varphi_{m,n}(f)) - \tau_2^{(n)}(\varphi_{m,n}(f))| = |\mathrm{tr}_{d_n}(f(\xi_i^{(m,n)}(x_1)) - f(\xi_i^{(m,n)}(x_2)))| < \epsilon.$$

Since any tracial state is a convex combination of extreme tracial states (Exercise 5.4.13), it follows that for sufficiently large N we can arrange

$$|\tau_1^{(n)}(\varphi_{m,n}(f)) - \tau_2^{(n)}(\varphi_{m,n}(f))| < \epsilon,$$

for any $\tau_1^{(n)}, \tau_2^{(n)} \in T(A_n)$.

Now let $\{\tau_n \in T(A_n)\}_{n \in \mathbb{N}}$ be a sequence of tracial states. For $m \in \mathbb{N}$, let

$$\theta_m(f) := \lim_{n \to \infty} \tau_n(\varphi_{m,n}(f)), \quad f \in A_m.$$

By the claim, θ_m is a well-defined tracial state on A_m which is moreover independent of the sequence τ_n. We also have $\theta_m = \theta_{m+1} \circ \varphi_m$, so the sequence θ_m passes to a well-defined tracial state θ on A. Since the maps in the inductive limit are unital and injective, for every $m \in \mathbb{N}$, any tracial state on A restricts to a tracial state on A_m. Since θ is independent of the initial sequence, it follows that θ is the unique tracial state on A.   □

15.2.7. Now we can put the results of this section together with continuity of K-theory (Theorems 12.2.4 and 12.4.8) to arrive at the theorem below. That the Jiang–Su algebra is nuclear comes from Exercise 15.5.4.

Theorem (Jiang–Su). *There exists an infinite-dimensional, simple, unital, nuclear C*-algebra $\mathcal{Z}$ with unique tracial state such that*

$$(K_0(\mathcal{Z}), K_0(\mathcal{Z})_+, [1_{\mathcal{Z}}]_0) \cong (K_0(\mathbb{C}), K_0(\mathbb{C})_+, [1_{\mathbb{C}}]) \cong (\mathbb{Z}, \mathbb{Z}_+, 1)$$

and

$$K_1(\mathcal{Z}) \cong K_1(\mathbb{C}) = 0.$$

In fact, Jiang and Su also show that $\mathcal{Z}$ is the unique such algebra. Proving that is beyond the scope of these notes, because we'd have to introduce KK-theory. However, the strategy of the proof is one which we have seen already, in Chapter 13: an approximate intertwining argument along an inductive limit, here with dimension-drop algebras.

15.2.8. Definition. The unique simple unital inductive limit of dimension-drop algebras with unique tracial state and K-theory isomorphic to the K-theory of $\mathbb{C}$ is called the *Jiang–Su algebra* and is denoted $\mathcal{Z}$.

We also state their classification theorem here without proof. This is an example of a classification theorem for a subclass of simple separable unital nuclear C*-algebras, which the reader may wish to compare to the classification of unital AF algebras, Theorem 13.1.8, and the classification theorem in the final section, Theorem 18.0.12.

15.2.9. Theorem (Jiang–Su). *Let A and B be simple unital infinite-dimensional inductive limits of finite direct sums of dimension-drop algebras. Suppose that there exists a homomorphism*

$$\varphi : \mathrm{Ell}(A) \to \mathrm{Ell}(B).$$

Then there exists a unital *-homomorphism $\Phi : A \to B$ inducing φ. Moreover, if φ is an isomorphism, then $\Phi : A \to B$ can be chosen to be a *-isomorphism.

15.2.10. Since the construction of $\mathcal{Z}$ works for arbitrary prime dimension-drop algebras, uniqueness of $\mathcal{Z}$ implies the following.

Corollary. *Every prime dimension-drop C*-algebra embeds unitally into $\mathcal{Z}$.*

15.2.11. Remark. Let $\mathcal{C}$ denote the class of C*-algebras covered by Theorem 15.2.9. Note that any simple unital AF algebra is in $\mathcal{C}$ as is the tensor product $A \otimes_{\min} B$ of any simple unital AF algebra A with any $B \in \mathcal{C}$. (Exercise).

15.3 $K_0(A \otimes \mathcal{Z})$ is weakly unperforated

The minimal tensor product of simple C*-algebras is again simple (Theorem 6.2.7), so if A is a simple separable unital nuclear C*-algebra, we can compute the Elliott invariant of $A \otimes \mathcal{Z}$. The fact that $\mathcal{Z}$ has a unique tracial state means that $T(A) \cong$

$T(A \otimes \mathcal{Z})$ (Exercise 7.3.13 (iv)). Moreover, since $K_0(\mathcal{Z})$ and $K_1(\mathcal{Z})$ are torsion free, we can apply Theorem 12.5.15 to get

$$K_0(A \otimes \mathcal{Z}) \cong K_0(A) \otimes \mathbb{Z} \oplus K_1(A) \otimes 0 \cong K_0(A),$$

and

$$K_1(A \otimes \mathcal{Z}) \cong K_1(A) \otimes \mathbb{Z} \oplus K_0(A) \otimes 0 \cong K_1(A).$$

Uniqueness of the traces also implies that the map pairing tracial states with state on K_0 will be the same for both A and $A \otimes \mathcal{Z}$.

This is *almost* enough to imply that $\mathrm{Ell}(A \otimes \mathcal{Z}) \cong \mathrm{Ell}(A)$, except for the important fact that we are not taking care of the order structure on K_0. In this section we will establish the fact, proved by Gong, Jiang and Su [53], that $K_0(A \otimes \mathcal{Z})$ is always weakly unperforated (Definition 13.3.7).

15.3.1. Let A be a unital C*-algebra and $p, q \in \mathbb{N}$ two positive integers. Define

$$I(p,q)(A) := \{f \in C([0,1], M_{pq}(A) \mid f(0) \in M_p(A) \otimes 1_q \text{ and } f(1) \in 1_p \otimes M_q(A)\}.$$

It follows easily from Exercise 6.5.5 (vi) that $I(p,q)(A) \cong A \otimes I(p,q)$. Let $\iota_{p,q} : A \hookrightarrow I(p,q)(A)$ denote the unital embedding

$$a \mapsto \begin{pmatrix} a & 0 & \cdots & 0 \\ 0 & a & \cdots & 0 \\ \vdots & & \ddots & \vdots \\ 0 & 0 & \cdots & a \end{pmatrix},$$

where a is copied pq-times down the diagonal.

15.3.2. Given what we know about the K_0-group of a dimension-drop algebra, the next lemma is not so surprising.

Lemma. *For any p, q with $I(p,q)$ a prime dimension-drop algebra, the map*

$$[\iota_{p,q}]_0 : K_0(A) \to K_0(I(p,q)(A)),$$

is a group isomorphism.

Proof. Let $\mathrm{ev}_0 : I(p,q)(A) \to M_p(A)$ denote evaluation at 0, that is, let $\mathrm{ev}_0(f) = a \in M_p(A)$ if $f(0) = a \otimes 1_{M_q(A)}$. Analogously, we define the map $\mathrm{ev}_1 : I(p,q)(A) \to M_q(A)$, the evaluation map at 1. Since p and q are relatively prime, there are integers m, n such that

$$np + mq = 1.$$

Set $\nu := n \cdot [\mathrm{ev}_0]_0 + m \cdot [\mathrm{ev}_1]_0$. Then, if $a \in M_r(A) \subset M_\infty(A)$ is a projection,

$$\nu \circ [\iota_{p,q}]_0([a]) = \nu(pq \cdot [a]) = n(p \cdot [a]) + m(q \cdot [a]) = [a],$$

so $\nu \circ [\iota_{p,q}]_0(x) = x$ for every $x \in K_0(A)$, which is to say, ν is an inverse of $[\iota_{p,q}]_0$. Thus $[\iota_{p,q}]_0 : K_0(A) \to K_0(I(p,q)(A))$ is group isomorphism. $\square$

Observe that by appealing to the Künneth Theorem for Tensor Products (Theorem 12.5.15), we already knew that $K_0(A) \cong K_0(I(p,q)(A))$ were isomorphic as abelian groups. However, the Künneth Theorem does not give us information on the order structure, while the explicit construction of maps in the previous lemma does. We will use this to determine the order structure on $K_0(A \otimes \mathcal{Z})$ when A is stably finite. First, a pair of lemmas.

15.3.3. Lemma. *Let A be a unital, simple, stably finite C*-algebra. Let $x \in K_0(A)$ and suppose that there exists $n \in \mathbb{N} \setminus \{0\}$ such that $nx \geq 0$. Then $kx \geq 0$ for all sufficiently large k, and if $nx > 0$ then $kx > 0$.*

Proof. Since A is a unital, simple, stably finite C*-algebra, $(K_0(A), K_0(A)_+)$ is a simple ordered abelian group (Proposition 12.1.18). Suppose that $x \in K_0(A)_+$ satisfies $nx \geq 0$ for some $n > 0$. Let $y := nx$. Then y is an order unit, so, for every $1 \leq m \leq n - 1$ there exists a positive integer n_m such that $n_m y \geq mx$. Let $N := n_1 \cdot n_2 \cdots n_m$. Then $Ny \geq mx$ for every $m = 1, \ldots, n-1$. Thus $(Nn-m)x \geq 0$. Let $k \geq Nn$. Then k can be written as the sum of $Nn - m$ for some $m = 1, \ldots, n - 1$ and a multiple of n. Since $nx \geq 0$ and $(Nn - m)x \geq 0$, it follows that also $kx \geq 0$. If we had $nx > 0$, then it is easy to see that also $kx > 0$. $\qquad\square$

15.3.4. In the next lemma, if $p \in M_n(A)$ is a projection, then $k \cdot p$ denotes the projection $p \oplus \cdots \oplus p \in M_{kn}(A)$, where $\oplus$ is as defined at the beginning of Section 12.1. Equivalently, $k \cdot p = p \otimes 1_k$ after identifying $M_{kn}(A)$ with $M_n(A) \otimes M_k$.

Lemma. *Let A be a unital, simple, stably finite C*-algebra. Suppose that $p, q \in M_\infty(A)$ are projections satisfying $p \oplus r \sim q \oplus r$ for some projection $r \in M_\infty(A)$. Then $k \cdot p \sim k \cdot q$ for all sufficiently large $k \in \mathbb{N}$.*

Proof. Let $p, q, r \in M_\infty(A)$ be projections satisfying the hypotheses be given. As in the previous lemma, the hypotheses on A mean that $(K_0(A), K_0(A)_+)$ is a simple ordered group. Thus both $[p]$ and $[q]$ are order units for $(K_0(A), K_0(A)_+)$. In particular, there are $m, n \in \mathbb{Z}$ such that $[r] \leq m[p] = [m \cdot p]$ and $[r] \leq n[q] = [n \cdot q]$. Assume that $p \in M_k(A)$. Then there exists $r' \in M_{mk}(A)$ such that $r \sim r'$ and $r' \leq m \cdot p$. Let $x = m \cdot p - r'$. Then $x \in M_{nk}(A)$ is a projection and $[r] + [x] = m[p]$. Similarly, we can find $y \in M_\infty(A)$ satisfying $[r] + [y] = n[q]$. It follows that

$$[p] + m[p] = [p] + [r] + [x] = [q] + [r] + [x] = [q] + m[p],$$

and similarly, $[q] + n[q] = [p] + n[q]$. Let $k \geq m + n$. Then

$$[k \cdot p] = k[p] = m[p] + n \cdot [p] + (k - m - n)[p]$$
$$= m[p] + n[q] + (k - m - n)[q]$$
$$= m[q] + n[q] + (k - m - n)[q] = k[q] = [k \cdot q].$$

Thus $k \cdot p \sim k \cdot q$. $\qquad\square$

15.3.5. Proposition. *Let A be a unital, simple, stably finite C*-algebra and let $\iota : A \to A \otimes \mathcal{Z}$ be the embedding obtained as the limit of the unital embeddings $\iota_{p,q} : A \hookrightarrow A \otimes I(p,q)(A)$. Let $x \in K_0(A)$. Then $\iota_0(x) > 0$ if and only if $nx > 0$ for some positive integer n.*

Proof. Let $x \in K_0(A)$ and suppose that $\iota_0(x) > 0$. By the construction of $\mathcal{Z}$, there is some prime dimension-drop algebra $I(p,q)$ and embedding $\iota_{p,q} : A \to A \otimes I(p,q)$ such that $[\iota_{p,q}]_0(x) \geq 0$. Let $m, n \in \mathbb{Z}$ be integers satisfying $np + mq = 1$, and define ev_0, ev_1 and ν as in the proof of Lemma 15.3.2. Then

$$px = [\mathrm{ev}_0]_0([pq \cdot x]) = [\mathrm{ev}_0]_0[\iota_{p,q}]_0(x) \geq 0,$$

and similarly $qx \geq 0$. Suppose that both $px = 0$ and $qx = 0$. Then $x = 1x = px + qx = 0$, which implies $\iota_0(x) = 0$, contradicting the fact that $\iota_0(x) > 0$. Thus either $px > 0$ or $qx > 0$.

Now suppose that $nx > 0$ for some $n > 0$. By Lemma 15.3.3, there is some $N > 0$ such that $mx > 0$ for every $m > N$. Choose $p, q \in \mathbb{N}$ such that $p, q > N$ and p and q are relatively prime. Since $px > 0$ and $qx > 0$, there are projections $e \in M_{jp}(A)$, $f \in M_{jq}(A)$ for some $j > 0$ such that

$$px = [e], \quad qx = [f].$$

Thus $q[e] = pq \cdot x = p[f]$. It follows from Lemma 15.3.4 that for all k sufficiently large, $kq \cdot e$ and $kp \cdot f$ are equivalent projections in $M_{jkpq}(A)$. In particular, we may find k such that $kq \cdot e$ and $kp \cdot f$ are equivalent and kp and q are relatively prime. Let $\tilde{e} = k \cdot e$. Then $\tilde{e} \in M_{kp}(M_j(A))$. Note that $q \cdot \tilde{e} = kq \cdot e \sim kp \cdot f$, so by taking $j \geq 1$ large enough, we may assume that there is a continuous path of projections r_t in $M_{jkpq}(A)$ (Proposition 12.1.5) such that

$$r_0 := q \cdot \tilde{e}, \quad r_1 := kp \cdot f.$$

Thus

$$r : [0,1] \to M_{jkpq}(A), \quad t \mapsto r_t$$

defines a projection in $I(kp, q)(M_j(A)) \cong M_j(I(kp, q)(A))$. Using the isomorphism

$$(\iota_{kp,q})_0 : K_0(A) \to K_0(I(kp,q)(A)),$$

we have $(\iota_{kp,q})_0(x) = [r]$. Finally, the unital embedding $j : I(kp,q) \hookrightarrow \mathcal{Z}$, gives us the commutative diagram

$$
\begin{array}{ccc}
A & \xrightarrow{\;\cong\;} & A \\
{\scriptstyle \iota_{kp,q}}\downarrow & & \downarrow{\scriptstyle \iota} \\
A \otimes I(kp,q) & \xrightarrow{\;\mathrm{id}_A \otimes j\;} & A \otimes \mathcal{Z}.
\end{array}
$$

Thus, $\iota_0(x) = (\mathrm{id}_A \otimes j)_0([r]) \geq 0$. Since $x \neq 0$ and ι_0 is injective, it follows that $\iota_0(x) > 0$. $\qquad\square$

15.3.6. Corollary. *Let A be a unital simple stably finite C*-algebra. Then $K_0(A \otimes \mathcal{Z})$ is weakly unperforated.*

15.3.7. Proposition. *Let A be a unital AF algebra. Then $K_0(A)$ is weakly unperforated.*

Proof. Exercise. $\qquad\square$

15.4 The Cuntz semigroup of $\mathcal{Z}$-stable C*-algebras

We saw in the previous section that if A is a simple unital C*-algebra then the ordered abelian group $K_0(A \otimes \mathcal{Z})$ is always weakly unperforated. Thus if $K_0(A)$ is not weakly unperforated, we cannot expect A to be $\mathcal{Z}$-stable. Is this the only obstruction to $\mathcal{Z}$-stability? Many of the early counterexamples to the Elliott conjecture suggested this could be the case. However, in [120] Toms constructed a simple separable unital nuclear C*-algebra A with weakly unperforated $K_0(A)$ which is not $\mathcal{Z}$-stable. In fact, he was able to construct a C*-algebra A that, from the point of view of any of the usual invariants (including $\mathrm{Ell}(\cdot)$, stable rank and real rank) appears to be the same as $A \otimes \mathcal{U}$ where $\mathcal{U}$ is a UHF algebra of infinite type, yet A is not isomorphic to $A \otimes \mathcal{U}$.

The difference between A and $A \otimes \mathcal{U}$ or indeed A and $A \otimes \mathcal{Z}$ can be seen in the order structure of their Cuntz semigroups.

In general, the Cuntz semigroup of a C*-algebra can be quite wild and intractable. However, for $\mathcal{Z}$-stable C*-algebras, we can at least see that the order structure is well behaved. The results in this section are due to Rørdam [102].

15.4.1. Lemma. *Let $n \in \mathbb{N} \setminus \{0\}$ and let A be the prime dimension-drop C*-algebra $A := I(n, n+1)$. For $m \in \{n, n+1\}$, let $e_{ij}^{(m)}$ denote the system of matrix units generating M_m. There exist pairwise orthogonal positive functions $f_1, f_2, \ldots, f_{n+1} \in A$ such that*

(i) $f_i(0) = \begin{cases} e_{ii}^{(n)} \otimes 1_{n+1} & i = 1, \ldots, n, \\ 0 & i = n+1, \end{cases}$

(ii) $f_i(t) = 1_n \otimes e_{ii}^{(n+1)}$ *for $t \in [1/2, 1]$,*

(iii) $\sum_{i=1}^{n+1} f_1 = 1_A$,

(iv) $f_{n+1} \precsim f_1 \sim f_2 \sim \cdots \sim f_n$.

Proof. Let

$$w := \sum_{i,j=1}^{n} e_{ij}^{(n)} \otimes e_{ji}^{(n+1)} + 1_n \otimes e_{n+1,n+1}^{(n+1)}.$$

Observe that $w^* = \sum_{i,j=1}^{n} e_{ji}^{(n)} \otimes e_{ij}^{(n+1)} + 1_n \otimes e_{n+1,n+1}^{(n+1)} = w$ and also that

$$w^2 = \sum_{i,j=1}^{n} e_{ij}^{(n)} e_{ji}^{(n)} \otimes e_{ji}^{(n+1)} e_{ij}^{(n+1)} + 1_n \otimes e_{n+1,n+1}^{(n+1)}$$

$$= \sum_{i=1}^{n} e_{ii}^{(n)} \otimes e_{ii}^{(n+1)} + 1_n \otimes e_{n+1,n+1}^{(n+1)}$$

$$= 1_A,$$

that is, $w \in M_n \otimes M_{n+1}$ is a self-adjoint unitary. Since the unitary group of a finite-dimensional C*-algebra is connected (Exercise 12.6.3), we can find a continuous path of unitaries $\{v_t\}_{t\in[0,1]} \subset M_n \otimes M_{n+1}$ such that $v_0 = 1$ and $v_t = w$ for $t \in [1/2, 1]$. Define $w_t := v_t w, \quad t \in [0,1]$ and let γ_t be a continuous path with $\gamma_0 = 0$ and $\gamma_t = 1$ for any $t \in [1/2, 1]$. Now, for $1 \leq i \leq n$, define $f_i : [0,1] \to M_n \otimes M_{n+1}$ by

$$f_i(t) := \gamma_t w_t (1_n \otimes e_{ii}^{(n+1)}) w_t^* + (1 - \gamma_t) v_t (e_{ii}^{(n)} \otimes 1_{n+1}) v_t^*,$$

and define $f_{n+1} : [0,1] \to M_n \otimes M_{n+1}$ by

$$f_{n+1}(t) := \gamma_t w_t (1_n \otimes e_{n+1,n+1}^{(n+1)}) w_t^*.$$

Observe that $f_1, \ldots, f_{n+1}$ are positive. For $1 \leq i \leq n$ we have

$$w(1_n \otimes e_{ii}^{(n+1)})w = \sum_{j=1}^{n} e_{ii}^{(n)} \otimes e_{jj}^{(n+1)} = e_{ii}^{(n)} \otimes (1_{n+1} - e_{n+1,n+1}^{(n+1)}) \leq e_{ii}^{(n)} \otimes 1_{n+1},$$

from which it follows that, when $1 \leq i \neq j \leq n$,

$$\begin{aligned}
f_i f_j(t) &= \gamma_t^2 w_t (1_n \otimes e_{ii}^{(n+1)})(1_n \otimes e_{jj}^{(n+1)}) w_t^* \\
&\quad + (\gamma_t - \gamma_t^2) v_t (e_{ii}^{(n)} \otimes 1_{n+1}) w (1_n \otimes e_{jj}^{(n+1)}) w_t^* \\
&\quad + (\gamma_t - \gamma_t^2) w_t (1_n \otimes e_{ii}^{(n+1)}) w (e_{jj}^{(n)} \otimes 1_{n+1}) v_t^* \\
&\quad + (1 - \gamma_t)^2 v_t (e_{ii}^{(n)} \otimes 1_{n+1})(e_{jj}^{(n)} \otimes 1_{n+1}) v_t^* \\
&\leq (\gamma_t - \gamma_t^2) v_t (e_{ii}^{(n)} \otimes 1_{n+1})(e_{jj}^{(n)} \otimes 1_{n+1}) v_t^* \\
&\quad + (\gamma_t - \gamma_t^2) v_t (e_{ii}^{(n)} \otimes 1_{n+1}) w (e_{jj}^{(n)} \otimes 1_{n+1}) v_t^* \\
&= 0.
\end{aligned}$$

A similar calculation shows that for $1 \leq i \leq n$ we have $f_i f_{n+1} = 0 = f_{n+1} f_i$, so that the functions $f_1, \ldots, f_{n+1}$ are pairwise orthogonal.

Now let us verify (i)–(iv). For (i), clearly $f_{n+1}(0) = 0$, and when $1 \leq i \leq n$, we have that $f_i(0) = 0 + v_0 (e_{ii}^{(n)} \otimes 1_{n+1}) v_0 = (e_{ii}^{(n)} \otimes 1_{n+1})$. For (ii), if $t \in [1/2, 1]$ we have $f_i(t) = w_t (1 \otimes e_{ii}^{(n+1)}) w_t + 0 = 1_n \otimes e_{ii}^{(n+1)}$ and $f_{n+1}(t) = 1_n \otimes e_{n+1,n+1}^{(n+1)}$. Notice that this also implies that each $f_i \in A$, since the boundary conditions defining the dimension-drop algebra are satisfied.

For (iii) we have

$$\begin{aligned}
\sum_{i=1}^{n+1} f_i(t) &= \sum_{i=1}^{n} f_i(t) + f_{n+1}(t) \\
&= \gamma_t w_t (1_n \otimes (1_{n+1} - e_{n+1,n+1}^{(n+1)})) w_t^* + (1 - \gamma_t) v_t (1_n \otimes 1_{n+1}) v_t^* \\
&\quad + \gamma_t w_t (1_n \otimes e_{n+1,n+1}^{(n+1)}) w_t^* \\
&= 1_n \otimes 1_{n+1} \\
&= 1_A.
\end{aligned}$$

Finally let us verify (iv). Let $s \in M_n$ be the matrix defined by $s_{i+1,i} = 1$ for $1 \le i \le n-1$, $s_{1,n} = 1$ and $s_{ij} = 0$ otherwise, that is,

$$s := \begin{pmatrix} 0 & 0 & \cdots & 0 & 1 \\ 1 & 0 & \cdots & 0 & 0 \\ 0 & 1 & & & \vdots \\ \vdots & & \ddots & & \vdots \\ 0 & 0 & \cdots & 1 & 0 \end{pmatrix}.$$

Then s is unitary and $s^{i-1}e_{11}^{(n)}s^{1-i} = e_{ii}^{(n)}$ for every $1 \le i \le n$. Define $d \in M_{n+1}$ analogously so that d is unitary and $d^{i-1}e_{11}^{(n+1)}d^{1-i} = e_{ii}^{(n+1)}$ for $1 \le i \le n+1$. Define

$$r_i(t) := v_t(s^{i-1} \otimes 1_{n+1})v_t^*, \quad t \in [0, 1/2].$$

Then $r_i(0) = s^{i-1} \otimes 1_{n+1}$. Moreover, when $t \in [0, 1/2]$, a calculation similar to those above shows that

$$r_i(t)f_1(t)r_i(t)^* = f_i(t),$$

and when $t = 1/2$, we have

$$\begin{aligned} r_i(1/2)f_1(1/2)r_i(1/2) &= w(s^{i-1} \otimes 1_{n+1})w(1_n \otimes e_{11}^{(n+1)})w(s^{i-1} \otimes 1_{n+1})w \\ &= w(e_{ii}^{(n)} \otimes 1_{n+1})w \\ &= 1_n \otimes e_{ii}^{(n+1)} \\ &= (1_n \otimes d^{i-1})(1_n \otimes e_{11}^{(n+1)})(1_n \otimes d^{1-i}). \end{aligned}$$

For $t \in [1/2, 1]$, put $r_i(t) := r(1/2)$. Then r_i is unitary and $r_i(0) \in M_n \otimes 1_{n+1}$ while $r_i(1) = 1_n \otimes d^{i-1} \in 1_n \otimes M_{n+1}$. Thus $r_i \in A$ and $r_i f_1 r_i^* = f_i$ for $1 \le i \le n$, which implies $f_1 \sim f_2 \sim \cdots \sim f_n$.

To show that $f_{n+1} \precsim f_1$, note that

$$r_i(t)(\gamma_t w_t(1_n \otimes e_{11}^{(n+1)})w_t^*)r_i(t)^* = \gamma_t w_t(1_n \otimes e_{ii}^{(n+1)})w_t^*$$

for every $1 \le i \le n+1$. It follows that

$$f_{n+1} = \gamma_t w_t(1_n \otimes e_{n+1,n+1}^{(n+1)})w_t^* \sim \gamma_t w_t(1_n \otimes e_{11}^{(n+1)})w_t^* \le f_1.$$

Thus $f_{n+1} \precsim f_1$, as required. $\qquad\square$

15.4.2. Lemma. *For every $n \in \mathbb{N} \setminus \{0\}$ there is a positive element $e_n \in \mathcal{Z}$ such that $n[e_n] \le [1_{\mathcal{Z}}] \le (n+1)[e_n]$ in $W(\mathcal{Z})$.*

Proof. Recall that by Corollary 15.2.10, every prime dimension-drop C*-algebra embeds unitally into $\mathcal{Z}$. In particular, for any $n \in \mathbb{N} \setminus \{0\}$, we may consider the prime dimension-drop C*-algebra $A := I(n, n+1)$ as a unital subalgebra of $\mathcal{Z}$. Thus, it is enough to show that we can find a positive element $e_n \in A$ such that $n[e_n] \le [1_A] \le (n+1)[e_n]$ in $W(A)$.

From the previous lemma, there are pairwise orthogonal positive elements $f_1, \ldots, f_{n+1} \in A$ such that $\sum_{i=1}^{n+1} f_i = 1$ and $f_{n+1} \precsim f_1 \sim f_2 \sim \cdots \sim f_n$. Let $e_n := f_1$. Then, since the f_i are pairwise orthogonal, Lemma 14.1.6 tells us that

$$n[e_n] \sim \sum_{i=1}^{n} [f_i] \sim [\oplus_{i=1}^n f_i] \precsim [\oplus_{i=1}^n f_i] + [f_{n+1}] \sim [1_A].$$

Also,

$$[1_A] = [\oplus_{i=1}^{n+1} f_{n+1}] \sim \sum_{i=1}^{n+1} [f_i] \sim \sum_{i=1}^{n} [f_1] + [f_{n+1}] \precsim (n+1)[f_1],$$

so $n[e_n] \leq [1_A] \leq (n+1)[e_n]$ in $W(A)$.   □

15.4.3. Lemma. *Let A and B be C*-algebras. Let $A \otimes B$ denote any C*-tensor product. For $a_1, a_2 \in A_+$, $b \in B_+$ and $m, n \in \mathbb{N}$, if $n[a_1] \leq m[a_2]$ in $W(A)$ then $n[a_1 \otimes b] \leq m[a_2 \otimes b]$ in $W(A \otimes B)$.*

Proof. Let $a_1, a_2 \in A_+$ and $b \in B_+$ and suppose there are $m, n \in \mathbb{N}$ such that $n[a_1] \leq m[a_2]$ in $W(A)$. Observe that, after identifying $M_n(A)$ with $A \otimes M_n$, we have $n[a_1] = [a_1 \oplus \cdots \oplus a_1] = [a_1 \otimes 1_n]$. Similarly, $m[a_2] = [a_2 \otimes 1_m]$. Thus, there exists a sequence $r_k = (r_{ij}^{(k)})_{ij} \subset M_{m,n}(A)$ such that $r_k^*(a_2 \otimes 1_m)r_k \to a_1 \otimes 1_n$ as $k \to \infty$. Let $(e_k)_{k \in \mathbb{N}}$ be a sequence of positive contractions such that $e_k b e_k \to b$ as $k \to \infty$ (for example, take an approximate unit for C*(b)). Define

$$s_{ij}^{(k)} := r_{ij}^{(k)} \otimes e_k, \quad \text{and let} \quad s_k := (s_{ij}^{(k)})_{ij} \in M_{m,n}(A \otimes B).$$

Then $s_k^*(a_2 \otimes 1_n \otimes b)s_k = r_k^*(a_2 \otimes 1_n)r_k \otimes e_k b e_k \to a_1 \otimes 1_m b$ as $k \to \infty$. Thus, after identifying $M_n(A \otimes B)$ with $M_n(A) \otimes B$, and similarly $M_m(A \otimes B)$ with $M_m(A) \otimes B$, we see that $n[a_1 \otimes b] \leq m[a_2 \otimes b]$.   □

Note that, by symmetry, the lemma we have just proved also shows that for positive elements $b_1, b_2 \in B_+$ with $n[b] \leq m[b]$ in $W(B)$ we have $n[a \otimes b_1] \leq m[a \otimes b_2]$ in $W(A \otimes B)$.

15.4.4. Lemma. *Let A be a C*-algebra and let $a, b \in A_+$ be positive elements satisfying $(n+1)[a] \leq n[b]$ in $W(A)$ for some $n \in \mathbb{N} \setminus \{0\}$. Then $[a \otimes 1_{\mathcal{Z}}] \leq [b \otimes 1_{\mathcal{Z}}]$ in $W(A \otimes \mathcal{Z})$.*

Proof. Let $e_n \in \mathcal{Z}$ be as in Lemma 15.4.2. By the previous lemma, we have

$$[a \otimes 1_{\mathcal{Z}}] \leq (n+1)[a \otimes e_n] \leq n[b \otimes e_n] \leq [b \otimes 1_{\mathcal{Z}}],$$

in $W(A \otimes \mathcal{Z})$, as required.   □

Now we are able to prove the main theorem of this section.

15.4.5. Theorem. *Let A be a $\mathcal{Z}$-stable C*-algebra. Then $W(A)$ is almost unperforated.*

Proof. Observe that it is enough to show for any $x, y \in W(A)$ and any natural number $n \in \mathbb{N} \setminus \{0\}$ whenever $(n+1)x \leq ny$ that $x \leq y$. So, let $a, b \in M_\infty(A)_+$. We may assume that both $a, b \in M_n(A)$ for some large enough $n \in \mathbb{N}$. Let $B := M_n(A) \cong M_n \otimes A$ and notice that B is also $\mathcal{Z}$-stable. From the previous lemma applied to B, we have $[a \otimes 1_{\mathcal{Z}}] \leq [b \otimes 1_{\mathcal{Z}}]$. Let $\epsilon > 0$ and let $r \in A \otimes \mathcal{Z}$ satisfy

$$\|r^*(b \otimes 1_{\mathcal{Z}})r - a \otimes 1_{\mathcal{Z}}\| < \epsilon/3.$$

By Proposition 16.2.2 there is a *-isomorphism $\varphi : A \otimes \mathcal{Z} \to A$ such that

$$\|\varphi(c \otimes 1_{\mathcal{Z}}) - c\| < \epsilon/3.$$

Let $s = \varphi(r) \in A$. Then

$$\begin{aligned}
\|s^*bs - a\| &= \|s^*bs - s^*\varphi(b \otimes 1_{\mathcal{Z}})s + s^*\varphi(b \otimes 1_{\mathcal{Z}})s - \varphi(a \otimes 1_{\mathcal{Z}}) + \varphi(a \otimes 1_{\mathcal{Z}}) - a\| \\
&\leq \|s^*bs - s^*\varphi(b \otimes 1_{\mathcal{Z}})s\| + \|\varphi(r^*(b \otimes 1_{\mathcal{Z}})r) - \varphi(a \otimes 1_{\mathcal{Z}})\| + \|\varphi(a \otimes 1_{\mathcal{Z}}) - a\| \\
&\leq \|b - \varphi(b \otimes 1_{\mathcal{Z}})\| + \|r(b \otimes 1_{\mathcal{Z}})r - a \otimes 1_{\mathcal{Z}}\| + \|\varphi(a \otimes 1_{\mathcal{Z}}) - a\| < \epsilon.
\end{aligned}$$

Since ϵ was arbitrary, it follows that $[a] \leq [b]$ in $W(A)$. Thus $W(A)$ is weakly unperforated. $\qquad\square$

15.4.6. Theorem (Rørdam)**.** *Let A be a simple separable unital exact $\mathcal{Z}$-stable C*-algebra. Then A has strict comparison of positive elements.*

Proof. Exercise. $\qquad\square$

In the same paper, Rørdam also proves the following [102, Theorem 6.7]. We omit the proof since we have not quite covered all the necessary material.

15.4.7. Theorem. *Every simple, unital, finite $\mathcal{Z}$-stable C*-algebra has stable rank one.*

15.4.8. Now we can put this together with the results of the last chapter.

Corollary. *Let A be a simple, separable, unital, exact C*-algebra. Suppose that A is $\mathcal{Z}$-stable and $W(A)$ is weakly divisible. Then*

$$W(A) \cong V(A) \sqcup \mathrm{LAff_b}(T(A))_{++}.$$

15.5 Exercises

15.5.1. Show that the dimension-drop algebra $I(p, pq, q)$ has no nontrivial projections if and only if p and q are relatively prime.

15.5.2. Let $I(p, n, q)$ be a dimension-drop algebra. Show that if $I \subset I(p, n, q)$ is an ideal, then there is a unique closed subset $E \subset [0, 1]$ such that

$$I = \{f \in I(p, n, q) \mid f(x) = 0 \text{ for all } x \in E_I\}.$$

15.5.3. Let $I = (p, d, q)$ be a dimension drop algebra. Show that a tracial state $\tau \in T(I(p, d, q))$ is extreme if and only if there exists $x \in [0, 1]$ such that $\tau(f) = \mathrm{tr}_d(f(x))$, where tr_d is the normalised trace on M_d. (Hint: See Exercise 7.3.13 (v).)

15.5.4. The Jiang–Su algebra is nuclear:

(i) Show that any dimension drop algebra is nuclear. (Hint: Use Theorem 6.4.3.)

(ii) Suppose that $(A_n, \varphi_n)_{n \in \mathbb{N}})$ is an inductive sequence of C*-algebras, each of which have the c.p.a.p. Show that $A = \varinjlim A_n$ has the c.p.a.p.

(iii) Conclude that $\mathcal{Z}$ is nuclear.

15.5.5. Let A be a simple separable unital AF algebra.

(i) By composing connecting maps with point evaluations, show that A can be written as an inductive limit of direct sums of C*-algebras of the form $C([0, 1], M_n)$.

(ii) Show that Theorem 15.2.9 covers simple unital AF algebras.

(iii) Let B be a simple unital infinite-dimensional inductive limit of dimension-drop algebras. Show that Theorem 15.2.9 covers tensor products of simple unital AF algebras with B.

(iv) Show that $A \otimes \mathcal{Z} \cong A$, that is, A is $\mathcal{Z}$-stable.

15.5.6. Show that the only projections in $\mathcal{Z}$ are 0 and 1. Deduce that $\mathcal{Z}$ is not an AF algebra.

15.5.7. Let A be a unital simple C*-algebra, and let $\iota : A \to A \otimes \mathcal{Z}$ be as in Proposition 15.3.5. Suppose that A is not stably finite. Let $x \in K_0(A)$. Show that $\iota_0(x) > 0$ if and only if there is some $n > 0$ such that $nx > 0$.

15.5.8. Let A be a simple separable unital $\mathcal{Z}$-stable C*-algebra. Show that A has strict comparison of positive elements.

15.5.9. We will show how almost unperforation (Definition 14.3.10) is related to weak unperforation. Say that a partially ordered abelian group (G, G_+) is almost unperforated if the positive semigroup G_+ is an almost unperforated semigroup.

(a) Suppose (G, G_+) is an almost unperforated abelian group. Show that for every $g \in G$ and $n \in \mathbb{N}$, if $ng, (n+1)g \in G_+$ we have $g \in G_+$.

(b) Let (G, G_+) be a simple ordered abelian group (Definition 12.1.17). Show that (G, G_+) is almost unperforated if and only if it is weakly unperforated.

(c) Let $G = \mathbb{Z}^2$ and let G_+ be generated by $(1, 0), (0, 1), (2, -2)$. Show that (G, G_+) is almost unperforated but not weakly unperforated.

16 Strongly self-absorbing algebras

In this chapter, we study certain unital C*-algebras which are *-isomorphic to their minimal tensor squares, that is, a C*-algebra A for which there exists a *-isomorphism $\varphi : A \to A \otimes_{\min} A$. Such C*-algebras are quite easy to find; indeed, just take your favourite unital C*-algebra A and consider the infinite minimal tensor product of A with itself, which is to say the inductive limit $A^{\otimes \infty} = \varinjlim(A^{\otimes n}, a_1 \otimes \cdots \otimes a_n \mapsto a_1 \otimes \cdots \otimes a_n \otimes 1_A)$. Here we look at a particularly nice subclass of such C*-algebras, those that are so-called *strongly self-absorbing*, which means that the *-isomorphism φ can be chosen to be approximately unitarily equivalent to the map

$$A \to A \otimes_{\min} A, \quad a \to a \otimes 1_A.$$

The strongly self-absorbing algebras have a number of interesting properties, for example, they are always nuclear and simple. The class is furthermore closed under tensor products and has a number of interesting implications for K-theory and classification. Most of this chapter is derived from [122].

In the first section, we define what it means to be strongly self-absorbing and show that any strongly self-absorbing C*-algebra is automatically simple and nuclear. We show that the class of strongly self-absorbing C*-algebras is closed under taking tensor products and provide conditions that are equivalent to being strongly self-absorbing. In the second section we look at the property of a C*-algebra A absorbing a strong self-absorbing C*-algebra $\mathcal{D}$ in the sense that $A \cong A \otimes \mathcal{D}$. This is property is called $\mathcal{D}$-*stability*. We investigate some permanence properties of $\mathcal{D}$-stability, for example, we show that it passes to inductive limits and to corners. Many of the arguments in this chapter use the *central sequence algebra* of a C*-algebra, introduced towards the end of Section 16.1.

16.1 Definition and characterisations

The property of being strongly self-absorbing had already been observed in various C*-algebras such as the Jiang–Su algebra of the last chapter, or the Cuntz algebras $\mathcal{O}_2$ and $\mathcal{O}_\infty$ of Chapter 10. In an effort to study these types of algebras together as a single class, Toms and Winter came up with the abstract definition below.

16.1.1. Definition (Toms–Winter [122]). We say that a separable, unital C*-algebra $\mathcal{D}$ is *strongly self-absorbing* if $\mathcal{D}$ is infinite-dimensional and there exist a *-isomorphism

$$\varphi : \mathcal{D} \to \mathcal{D} \otimes_{\min} \mathcal{D}$$

and a sequence of unitaries $(u_n)_{n \in \mathbb{N}}$ in $\mathcal{D} \otimes_{\min} \mathcal{D}$ satisfying

$$\|u_n^* \varphi(a) u_n - a \otimes 1_{\mathcal{D}}\| \to 0 \text{ as } n \to \infty.$$

Recall from Definition 13.2.1 that unital *-homomorphisms $\varphi, \psi : A \to B$ between unital C*-algebras are approximately unitarily equivalent if there is a sequence of unitaries $(u_n)_{n \in \mathbb{N}}$ in B such that

$$\lim_{n \to \infty} \|u_n \varphi(a) u_n^* - \psi(a)\| = 0, \text{ for every } a \in A,$$

and in such a case we write $\varphi \approx_{a.u.} \psi$.

16.1.2. Let A be a separable unital C*-algebra and define the *flip* map

$$\sigma : A \otimes_{\min} A \to A \otimes_{\min} A,$$

to be the automorphism induced by the map defined on simple tensors by

$$a \otimes b \mapsto b \otimes a.$$

We say that A has *approximately inner flip* if the flip map is approximately unitarily equivalent to $\mathrm{id}_A \otimes \mathrm{id}_A$,

$$\sigma \approx_{a.u.} \mathrm{id}_A \otimes \mathrm{id}_A.$$

We say A has *approximately inner half-flip* if the first- and second-factor embeddings

$$\iota_1 : A \hookrightarrow A \otimes_{\min} A, \ a \mapsto a \otimes 1_A, \qquad \iota_2 : A \hookrightarrow A \otimes_{\min} A, \ a \mapsto 1_A \otimes a,$$

are approximately unitarily equivalent.

16.1.3. Suppose that A has approximately inner flip. Let $n \in \mathbb{N} \setminus \{0\}$ and i, j with $1 \leq i < j \leq n$. Define a *-homomorphism

$$\sigma_{i,j} : A^{\otimes n} \to A^{\otimes n}$$

on simple tensors by

$$a_1 \otimes \cdots \otimes a_i \otimes \cdots \otimes a_j \otimes \cdots \otimes a_n \mapsto a_1 \otimes \cdots \otimes a_j \otimes \cdots \otimes a_i \otimes \cdots \otimes a_n.$$

Then $\sigma_{i,j}$ is approximately inner (exercise). Similarly, define

$$\iota_i : A \hookrightarrow A^{\otimes n}, \qquad a \mapsto 1_A \otimes \cdots \otimes 1_A \otimes a \otimes 1_A \otimes \cdots \otimes 1_A,$$

where a is in the ith tensor factor. Then if A has approximately inner half-flip, ι_i is approximately unitarily equivalent to ι_j (exercise).

Let A and B be C*-algebras with B unital. Suppose that $\varphi : A \to B$ is a *-homomorphism. We denote by

$$\varphi \otimes 1_B : A \to A \otimes B$$

the *-homomorphism mapping $a \mapsto \varphi(a) \otimes 1_B$.

16.1.4. Proposition. *Let $\mathcal{D}$ be a strongly self-absorbing C*-algebra. Then $\mathcal{D}$ has approximately inner half-flip.*

Proof. Let $\varphi : \mathcal{D} \to \mathcal{D} \otimes_{\min} \mathcal{D}$ be a *-isomorphism that is approximately unitarily equivalent to the first factor embedding ι_1, that is, $\varphi \approx_{a.u.} \iota_1$. Let $\psi := \varphi^{-1} \circ \iota_2$. By Proposition 13.2.2 (ii), the property $\varphi \approx_{a.u.} \iota_1$ implies $\varphi \circ \psi \approx_{a.u.} \iota_1 \circ \psi$, which is to say

$$\iota_2 \approx_{a.u.} \iota_1 \circ \psi = \psi \otimes 1_{\mathcal{D}}.$$

Let $\sigma : \mathcal{D} \otimes_{\min} \mathcal{D} \to \mathcal{D} \otimes_{\min} \mathcal{D}$ denote the flip automorphism. Again by Proposition 13.2.2 (ii), we get

$$\iota_1 = \sigma \circ \iota_2 \approx_{a.u.} \sigma \circ (\psi \otimes 1_{\mathcal{D}}) = 1_{\mathcal{D}} \otimes \psi,$$

and similarly $\mathrm{id}_{\mathcal{D}} \otimes 1_{\mathcal{D}} \otimes 1_{\mathcal{D}} \approx_{a.u.} 1_{\mathcal{D}} \otimes 1_{\mathcal{D}} \otimes \psi$ by applying the automorphism which flips the first and third tensor factors. Then,

$$
\begin{aligned}
\psi \otimes 1_{\mathcal{D}} = \quad & \psi \otimes (\varphi^{-1}(1_{\mathcal{D}} \otimes 1_{\mathcal{D}})) \\
= \quad & (\mathrm{id}_{\mathcal{D}} \otimes \varphi^{-1}) \circ (\psi \otimes 1_{\mathcal{D}} \otimes 1_{\mathcal{D}}) \\
\approx_{a.u.} & (\mathrm{id}_{\mathcal{D}} \otimes \varphi^{-1}) \circ (\iota_2 \otimes 1_{\mathcal{D}}) \\
\approx_{a.u.} & (\mathrm{id}_{\mathcal{D}} \otimes \varphi^{-1}) \circ (1_{\mathcal{D}} \otimes \iota_1) \\
\approx_{a.u.} & (\mathrm{id}_{\mathcal{D}} \otimes \varphi^{-1}) \circ (1_{\mathcal{D}} \otimes 1_{\mathcal{D}} \otimes \psi) \\
\approx_{a.u.} & (\mathrm{id}_{\mathcal{D}} \otimes \varphi^{-1}) \circ (\mathrm{id}_{\mathcal{D}} \otimes 1_{\mathcal{D}} \otimes 1_{\mathcal{D}}) \\
= \quad & \mathrm{id}_{\mathcal{D}} \otimes 1_{\mathcal{D}} = \iota_1.
\end{aligned}
$$

Thus $\iota_1 \approx_{a.u.} \psi \otimes 1_{\mathcal{D}}$ and $\psi \otimes 1_{\mathcal{D}} \approx_{a.u.} \iota_2$, so $\iota_1 \approx_{a.u.} \iota_2$ by Proposition 13.2.2 (i). $\square$

16.1.5. Theorem (Kirchberg–Phillips [67]). *Let A be a separable unital C*-algebra with approximately inner half-flip. Then A is simple.*

Proof. Suppose that A has a nontrivial proper ideal J. Then $J \otimes_{\min} A$ and $A \otimes_{\min} J$ are ideals in $A \otimes_{\min} A$. Let $a \in J$. If J is nontrivial then $1_A \notin J$ so $a \otimes 1_A \in J \otimes A$ but $a \otimes 1_A \notin A \otimes_{\min} J$. Since A has approximately inner half-flip, there is a sequence of unitaries $(u_n)_{n \in \mathbb{N}} \subset A \otimes_{\min} A$ such that

$$\|u_n(1_A \otimes a)u_n^* - a \otimes 1_A\| \to 0 \text{ as } n \to \infty.$$

But then for every n, $u_n(1_A \otimes a)u_n^* \in A \otimes_{\min} J$, so $a \otimes 1_A \in A \otimes_{\min} J$, a contradiction. This shows that A must be simple. $\square$

16.1.6. For the next lemma, we require the notion of a conditional expectation, the definition of which was given in Definition 9.5.1.

Lemma. *Let A and B be unital C*-algebras and let ϕ be a state on A. Then the (right) slice map, defined on the algebraic tensor product as*

$$r_\phi : A \odot B \to B, \quad \sum_{i=1}^{n} a_i \otimes b_i \mapsto \sum_{i=1}^{n} \phi(a_i)b_i,$$

extends to a unital completely positive map

$$R_\phi : A \otimes_{\min} B \to B.$$

In fact, identifying $B \cong \mathbb{C} \otimes B$ as a subalgebra of $A \otimes_{\min} B$, the map R_ϕ is a conditional expectation.

Proof. Identifying B with $\mathbb{C} \otimes B$ we have $r_\phi = \phi \otimes \mathrm{id}_B : A \otimes_{\min} B \to \mathbb{C} \otimes B$. Since ϕ and id are both completely positive contractive maps, it follows from Theorem 7.1.6 that r_ϕ extends to a well-defined c.p.c. map $R_\phi : A \otimes_{\min} B \to \mathbb{C} \otimes B$.

To see that R_ϕ is a conditional expectation, we need to show first that $R_\phi(c) = c$ for any $c \in \mathbb{C} \otimes B$ and second, that $R_\phi(c_1 d c_2) = c_1 R_\phi(d) c_2$ for every $c_1, c_2 \in \mathbb{C} \otimes B$ and $d \in A \otimes_{\min} B$. Since $\mathbb{C} \otimes B$ consists of finite sums of simple tensors, both of these are easy to check. $\qquad\square$

16.1.7. Theorem. *Let A be a separable unital C*-algebra with an approximately inner half-flip. Then A is nuclear.*

Proof. We will show that A has the completely positive approximation property. By Theorem 7.2.2, this implies that A is nuclear. Let $\epsilon > 0$ and $\mathcal{F} \subset A$ be a finite subset. Let $u \in A \otimes_{\min} A$ be a unitary such that

$$\|u(a \otimes 1_A)u^* - 1_A \otimes a\| \leq \epsilon/4, \quad \text{for every } a \in \mathcal{F}.$$

Since the algebraic tensor product $A \odot A$ is dense in $A \otimes_{\min} A$, there is a $v \in A \odot A$ such that $\|u - v\| < \epsilon/4$. Let ϕ be a state on A and R_ϕ the right slice map on $A \otimes_{\min} A$. Define a map $T : A \to A$ by

$$T(a) := R_\phi(v(a \otimes 1_A)v^*).$$

The fact that $\|v\| \leq 1$ together with the previous lemma implies T is a completely positive contractive map. Now, v is in the algebraic tensor product, so there are $n \in \mathbb{N}$ and $x_i, y_i \in B$, $1 \leq i \leq n$ with $v = \sum_{i=1}^{n} x_i \otimes y_i$. Consequently, we can rewrite the map T as

$$T(a) = \sum_{i,j=1}^{n} \phi(x_i a x_j^*) y_i y_j^*.$$

Define $\Psi : A \to M_n$ by $\Psi(a) = \sum_{i,j=1}^{n} \phi(x_i a x_j^*) e_{ij}$, where e_{ij} are the matrix units generating M_n, that is, e_{ij} is the $n \times n$ matrix with 1 in the (i,j)th entry and zeros elsewhere. Define $\Phi : M_n \to A$ by mapping the generators $e_{ij} \mapsto y_i y_j^*$. It is straightforward to check that Ψ and Φ are completely positive contractive maps and that $\Phi \circ \Psi = T$. So it is enough to show that T approximates the identity up to ϵ on the given finite subset $\mathcal{F}$. First, observe that $(\phi \otimes \mathrm{id}_A)(1_A \otimes a) = a$, since ϕ is a state. Then, for every $a \in \mathcal{F}$, we have

$$
\begin{aligned}
\|T(a) - a\| &= \|R_\phi(v(a \otimes 1_A)v^*) - (\phi \otimes \mathrm{id}_A)(1_A \otimes a)\| \\
&= \|(\phi \otimes \mathrm{id}_A)(v(a \otimes 1_A)v^*) - (\phi \otimes \mathrm{id}_A)(1_A \otimes a)\| \\
&\leq \|v(a \otimes 1_A)v^* - 1_A \otimes a\| \\
&\leq \|u(a \otimes 1_A)u^* - 1_A \otimes a\| + \epsilon/2 \\
&\leq \epsilon/4 + \epsilon/2 < \epsilon.
\end{aligned}
$$

So A has the completely positive approximation property. $\qquad\square$

16.1.8. When A and B are both separable and unital, the properties of being strongly self-absorbing, having approximately inner half-flip, and approximately inner flip, are preserved under taking tensor products. Observe that the previous theorem means we can drop the subscript min on the tensor products.

Proposition. *Suppose that A and B are separable and unital.*

(i) *If A and B have approximately inner flip, then $A \otimes B$ has approximately inner flip.*

(ii) *If A and B have approximately inner half-flip, then $A \otimes B$ has approximately inner half-flip.*

(iii) *If A and B are strongly self-absorbing, then $A \otimes B$ is strongly self-absorbing.*

Proof. Suppose that A and B have approximately inner flips σ_A and σ_B. Then, by Exercise 13.4.1, their tensor product $\sigma_A \otimes \sigma_B : A \otimes A \otimes B \otimes B \to A \otimes A \otimes B \otimes B$ is also approximately inner. Let $\sigma_{A,B}$ denote the flip map on $A \otimes B$. Suppose that $u \in A \otimes A$ and $v \in B \otimes B$ are unitaries. Put

$$
w := (1_A \otimes \sigma_{A,B} \otimes 1_B)(u \otimes v).
$$

Since $\sigma_{A,B}$ is a *-homomorphism, w is a unitary in $A \otimes B \otimes A \otimes B$. It now follows easily that since $\sigma_A \otimes \sigma_B$ is approximately inner, so too is $\sigma_{A \otimes B}$.

The proof for the half-flip and for being strongly self-absorbing follows similarly from Exercise 13.4.1, so we leave the details as an exercise. $\qquad\square$

16.1.9. For a separable unital nuclear C*-algebra we denote

$$
A^{\otimes\infty} := \varinjlim(A^{\otimes n}, a \mapsto a \otimes 1_A)_{n \in \mathbb{N}}.
$$

Proposition. *Let A be a unital separable C*-algebra with approximately inner half-flip. Then $A^{\otimes\infty}$ has approximately inner flip.*

Proof. For $k \in \mathbb{N}$, let $\sigma_{A^{\otimes k}}$ denote the flip $A^{\otimes k} \otimes A^{\otimes k} \to A^{\otimes k} \otimes A^{\otimes k}$. Define a map

$$\lambda_k : A^{\otimes k} \otimes A^{\otimes k} \to A^{\otimes 2k} \otimes A^{\otimes 2k}$$

by setting, for $a, b \in A^{\otimes k}$,

$$\lambda_k(a \otimes b) := a \otimes 1_{A^{\otimes k}} \otimes b \otimes 1_{A^{\otimes k}}.$$

Since $A^{\otimes \infty}$ is the inductive limit $\varinjlim A^{\otimes n}$, to show that $A^{\otimes \infty}$ has approximately inner flip, it suffices to prove that for any $k \in \mathbb{N}$,

$$\lambda_k \approx_{a.u.} \lambda_k \circ \sigma_{A^{\otimes k}}.$$

For $i \in \{1, 2, 3, 4\}$, let $\iota_k^{(i)} : A^{\otimes k} \to (A^{\otimes k})^{\otimes 4}$ denote the embedding of $A^{\otimes k}$ into the ith factor. Observe that if $i \neq j \in \{1, 2, 3, 4\}$, then $\iota_k^{(i)}(A^{\otimes k})$ and $\iota_k^{(j)}(A^{\otimes k})$ commute. Thus we may define, for $i \neq j \in \{1, 2, 3, 4\}$, *-homomorphisms

$$\iota_k^{(i,j)} : (A^{\otimes k})^{\otimes 2} \to (A^{\otimes k})^{\otimes 4}$$

satisfying

$$\iota_k^{(i,j)}|_{(A^{\otimes k}) \otimes (1_{A^{\otimes k}})} = \iota^{(i)} \quad \text{and} \quad \iota_k^{(i,j)}|_{(1_{A^{\otimes k}}) \otimes (A^{\otimes k})} = \iota^{(j)}.$$

Identify $A^{\otimes 2k} \otimes A^{\otimes 2k}$ with $(A^{\otimes k})^{\otimes 4}$. Then, it easy to check that

$$\lambda_k = \iota_k^{(1,3)},$$

while

$$\lambda_k \otimes \sigma_{A^{\otimes k}} = \iota_k^{(3,1)}.$$

By Proposition 16.1.8, since A has approximately inner half-flip, so does $A^{\otimes k}$. Thus the first- and second-factor embeddings

$$A^{\otimes k} \hookrightarrow A^{\otimes k} \otimes A^{\otimes k}, \quad a \mapsto a \otimes 1_{A^{\otimes k}},$$

and

$$A^{\otimes k} \hookrightarrow A^{\otimes k} \otimes A^{\otimes k}, \quad a \mapsto 1_{A^{\otimes k}} \otimes a,$$

are approximately unitarily equivalent.

Let $(v_m)_{m \in \mathbb{N}} \subset A^{\otimes k} \otimes A^{\otimes k}$ denote a sequence of unitaries implementing this equivalence. Then $\iota_k^{(j)}$ and $\iota_k^{(l)}$ are approximately unitarily equivalent, as may be seen by using the sequence of unitaries $(\iota_k^{(j,l)}(v_m))_{m \in \mathbb{N}} \subset (A^{\otimes k})^{\otimes 4}$. Furthermore, since each $\iota_k^{(j,l)}(v_m)$, $m \in \mathbb{N}$, commutes with $\iota_k^{(i)}(A^{\otimes k})$, the unitaries $(\iota_k^{(j,l)}(v_m))_{m \in \mathbb{N}}$ also give the approximate unitary equivalence of $\iota_k^{(i,j)}$ and $\iota^{(i,l)}$. Thus

$$\iota_k^{(1,3)} \approx_{a.u.} \iota_2^{(1,2)} \approx_{a.u.} \iota_k^{(3,2)} \approx_{a.u.} \iota_k^{(3,1)}.$$

Finally,

$$\lambda_k = \iota^{(1,3)} \approx_{a.u.} \iota^{(3,1)} = \lambda_k \circ \sigma_{A^{\otimes k}},$$

as required. $\qquad\square$

16.1.10. We don't have a converse for Proposition 16.1.4 – a separable, unital C*-algebra with approximately inner half-flip need not be strongly self-absorbing. However, it is always the case that its infinite tensor product will be strongly self-absorbing.

Proposition. *Let A be a unital, separable C*-algebra with approximately inner half-flip. Then $A^{\otimes\infty}$ is strongly self-absorbing.*

Proof. For $k \in \mathbb{N}$ let us denote the connecting map $A^{\otimes k} \to A^{\otimes k+1}$ by α_k. Note that $\alpha_k = \mathrm{id}_{A^{\otimes k}} \otimes 1_A$. Then

$$A^{\otimes\infty} \otimes A^{\otimes\infty} = \varinjlim(A^{\otimes k} \otimes A^{\otimes k}, \alpha_k \otimes \alpha_k),$$

and also

$$A^{\otimes\infty} = \varinjlim(A^{\otimes 2k}, \alpha_{2k+1} \circ \alpha_{2k}).$$

Let

$$\beta_k := \alpha_k \otimes \alpha_k, \quad \gamma_k := \alpha_{2k+1} \circ \alpha_{2k},$$

and as in 8.2.5, we let

$$\beta^{(k)} : A^{\otimes k} \otimes A^{\otimes k} \to A^{\otimes\infty} \otimes A^{\otimes\infty}, \quad \gamma^{(k)} : A^{\otimes 2k} \to A^{\otimes\infty}$$

denote the maps induced by β_k and γ_k.

The map $\psi_k : A^{\otimes 2k} \to A^{\otimes k} \otimes A^{\otimes k}$ defined by

$$\psi_k(a_1 \otimes b_1 \otimes \cdots \otimes a_k \otimes b_k) = a_1 \otimes \cdots \otimes a_k \otimes b_1 \otimes \cdots \otimes b_k,$$

for $a_1, b_1, \ldots, a_k, b_k \in A$, is evidently a *-isomorphism.

Observe that

$$\psi_{k+1} \circ \gamma_k = \beta_k \circ \psi_k.$$

Thus the diagram

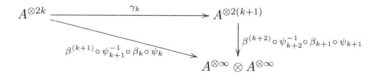

commutes. By the universal property of inductive limits (Theorem 8.2.6) applied to $A^{\otimes\infty}$, we get a map $\psi : A^{\otimes\infty} \to A^{\otimes\infty} \otimes A^{\otimes\infty}$ such that the diagram

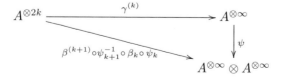

commutes. Using the universal property again, this time for $A^{\otimes\infty} \otimes A^{\otimes\infty}$, we also obtain a map $\varphi : A^{\otimes\infty} \otimes A^{\otimes\infty} \to A^{\otimes\infty}$ satisfying

$$
\begin{array}{ccc}
A^{\otimes k} \otimes A^{\otimes k} & \xrightarrow{\ \beta^{(k)}\ } & A^{\otimes\infty} \otimes A^{\otimes\infty} \\
& \searrow{\scriptstyle \gamma^{(k)} \circ \psi_k^{-1}} & \downarrow{\scriptstyle \varphi} \\
& & A^{\otimes\infty}
\end{array}
$$

and it is straightforward to check that φ and ψ are mutual inverses.

Now we would like to show that ψ is approximately unitarily equivalent to $\mathrm{id}_{A^{\otimes\infty}} \otimes 1_{A^{\otimes\infty}}$.

We can write the inductive limits as

$$
A^{\otimes\infty} \otimes A^{\otimes\infty} = \varinjlim \left(A^{\otimes k} \otimes A^{\otimes k}, \lambda_k := (\mathrm{id}_{A^{\otimes k}} \otimes 1_{A^{\otimes k}}) \otimes (\mathrm{id}_{A^{\otimes k}} \otimes 1_{A^{\otimes k}}) \right)
$$

and

$$
A^{\otimes\infty} = \varinjlim (A^{\otimes 2^m}, \mathrm{id}_{A^{\otimes 2^m}} \otimes 1_{A^{\otimes 2^m}}),
$$

giving the diagram

$$
\begin{array}{ccccccccc}
\cdots \longrightarrow & A^{\otimes 2^m} \otimes A^{\otimes 2^m} & \xrightarrow{\lambda_{2^m}} & A^{\otimes 2^{m+1}} \otimes A^{\otimes 2^{m+1}} & \xrightarrow{\lambda_{2^{m+1}}} & \cdots \longrightarrow & A^{\otimes\infty} \otimes A^{\otimes\infty} \\
& {\scriptstyle \psi_{2^m}}\uparrow\ \uparrow{\scriptstyle \mathrm{id}_{A^{\otimes 2^m}} \otimes 1} & & {\scriptstyle \psi_{2^{m+1}}}\uparrow\ \uparrow{\scriptstyle \mathrm{id}_{A^{\otimes 2^{m+1}}} \otimes 1} & & & \uparrow{\scriptstyle \psi} \\
\cdots \longrightarrow & A^{\otimes 2^m} & \xrightarrow[\mathrm{id}_{A^{\otimes 2^m}} \otimes 1]{} & A^{\otimes 2^{m+1}} & \xrightarrow[\mathrm{id}_{A^{\otimes 2^{m+1}}} \otimes 1]{} & \cdots \longrightarrow & A^{\otimes\infty}.
\end{array}
$$

Thus it is enough to show that $\lambda_{2^m} \circ \psi_{2^m} \circ (\mathrm{id}_{A^{\otimes 2^m}} \otimes 1_{A^{\otimes 2^m}})$ is approximately unitarily equivalent to $\lambda_{2^m} \circ \mathrm{id}_{A^{\otimes 2^m}} \otimes 1_{A^{\otimes 2^m}}$ for every $m \in \mathbb{N}$.

Let $k \in \{2^m \mid m \in \mathbb{N}\}$. Define a *-isomorphism

$$
\rho_k : (A^{\otimes k})^4 \to (A^{\otimes k})
$$

by setting

$$
\rho_k := \mathrm{id}_{A^{\otimes k}} \otimes \sigma_{A^{\otimes k}} \otimes \mathrm{id}_{A^{\otimes k}} .
$$

Identify $A^{\otimes 2k} \otimes A^{\otimes 2k}$ with $(A^{\otimes k})^4$ and define $\iota_k^{(i)}$, $i \in \{1,2,3,4\}$, as in the previous proof, to be the embedding of $A^{\otimes k}$ into the ith position of $A^{\otimes k} \otimes A^{\otimes k} \otimes A^{\otimes k} \otimes A^{\otimes k}$. Then, for $i = 1, 2$ we have *-homomorphisms

$$
\rho_k \circ (\psi_k \otimes \mathrm{id}_{D^{\otimes 2k}}) \circ \iota_k^{(i)} : A^{\otimes k} \to A^{\otimes 2k} \otimes A^{\otimes 2k},
$$

satisfying

$$
\rho_k \circ (\psi_k \otimes \mathrm{id}_{D^{\otimes 2k}}) \circ \iota_k^{(1)} = \lambda_k \otimes \psi_k \otimes (\mathrm{id}_{D^{\otimes k}} \otimes 1_{D^{\otimes k}})
$$

and

$$
\iota_k^{(2)} = \rho_k \circ (\psi_k \otimes \mathrm{id}_{A^{\otimes 2}}) \circ \iota_k^{(3)}.
$$

Now $\iota_k^{(i)} \approx_{a.u.} \iota_k^{(j)}$. So, using Proposition 13.2.2 (ii), we have

$$
\begin{aligned}
\lambda_k \circ \psi_k \circ (\mathrm{id}_{A^{\otimes k}} \otimes 1_{A^{\otimes k}}) &= \rho_k \circ (\psi_k \otimes \mathrm{id}_{A^{\otimes 2k}}) \circ \iota_k^{(1)} \\
&\approx_{a.u.} \rho_k \circ (\psi_k \otimes \mathrm{id}_{A^{\otimes 2k}}) \circ \iota_k^{(3)} \\
&= \iota_k^{(2)} \\
&\approx_{a.u.} \iota_k^{(1)} \\
&= \lambda_k \otimes (\mathrm{id}_{A^{\otimes}} \otimes 1_{A^{\otimes k}}).
\end{aligned}
$$

This shows $\lambda_{2^m} \circ \psi_{2^m} \circ (\mathrm{id}_{A^{\otimes 2^m}} \otimes 1_{A^{\otimes 2^m}})$ is approximately unitarily equivalent to $\lambda_{2^m} \circ \mathrm{id}_{A^{\otimes 2^m}} \otimes 1_{A^{\otimes 2^m}}$ for every $m \in \mathbb{N}$, proving the proposition. $\square$

16.1.11. Proposition. *Let A be a separable unital C*-algebra with approximately inner half-flip. Then there is a sequence of *-homomorphisms*

$$
(\varphi_n : A^{\otimes \infty} \otimes A^{\otimes \infty} \to A^{\otimes \infty})_{n \in \mathbb{N}}
$$

such that

$$
\|\varphi_n(a \otimes 1_{A^{\otimes \infty}}) - a\| \to 0, \quad \text{as } n \to \infty,
$$

for every $a \in A^{\otimes \infty}$.

Proof. By the previous theorem, $A^{\otimes \infty}$ is strongly self-absorbing, so there is a *-isomorphism $\psi : A^{\otimes \infty} \to A^{\otimes \infty} \otimes A^{\otimes \infty}$ and a sequence of unitaries $(u_n)_{n \in \mathbb{N}}$ in $A^{\otimes \infty} \otimes A^{\otimes \infty}$ such that

$$
\|u_n^* \psi(a) u_n - a \otimes 1_{A^{\otimes \infty}}\| \to 0 \text{ as } n \to \infty, \text{ for every } a \in A^{\otimes \infty}.
$$

Then

$$
\begin{aligned}
\|\psi^{-1}(u_n(a \otimes 1_{A^{\otimes \infty}})u_n^*) - a\| &= \|u_n(a \otimes 1_{A^{\otimes \infty}})u_n^* - \psi(a)\| \\
&= \|a \otimes 1_{A^{\otimes \infty}} - u_n^* \psi(a) u_n\| \to 0 \text{ as } n \to \infty.
\end{aligned}
$$

For $n \in \mathbb{N}$, define $\varphi_n : A^{\otimes \infty} \otimes A^{\otimes \infty} \to A^{\otimes \infty}$ by

$$
\varphi_n(a \otimes b) := \psi^{-1}(u_n(a \otimes b)u_n^*),
$$

for $a, b \in A^{\otimes \infty}$. Then the sequence $(\varphi_n)_{n \in \mathbb{N}}$ satisfies the requirements. $\square$

16.1.12. Recall that the commutant A' of $A \subset \mathcal{B}(H)$ is the set of all operators in $\mathcal{B}(H)$ that commute with every operator in A (5.1.5). More generally, if $B \subset A$ is a C*-subalgebra, we define the *relative commutant* by

$$
A \cap B' = \{a \in A \mid ab = ba \text{ for all } b \in B\}.
$$

Let A be a unital separable C*-algebra. We denote by $\prod_{n\in\mathbb{N}} A$ the C*-algebra of bounded sequences in A. Let $\bigoplus_{n\in\mathbb{N}} A$ denote those sequences that converge to zero, which is easily seen to be an ideal in $\Pi_{n\in\mathbb{N}}A$. Let

$$A^\infty := \prod_{n\in\mathbb{N}} A \Big/ \bigoplus_{n\in\mathbb{N}} A.$$

There is a canonical embedding $\iota : A \hookrightarrow \prod_{n\in\mathbb{N}} A$ which sends an element $a \in A$ to the constant sequence $(a_n = a)_{n\in\mathbb{N}}$. This embedding passes to an embedding of A into A^∞, and we will identify A with its image in A^∞ under this map.

16.1.13. Definition. For a unital separable C*-algebra A its *central sequence algebra* is the relative commutant of A in A^∞, that is, the C*-algebra

$$A^\infty \cap A' = \{x \in A^\infty \mid xa = ax \text{ for every } a \in A\}.$$

Note that $A^\infty \cap A'$ consists of sequences $(x_n)_{n\in\mathbb{N}} \subset A$, such that, for all $a \in A$, one has $\|x_n a - a x_n\| \to 0$ as $n \to \infty$ and $(x_n)_{n\in\mathbb{N}} = (y_n)_{n\in\mathbb{N}}$ if $\|x_n - y_n\| \to 0$ as $n \to \infty$. Such a sequence $(x_n)_{n\in\mathbb{N}} \subset A$ is called an *approximately central sequence*.

16.1.14. A sequence of *-homomorphisms $(\varphi_n : A \to B)_{n\in\mathbb{N}}$ is called *approximately central* if for every $a \in A$ and $b \in B$ we have $\|\varphi_n(a)b - b\varphi_n(a)\| \to 0$ as $n \to \infty$. Any approximately central sequence of *-homomorphisms $(\varphi_n : A \to B)_{n\in\mathbb{N}}$ induces a *-homomorphism $\varphi : A^\infty \cap A' \to B$. (Exercise.)

16.1.15. Lemma. *Let A be a unital C*-algebra, and let $\pi : \prod_{n\in\mathbb{N}} A \twoheadrightarrow A^\infty$ denote the quotient map. Suppose that $u \in A^\infty$ is a unitary. Then there is a sequence of unitaries $(u_n)_{n\in\mathbb{N}}$ in A such that $\pi((u_n)_{n\in\mathbb{N}}) = u$.*

Proof. Let $u \in A^\infty$ be a unitary and find $(a_n)_{n\in\mathbb{N}}$ satisfying $\pi((a_n)_{n\in\mathbb{N}}) = u$. Then $\|a_n a_n^* - 1\| \to 0$ and $\|a_n^* a_n - 1\| \to 0$ as $n \to \infty$. Let N be large enough so that

$$\|a_n a_n^* - 1_A\| < 1, \quad \|a_n^* a_n - 1_A\| < 1,$$

for every $n > N$. Then both $a_n a_n^*$ and $a_n^* a_n$ are invertible as long as $n > N$ (Theorem 1.2.4), which implies a_n is invertible. Let $a_n = u_n |a_n|$ be the polar decomposition of a_n. Since a_n is invertible, the isometry u_n is a unitary in A (Exercise 5.4.12). By the above, $\||a_n| - 1_A\| \to 0$ so $\|a_n - u_n\| \to 0$. For $n \leq N$, let $u_n = 1_A$. Then $(u_n)_{n\in\mathbb{N}}$ satisfies our requirements. $\square$

16.1.16. Lemma. *Let A and B be separable C*-algebras and $\varphi : A \to B$ an injective *-homomorphism. If there exist unitaries $u_n \in B^\infty \cap \varphi(A)'$, $n \in \mathbb{N}$, such that*

$$\lim_{n\to\infty} \operatorname{dist}(u_n^* b u_n, \varphi(A)^\infty) = 0,$$

for every $b \in B$, then there is an isomorphism $\psi : A \to B$ which is approximately unitarily equivalent to φ.

Proof. Since B is separable, there exists a sequence which is dense in B. Let $(b_n)_{n\in\mathbb{N}}$ be a such a sequence. For $m \in \mathbb{N}$, let k be sufficiently large so that we can choose $c_1,\ldots,c_m \in \varphi(A)^\infty$ with $\|u_k^* b_j u_k - c_j\| < 1/(2m)$ for every $j = 1,\ldots,m$. Since $c_j \in \varphi(A)^\infty$, there is a bounded sequence $(a_{j,n})_{n\in\mathbb{N}}$ in A such that $\pi((a_{j,n})_{n\in\mathbb{N}}) = c_j$ and, by the previous lemma, unitaries $v_{k,n} \in B$ such that $\pi((v_{k,n})_{n\in\mathbb{N}}) = u_k$, where $\pi : \prod \varphi(A) \to \varphi(A)^\infty \subset B^\infty$ is the quotient map. Since $u_k \in \varphi(A)'$, there is a large enough n_m such that, for every $n \geq n_m$,

$$\|v_{k,n}\varphi(\bar{a}) - \varphi(a)v_{k,n}\| < 1/m,$$

for every $a \in A$ and since $\|u_k^* b_j u_k - c_j\| < 1/(2m)$, we can also assume that

$$\|v_{k,n}^* b_j v_{k,n} - \varphi(a_{n,j})\| < 1/m,$$

for every $1 \leq j \leq m$ and every $n \geq n_m$. Let $w_m := v_{k,n_m}$. Then

$$\lim_{m\to\infty} \|w_m\varphi(a) - \varphi(a)w_m\| = 0.$$

Furthermore, since $(b_n)_{n\in\mathbb{N}}$ is dense, it follows that

$$\lim_{m\to\infty} \text{dist}(w_m^* b w_m, \varphi(A)) = 0,$$

for every $b \in B$. Thus by Proposition 13.2.3 there is an isomorphism $\psi : A \to B$ that is approximately unitarily equivalent to φ. $\qquad\square$

16.1.17. Proposition. *Let $\mathcal{D}$ be a unital separable C^*-algebra with approximately inner half-flip. The following are equivalent:*

(i) *$\mathcal{D}$ is strongly self-absorbing,*

(ii) *there exists a unital *-homomorphism $\gamma : \mathcal{D} \otimes \mathcal{D} \to \mathcal{D}$ such that $\gamma \circ \iota_1$ is approximately unitarily equivalent to the identity $\text{id}_{\mathcal{D}}$,*

(iii) *there is a unital *-homomorphism $\gamma : \mathcal{D} \otimes \mathcal{D} \to \mathcal{D}$ and an approximately central sequence $(\varphi_n)_{n\in\mathbb{N}}$ of unital *-endomorphisms of $\mathcal{D}$,*

(iv) *there exists an approximately central sequence of unital *-homomorphisms $\mathcal{D}^{\otimes\infty} \to \mathcal{D}$,*

(v) *there exists an *-isomorphism $\mathcal{D} \to \mathcal{D}^{\otimes\infty}$.*

Proof. For (i) implies (ii), assume that $\mathcal{D}$ is strongly self-absorbing. Then there are a *-isomorphism $\varphi : \mathcal{D} \to \mathcal{D} \otimes \mathcal{D}$ and a sequence of unitaries $(u_n)_{n\in\mathbb{N}} \subset \mathcal{D} \otimes \mathcal{D}$ such that

$$\|u_n^*\varphi(a)u_n - a \otimes 1_{\mathcal{D}}\| \to 0 \text{ as } n \to \infty, \text{ for every } a \in \mathcal{D}.$$

Let $\gamma := \varphi^{-1}$. Then $(\gamma(u_n))_{n\in\mathbb{N}}$ is a sequence of unitaries such that

$$\|\gamma(u_n)^* a \gamma(u_n) - \gamma(a \otimes 1_{\mathcal{D}})\| \to 0 \text{ as } n \to \infty,$$

that is $\gamma \circ (\text{id}_{\mathcal{D}} \otimes 1_{\mathcal{D}}) \approx_{a.u.} \text{id}_{\mathcal{D}}$, showing (ii).

Now if (ii) holds, then there is a *-homomorphism $\gamma : \mathcal{D} \otimes \mathcal{D} \to \mathcal{D}$ such that $\gamma \circ \iota_1 \approx_{a.u.} \mathrm{id}_{\mathcal{D}}$. Let $(v_n)_{n \in \mathbb{N}} \subset \mathcal{D}$ be a sequence of unitaries implementing the equivalence, that is,

$$\|v_n^* \gamma(a \otimes 1_{\mathcal{D}}) v_n - a\| \to 0 \text{ as } n \to \infty \text{ for every } a \in \mathcal{D}.$$

Define a *-homomorphism $\varphi_n : \mathcal{D} \to \mathcal{D}$ by

$$\varphi_n(a) := v_n^* \gamma(1_{\mathcal{D}} \otimes a) v_n.$$

Then, for each n, φ_n is a unital *-endomorphism of $\mathcal{D}$. Observe that for any $a, b \in \mathcal{D}$, $\varphi_n(a)$ commutes with $v_n^* \gamma(b \otimes 1_{\mathcal{D}}) v_n$. Thus, for any $a, b \in \mathcal{D}$ we have

$$\|\varphi_n(a)b - b\varphi_n(a)\|$$
$$= \|\varphi_n(a)b - \varphi_n(a)v_n^* \gamma(b \otimes 1_{\mathcal{D}}) v_n + v_n^* \gamma(b \otimes 1_{\mathcal{D}}) v_n \varphi_n(a) - b\varphi_n(a)\|$$
$$\leq 2\|b - v_n^* \gamma(b \otimes 1_{\mathcal{D}}) v_n\| \to 0,$$

as $n \to \infty$. Thus the sequence $(\varphi_n)_{n \in \mathbb{N}}$ is approximately central, showing (iii).

Suppose that (iii) holds. Let $\gamma : \mathcal{D} \otimes \mathcal{D} \to \mathcal{D}$ be a unital *-homomorphism and $(\varphi_n : \mathcal{D} \to \mathcal{D})_{n \in \mathbb{N}}$ an approximately central sequence of unital *-endomorphisms. To show (vi), observe that it is enough to construct a unital *-homomorphism $\psi : \mathcal{D}^{\otimes \infty} \to \mathcal{D}$, since then the sequence $(\varphi_n \circ \psi)_{n \in \mathbb{N}}$ will satisfy the requirements. For $k \in \mathbb{N}$, define unital *-homomorphisms $\rho_k : \mathcal{D}^{\otimes (k+1)} \to \mathcal{D}^{\otimes k}$ by setting

$$\rho_k := \mathrm{id}_{\mathcal{D}^{\otimes (k-1)}} \otimes \gamma,$$

and $\psi_k : \mathcal{D}^{\otimes k} \to \mathcal{D}$ by

$$\psi_k := \rho_1 \circ \rho_2 \circ \cdots \circ \rho_{k+1} \circ (\mathrm{id}_{\mathcal{D}^{\otimes k}} \otimes 1_{\mathcal{D}^{\otimes 2}}).$$

Then, for every $k \in \mathbb{N}$ the maps ψ_k are each unital and satisfy

$$\psi_k = \rho_1 \circ \cdots \rho_{k+1} \circ \rho_{k+2} \circ (\mathrm{id}_{\mathcal{D}^{\otimes k}} \otimes 1_{\mathcal{D}^{\otimes 3}}) = \psi_{k+1} \circ (\mathrm{id}_{\mathcal{D}^{\otimes k}} \otimes 1_{\mathcal{D}}).$$

That is to say, the diagram

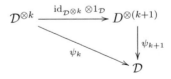

commutes, so by the universal property of the inductive limit, we have an induced unital *-homomorphism $\psi : \mathcal{D}^{\otimes \infty} \to \mathcal{D}$, showing (iv).

Suppose now that (iv) holds. Since $\mathcal{D}$ has approximately inner half-flip, $\mathcal{D}^{\otimes \infty}$ is strongly self-absorbing by Proposition 16.1.10. In particular $\mathcal{D}^{\otimes \infty}$ is simple and nuclear. Define

$$\varphi : \mathcal{D} \to \mathcal{D} \otimes \mathcal{D}^{\otimes \infty}, \quad d \mapsto d \otimes 1_{\mathcal{D}^{\otimes \infty}},$$

which is evidently an injective *-homomorphism. Thus, there exist unital *-homo-morphisms $\alpha_n : \mathcal{D}^{\otimes\infty} \to \varphi(\mathcal{D}) \subset \mathcal{D} \otimes \mathcal{D}^{\otimes\infty}$ such that

$$\|\alpha_n(a)\varphi(d) - \varphi(d)\alpha_n(a)\| \to 0 \text{ as } n \to \infty.$$

The asymptotic centrality of the sequence $(\alpha_n)_{n\in\mathbb{N}}$ induces a *-homomorphism

$$\alpha : \mathcal{D}^{\otimes\infty} \to \varphi(\mathcal{D})^\infty \cap \varphi(\mathcal{D})', \quad a \mapsto (\alpha_n(a))_{n\in\mathbb{N}}.$$

Define

$$\beta : \mathcal{D}^{\otimes\infty} \to (\mathcal{D} \otimes \mathcal{D}^{\otimes\infty})^\infty \cap \varphi(\mathcal{D})', \quad d \mapsto 1_\mathcal{D} \otimes d,$$

where we consider the $1_\mathcal{D} \otimes d$ as a constant sequence in $(\mathcal{D} \otimes \mathcal{D}^{\otimes\infty})^\infty$. Observe that the images of α and β commute in $(\mathcal{D} \otimes \mathcal{D}^{\otimes\infty})^\infty \cap \varphi(\mathcal{D})'$. Thus, by Theorem 6.3.2 they induce a map

$$\alpha \otimes \beta : \mathcal{D}^{\otimes\infty} \otimes D^{\otimes\infty} \to (\mathcal{D} \otimes D^{\otimes\infty})^\infty \cap \varphi(\mathcal{D})'.$$

It follows that

$$\mathcal{D}^{\otimes\infty} \otimes \mathcal{D}^{\otimes\infty} \cong \mathrm{C}^*(\alpha(\mathcal{D}^{\otimes\infty}), \beta(\mathcal{D}^{\otimes\infty})) \subset (\mathcal{D} \otimes \mathcal{D}^{\otimes\infty})^\infty \cap \varphi(\mathcal{D})'.$$

Since $\mathcal{D}^{\otimes\infty}$ has approximately inner half-flip, there is a sequence of unitaries $(u_n)_{n\in\mathbb{N}} \subset \mathrm{C}^*(\alpha(\mathcal{D}^{\otimes\infty}), \beta(\mathcal{D}^{\otimes\infty}))$, such that $u_n^*\beta(d)u_n \to \alpha(d)$ as $n \to \infty$ for every $d \in \mathcal{D}^{\otimes\infty}$.

Let $a \in \mathcal{D}$ and $d \in \mathcal{D}^{\otimes\infty}$. We have

$$\lim_{n\to\infty} u_n^*(a \otimes d)u_n = \lim_{n\to\infty} u_n^*\varphi(a)\beta(d)u_n.$$

Since $\mathrm{C}^*(\alpha(\mathcal{D}^{\otimes\infty}), \beta(\mathcal{D}^{\otimes\infty}))$ commutes with $\varphi(\mathcal{D})$, we further compute

$$\lim_{n\to\infty} u_n^*\varphi(a)\beta(d)u_n = \varphi(a)u_n^*\beta(d)u_n = \varphi(a)\alpha(d) = \alpha(d)\varphi(a).$$

Thus, for every $c \in \mathcal{D} \otimes \mathcal{D}^{\otimes\infty}$ we have $\lim_{n\to\infty} u_n^*cu_n \in \varphi(\mathcal{D})^\infty$. So

$$\mathrm{dist}(u_n^*cu_n, \varphi(\mathcal{D})^\infty) \to 0 \text{ as } n \to \infty.$$

Now by Lemma 16.1.16, there exists an isomorphism $\mathcal{D} \to \mathcal{D}^{\otimes\infty}$, showing (iv) implies (v). Finally, that (v) implies (i) is obvious since $\mathcal{D}^{\otimes\infty}$ is strongly self-absorbing. $\square$

16.1.18. Corollary. *Let A and $\mathcal{D}$ be separable unital C^*-algebras and suppose that $\mathcal{D}$ is strongly self-absorbing. Then any two unital *-homomorphisms*

$$\alpha, \beta : \mathcal{D} \to A \otimes \mathcal{D}$$

are approximately unitarily equivalent.

Proof. Since $\mathcal{D}$ is strongly self-absorbing, by Proposition 16.1.11 there is sequence of *-homomorphisms $\varphi_n : \mathcal{D}^{\otimes\infty} \otimes \mathcal{D}^{\otimes\infty} \to \mathcal{D}^{\otimes\infty}$ satisfying $\|\varphi_n(d \otimes 1_{\mathcal{D}^{\otimes\infty}}) - d\| \to 0$ as $n \to \infty$, for every $d \in \mathcal{D}^{\otimes\infty}$. Putting this together with the existence of *-isomorphisms $\mathcal{D} \to \mathcal{D}^{\otimes\infty}$ and $\mathcal{D}^{\otimes\infty} \to \mathcal{D}$ from Proposition 16.1.17 (iv) and (v), we get a sequence $\psi_n : \mathcal{D} \otimes \mathcal{D} \to \mathcal{D}$ such that

$$\lim_{n\to\infty} \|\psi_n \circ (\mathrm{id}_{\mathcal{D}} \otimes 1_{\mathcal{D}})(d) - d\| = 0,$$

for every $d \in \mathcal{D}$. For $n \in \mathbb{N}$ define a unital *-homomorphism

$$\tilde{\alpha}_n := (\mathrm{id}_A \otimes \psi_n) \circ (\alpha \otimes \mathrm{id}_{\mathcal{D}}) \circ (\mathrm{id}_{\mathcal{D}} \otimes 1_{\mathcal{D}}),$$

and similarly, for $n \in \mathbb{N}$ define

$$\tilde{\beta}_n := (\mathrm{id}_A \otimes \psi_n) \circ (\beta \otimes \mathrm{id}_{\mathcal{D}}) \circ (\mathrm{id}_{\mathcal{D}} \otimes 1_{\mathcal{D}}).$$

Observe that $\tilde{\alpha}_n \to \alpha$ and $\tilde{\beta}_n \to \beta$ pointwise.

Since $\mathcal{D}$ has approximately inner half-flip, and both α and β are unital, Proposition 13.2.2 (ii) implies

$$\begin{aligned}
\tilde{\alpha}_n &\approx_{a.u.} (\mathrm{id}_A \otimes \psi_n) \circ (\alpha \otimes \mathrm{id}_{\mathcal{D}}) \circ (1_{\mathcal{D}} \otimes \mathrm{id}_{\mathcal{D}}) \\
&= (\mathrm{id}_A \otimes \psi_n) \circ (1_A \otimes 1_{\mathcal{D}} \otimes \mathrm{id}_{\mathcal{D}}) \\
&= (\mathrm{id}_A \otimes \psi_n) \circ (\beta \otimes \mathrm{id}_{\mathcal{D}}) \circ (1_{\mathcal{D}} \otimes \mathrm{id}_{\mathcal{D}}) \\
&\approx_{a.u.} \tilde{\beta}_n.
\end{aligned}$$

Then $\tilde{\alpha}_n \approx_{a.u.} \tilde{\beta}_n$ for every $n \in \mathbb{N}$ by Proposition 13.2.2 (i), and by Proposition 13.2.2 $\alpha \approx_{a.u.} \beta$. $\qquad\square$

16.1.19. Let $\mathcal{Z}$ denote the Jiang–Su algebra (Definition 15.2.8). In addition to the construction of $\mathcal{Z}$ and classification of simple unital inductive limits of dimension-drop algebras, in [62] Jiang and Su also prove a number of technical properties enjoyed by $\mathcal{Z}$. Since the proofs use some things we have not introduced, we simply state them here. Together they imply Corollary 16.1.20 below, that $\mathcal{Z}$ is strongly self-absorbing.

Proposition. *Let $\mathcal{Z}$ be the Jiang–Su algebra. Then*

(i) *$\mathcal{Z}$ has approximately inner half-flip.*

(ii) *There exists a unital *-homomorphism $\psi : \mathcal{Z} \otimes \mathcal{Z} \to \mathcal{Z}$.*

(iii) *There exists an approximately central sequence of inner automorphisms $\mathcal{Z} \to \mathcal{Z}$.*

16.1.20. Corollary. *$\mathcal{Z}$ is strongly self-absorbing.*

Proof. By Proposition 16.1.19 (i), $\mathcal{Z}$ is a separable unital C*-algebra with approximately inner half-flip. By Proposition 16.1.17, the property that $\mathcal{Z}$ is strongly self-absorbing is equivalent to the existence of a unital *-homomorphism $\psi : \mathcal{Z} \otimes \mathcal{Z} \to \mathcal{Z}$ and an approximately central sequence of unital endomorphisms. Thus, by Proposition 16.1.19 (ii) and (iii), $\mathcal{Z}$ is strongly self-absorbing. $\qquad\square$

16.1.21. We have already met some other strongly self-absorbing C*-algebras, namely the UHF algebras of infinite type (exercise!) and the Cuntz algebras $\mathcal{O}_2$ and $\mathcal{O}_\infty$ [122, Examples 1.14 (ii)]. Let $\mathcal{U}$ denote any UHF algebra of infinite type. Then, by Proposition 16.1.8, the tensor products $\mathcal{U} \otimes \mathcal{O}_2, \mathcal{U} \otimes \mathcal{O}_\infty$, and $\mathcal{O}_\infty \otimes \mathcal{O}_2$ are all strongly self-absorbing. However, if A is simple, separable, unital and nuclear, Kirchberg and Phillips proved that $A \otimes \mathcal{O}_2 \cong \mathcal{O}_2$ [67, Theorem 3.8]. So the tensor products $\mathcal{U} \otimes \mathcal{O}_2$ and $\mathcal{O}_\infty \otimes \mathcal{O}_2$ do not give us anything new. If we take the tensor product of $\mathcal{D} \otimes \mathcal{Z}$, where $\mathcal{D}$ is any strongly self-absorbing C*-algebra, then $\mathcal{D} \otimes \mathcal{Z} \cong \mathcal{D}$ [135], so once again we don't get anything new upon taking tensor products. The UCT (see 12.5.14) is satisfied by each of $\mathcal{Z}, \mathcal{U}, \mathcal{O}_\infty, \mathcal{O}_\infty \otimes \mathcal{U}$ and $\mathcal{O}_2$ (where $\mathcal{U}$ is any UHF algebra of infinite type). In fact, these are the *only* strongly self-absorbing C*-algebras satisfying the UCT [118, Corollary 6.7].

16.2 $\mathcal{D}$-stability

Let A be a simple separable unital nuclear C*-algebra. Recall from Chapter 15 that, provided $(K_0(A), K_0(A)_+)$ is weakly unperforated, the Elliott invariant cannot tell the difference between A and $A \otimes \mathcal{Z}$. We would like to lift isomorphisms of Elliott invariants to *-isomorphisms of C*-algebras, so we would like to know when $A \cong A \otimes \mathcal{Z}$, that is, when A is $\mathcal{Z}$-stable. In the purely infinite case, one can ask similar questions about $\mathcal{O}_\infty$ since the Kirchberg–Phillips classification tells us that a simple, separable, unital purely infinite C*-algebra which satisfies the UCT is $\mathcal{O}_\infty$-stable [65, 93]. Although this book focuses more on the stably finite case, this next section considers the general case of $\mathcal{D}$-stability, where $\mathcal{D}$ can be any strongly self-absorbing C*-algebra.

16.2.1. Definition. Let $\mathcal{D}$ be a strongly self-absorbing C*-algebra. We say that a C*-algebra A is $\mathcal{D}$-*stable*, or $\mathcal{D}$-*absorbing*, if

$$A \otimes \mathcal{D} \cong A.$$

16.2.2. Proposition ([122, Theorem 2.3]). *For any $\mathcal{D}$-stable C*-algebra A there is a sequence of *-isomorphisms $(\varphi_n : A \otimes \mathcal{D} \to A)_{n \in \mathbb{N}}$ such that*

$$\lim_{n \to \infty} \|\varphi_n(a \otimes 1_\mathcal{D}) - a\| = 0$$

for every $a \in A$.

Proof. Since $\mathcal{D}$ is strongly self-absorbing, it follows from Proposition 16.1.17 that $\mathcal{D} \cong \mathcal{D}^{\otimes\infty}$. Since A is $\mathcal{D}$-stable, we may identify A with $A \otimes \mathcal{D}^{\otimes\infty}$. Define

$$\varphi_n : A \otimes \mathcal{D}^{\otimes\infty} \otimes \mathcal{D} \to A \otimes \mathcal{D}^{\otimes\infty}$$

to be the *-isomorphism that fixes A and the first n copies of $\mathcal{D}$ in $\mathcal{D}^{\otimes\infty}$, sends the last copy of $\mathcal{D}$ to the $n+1$ position in $\mathcal{D}^{\otimes\infty}$, and, for $m > n$, shifts the mth copy of $\mathcal{D}$ to the $m + 1$ position. Then, an elementary argument shows that the sequence $(\varphi_n)_{n \in \mathbb{N}}$ satisfies

$$\lim_{n \to \infty} \|\varphi_n(a \otimes 1_{\mathcal{D}}) - a\| = 0$$

for every $a \in A$ (exercise). $\square$

16.2.3. Theorem. *Let A and $\mathcal{D}$ be unital separable C*-algebras, with $\mathcal{D}$ strongly self-absorbing. Then there is a *-isomorphism $\varphi : A \to A \otimes \mathcal{D}$ if and only if there is a unital *-homomorphism*

$$\psi : \mathcal{D} \to A^\infty \cap A'.$$

In this case, φ and

$$\iota : A \to A \otimes \mathcal{D}, \quad a \mapsto a \otimes 1_{\mathcal{D}}$$

are approximately unitarily equivalent.

Proof. Suppose there is a *-isomorphism $\varphi : A \to A \otimes \mathcal{D}$. By Proposition 16.1.17 (iii) we can find an approximately central sequence of unital *-endomorphisms of $\mathcal{D}$, say $\rho_n : \mathcal{D} \to \mathcal{D}$, $n \in \mathbb{N}$. Define

$$\psi_n : \mathcal{D} \to A, \quad d \mapsto \varphi^{-1}(1_A \otimes \rho_n(d)).$$

Note that each ψ_n is unital. We will show that the sequence ψ_n is approximately central in A. Fix $d \in \mathcal{D}$ and $a \in A$. For any $\epsilon > 0$ there are $m \in \mathbb{N}$, $a_i \in A$, $d_i \in \mathcal{D}$, $i = 1, \ldots, m$ such that $\|\sum_{i=1}^m a_i \otimes d_i - \varphi(a)\| < \epsilon$. Then

$$\|\varphi^{-1}(1_A \otimes \rho_n(d))a - a\,\varphi^{-1}(1_A \otimes \rho_n(d))\|$$
$$= \|(1_A \otimes \rho_n(d))\varphi(a) - \varphi(a)(1_A \otimes \rho_n(d))\|$$
$$= 2\epsilon + \left\|(1_A \otimes \rho_n(d))\left(\sum_{i=1}^m a_i \otimes d_i\right) - \left(\sum_{i=1}^m a_i \otimes d_i\right)(1_A \otimes \rho_n(d))\right\|$$
$$= 2\epsilon + \left\|\sum_{i=1}^m a_i \otimes \rho_n(d)d_i - \sum_{i=1}^m a_i \otimes d_i\rho_n(d)\right\| \to 2\epsilon,$$

as $n \to \infty$. Since ϵ was arbitrary, this shows that $(\psi_n)_{n \in \mathbb{N}}$ is an approximately central sequence of unital *-homomorphisms. Thus,

$$\psi : \mathcal{D} \to A^\infty \cap A', \quad d \mapsto (\psi_n(d))_{n \in \mathbb{N}},$$

is a well-defined unital *-homomorphism satisfying our requirements.

Now assume that we have a unital *-homomorphism $\psi : \mathcal{D} \to A^\infty \cap A'$. Note that $\iota : A \to A \otimes \mathcal{D}$ induces a map $A^\infty \cap A' \to \iota(A)^\infty \cap \iota(A)'$. Let

$$\alpha : \mathcal{D} \to \iota(A)^\infty \cap \iota(A)' \subset (A \otimes \mathcal{D})_\infty \cap \iota(A)'$$

denote the composition of ψ with the map induced by ι. For $d \in \mathcal{D}$, let $\beta(d) = 1_A \otimes d$, considered as a constant sequence embedded in $(A \otimes \mathcal{D})_\infty$. Then $\beta(d)$ commutes with $\iota(A)$, so in fact we have

$$\beta : \mathcal{D} \to (A \otimes \mathcal{D})_\infty \cap \iota(A)'.$$

Observe that the images of α and β commute. Thus by Theorem 6.3.2 they induce an injective *-homomorphism

$$\mathcal{D} \otimes \mathcal{D} \to (A \otimes \mathcal{D})_\infty \cap \iota(A)',$$

and under the identification

$$\mathcal{D} \otimes \mathcal{D} \cong \mathrm{C}^*(\alpha(\mathcal{D}), \beta(\mathcal{D})) \subset (A \otimes \mathcal{D})_\infty \cap \iota(A)',$$

α and β are simply the first and second factor embeddings, respectively. Since $\mathcal{D}$ is strongly self-absorbing and therefore has an approximately inner half-flip, there are unitaries $(u_n)_{n \in \mathbb{N}} \subset \mathrm{C}^*(\alpha(\mathcal{D}), \beta(\mathcal{D}))$ such that

$$\|u_n^* \beta(d)u_n - \alpha(d)\| \to 0 \text{ as } n \to \infty$$

for every $d \in \mathcal{D}$. Thus, for every $a \in A$ and $d \in \mathcal{D}$ we have

$$\lim_{n \to \infty} u_n^*(a \otimes b)u_n = \lim_{n \to \infty} u_n^* \iota(a)\beta(d)u_n = \lim_{n \to \infty} \iota(a)\alpha(b).$$

Since $\iota(a)\alpha(b) \in \iota(A)^\infty$, we have

$$\lim_{n \to \infty} \mathrm{dist}(u_n^* x u_n, \iota(A)^\infty) = 0.$$

Thus, by Lemma 16.1.16, there is a *-isomorphism $\varphi : A \to A \otimes \mathcal{D}$ which is approximately unitarily equivalent to ι. $\square$

16.2.4. The corollary is a slight reformulation of the previous theorem, which will be more suitable for our purposes.

Corollary. *Let A and $\mathcal{D}$ be unital separable C*-algebras with $\mathcal{D}$ strongly self-absorbing. Then there is an isomorphism $\varphi : A \to A \otimes \mathcal{D}$ if and only if there is a unital *-homomorphism*

$$\psi : A \otimes \mathcal{D} \to A^\infty,$$

satisfying $\psi(a \otimes 1_\mathcal{D}) = a$ for every $a \in A$. In this case, the maps φ and $\mathrm{id}_A \otimes 1_\mathcal{D}$ are approximately unitarily equivalent.

Proof. Exercise. $\square$

16.2.5. Theorem. *Let $\mathcal{D}$ be a unital strongly self-absorbing C*-algebra and let A be unital, separable and $\mathcal{D}$-stable. Suppose that $p \in A$ is a projection. Then pAp is $\mathcal{D}$-stable.*

Proof. Let $B := pAp$. Identifying p with its image in $B^\infty \subset A^\infty$, the map

$$\beta : A^\infty \to B^\infty, \quad x \mapsto pxp,$$

is c.p.c. (Exercise 3.4.5) and $\beta(b) = b$ for every $b \in B$. Since A is $\mathcal{D}$-stable, we have, using Corollary 16.2.4, a *-homomorphism $\psi : A \otimes \mathcal{D} \to A^\infty$ satisfying $\psi(a \otimes 1_\mathcal{D}) = a$ for every $a \in A$. Let

$$\tilde{\psi} : B \otimes \mathcal{D} \to B^\infty$$

be the c.p.c. map given by the composition $\tilde{\psi} := \beta \circ \psi \circ (\iota \otimes \mathrm{id}_\mathcal{D})$, where $\iota : B \to A$ is the inclusion map. For $b \in B_+$ we have

$$\tilde{\psi}(b \otimes 1_\mathcal{D}) = \beta \circ \psi(b \otimes 1_\mathcal{D}) = \beta(b) = b,$$

and for $d \in \mathcal{D}_+$ we have

$$\begin{aligned}
\tilde{\psi}(b \otimes d) &= \beta(\psi(b^{1/4} \otimes 1_\mathcal{D})\psi(b^{1/2} \otimes 1_\mathcal{D})\psi(b^{1/4} \otimes 1_\mathcal{D})) \\
&= p(b^{1/4} \otimes 1_\mathcal{D})(\psi(b^{1/2} \otimes 1_\mathcal{D}))(b^{1/4} \otimes 1_\mathcal{D})p \\
&= (b^{1/4} \otimes 1_\mathcal{D})(\psi(b^{1/2} \otimes 1_\mathcal{D}))(b^{1/4} \otimes 1_\mathcal{D}) \\
&= \psi(b \otimes d).
\end{aligned}$$

This shows that $\psi(B \otimes \mathcal{D}) \subset B^\infty \subset A^\infty$ and that $\tilde{\psi} = \psi$ is multiplicative, since ψ is. Thus, $\tilde{\psi}$ is a *-homomorphism satisfying the conditions of Corollary 16.2.4, showing that B is $\mathcal{D}$-stable. $\qquad\square$

16.2.6. Remark. In [122], Toms and Winter show that $\mathcal{D}$-stability passes not only to corners, but to any hereditary C*-subalgebra. To show that, we'd need Corollary 16.2.4 for A not necessarily unital, which is quite a bit trickier to prove. For the interested reader, the details can be found in their paper.

16.2.7. Theorem. *Let $\mathcal{D}$ be a unital strongly self-absorbing C*-algebra and let $A = \lim(A_n, \varphi_n)$ be a unital inductive limit of separable and $\mathcal{D}$-stable C*-algebras A_n with injective connecting maps $\varphi_n : A_n \to A_{n+1}$. Then A is $\mathcal{D}$-stable.*

Proof. For every $n \in \mathbb{N}$ let $\varphi^{(n)} : A_n \to A$ and, for $n < m$, $\varphi_{n,m} : A_n \to A_m$ denote the induced maps as defined in 8.2.5. Since the maps φ_n are injective, we have $\varphi^{(n)}(A_n) \cong A_n$. By Proposition 16.1.11, there are *-homomorphisms $\psi_m : \mathcal{D} \otimes \mathcal{D} \to \mathcal{D}$, $m \in \mathbb{N}$ such that $\|\psi_m(d \otimes 1_\mathcal{D}) - d\| \to 0$ as $m \to \infty$. For each natural number n, define $\rho_m^{(n)} : \varphi^{(n)}(A_n) \otimes \mathcal{D} \otimes \mathcal{D} \to \varphi^{(n)}(A) \otimes \mathcal{D}$ by

$$\rho_m^{(n)} := \mathrm{id}_{\varphi^{(n)}(A_n)} \otimes \psi_m.$$

Then, for $a \in \varphi^{(n)}(A_n)$,

$$\|\rho_n^{(m)}(a \otimes d \otimes 1_\mathcal{D}) - a \otimes d \otimes 1_\mathcal{D}\| \to 0, \quad \text{as } m \to \infty.$$

Since $A_n \otimes \mathcal{D} \cong A_n$ and hence also $\varphi^{(n)}(A) \otimes \mathcal{D} \cong \varphi^{(n)}(A)$, this gives us maps, which by abuse of notation we still denote $\rho_m^{(n)}$,

$$\rho_m^{(n)} : \varphi^{(n)}(A_n) \otimes \mathcal{D} \to \varphi^{(n)}(A_n) \subset A,$$

satisfying $\|\rho_m^{(n)}(a \otimes 1_\mathcal{D}) - a\| \to 0$ as $m \to \infty$, for every $a \in \varphi^{(n)}(A_n)$.

Since A_n is separable, we can fix a dense subsequence $(a_i)_{i \in \mathbb{N}}$. For every $i \in \mathbb{N}$, and $N \geq n$, there exists m_N such that

$$\|\rho_{m_N}^{(N)}(\varphi^{(n)}(a_i) \otimes 1_\mathcal{D}) - \varphi^{(n)}(a_i)\|$$
$$= \|\rho_{m_N}^{(N)}(\varphi^{(N)} \circ \varphi_{n,N}(a_i) \otimes 1_\mathcal{D}) - \varphi^{(N)} \circ \varphi_{n,N}(a_i)\| < 1/(3N)$$

Let $a \in A_n$ and choose $i \in \mathbb{N}$ such that $\|a - a_i\| < 1/(3N)$. Then

$$\|\rho_{m_N}^{(N)}(\varphi^{(n)}(a) \otimes 1_\mathcal{D}) - \varphi^{(n)}(a)\| = \|\rho_{m_N}^{(N)}(\varphi^{(n)}(a) \otimes 1_\mathcal{D}) - \rho_{m_N}^{(N)}(\varphi^{(n)}(a_i) \otimes 1_\mathcal{D})\|$$
$$+ \|\rho_{m_N}^{(N)}(\varphi^{(n)}(a_i) \otimes 1_\mathcal{D}) - \varphi^{(n)}(a_i)\|$$
$$+ \|\varphi^{(n)}(a_i) - \varphi^{(n)}(a)\|$$
$$< 1/N.$$

Thus we can find a sequence $(i_n)_{n \in \mathbb{N}} \subset \mathbb{N}$ such that,

$$\rho_{i_n}^{(n)}(a \otimes 1_\mathcal{D}) \to a, \quad n \to \infty,$$

for every $a \in \varphi^{(m)}(A_m)$. Define

$$\tilde{\rho} : \bigcup_{n \in \mathbb{N}} \varphi^{(n)}(A_n) \otimes \mathcal{D} \to \prod_{n \in \mathbb{N}} \varphi^{(n)}(A_n) \subset \prod_{n \in \mathbb{N}} A$$

by defining $\tilde{\rho}(x)$ to be the sequence with nth coordinate entry $(\tilde{\rho}(x))_n$ given by

$$(\tilde{\rho}(x))_n := \begin{cases} \rho_{i_n}^{(n)}(x) & x \in \varphi^{(n)}(A_n) \otimes \mathcal{D}, \\ 0 & \text{otherwise}. \end{cases}$$

Then $\tilde{\rho}$ induces a map

$$\hat{\rho} : \bigcup_{n \in \mathbb{N}} \varphi^{(n)}(A_n) \otimes \mathcal{D} \to A^\infty,$$

which is multiplicative and $*$-preserving. Furthermore, $\hat{\rho}$ satisfies $\hat{\rho}(a \otimes 1_\mathcal{D}) = a$ for every $a \in A$. Since $\hat{\rho}$ is a $*$-homomorphism on $\varphi^{(n)}(A_n) \otimes \mathcal{D}$, it is norm-decreasing on $\bigcup_{n \in \mathbb{N}} \varphi^{(n)}(A_n) \otimes \mathcal{D}$, which is a dense subalgebra of $A \otimes \mathcal{D}$. Thus $\hat{\rho}$ extends to a $*$-homomorphism

$$\rho : A \otimes \mathcal{D} \to A^\infty,$$

that satisfies $\rho(a \otimes 1_\mathcal{D}) = a$ for every $a \in A$. Thus by Corollary 16.2.4, A is $\mathcal{D}$-stable. $\qquad \square$

To end the chapter, let us return to the Cuntz semigroup of a $\mathcal{Z}$-stable C*-algebra. Let A be a separable, unital C*-algebra. We saw in Theorem 15.4.5 at the end of the previous chapter that the Cuntz semigroup of $A \otimes \mathcal{Z}$ is always almost unperforated. Using Corollary 16.1.20, we are able to show that for any separable unital C*-algebra A, if A is $\mathcal{Z}$-stable then $W(A)$ is weakly divisible (Definition 14.3.35). In particular, we see that asking that $W(A)$ is weakly divisible in Corollary 15.4.8 was in fact redundant.

16.2.8. Theorem. *Suppose that A is a separable, unital, $\mathcal{Z}$-stable C*-algebra. Then $W(A)$ is weakly divisible.*

Proof. We need to show that for every $x \in W(A)_+$ and $n \in \mathbb{N}$, there exists $y \in W(A)_+$ such that $ny \le x \le (n+1)y$. We claim it is enough to assume that $a = b \otimes 1_{\mathcal{Z}}$ for some $b \in A_+$. So let $a \in A \otimes \mathcal{Z}$ be a positive element and choose a *-isomorphism $\varphi : \mathcal{Z} \otimes \mathcal{Z} \to \mathcal{Z}$. Let

$$\psi := (\mathrm{id}_{\mathcal{Z}} \otimes 1_{\mathcal{Z}}) \circ \varphi : \mathcal{Z} \otimes \mathcal{Z} \otimes 1_{\mathcal{Z}} \to \mathcal{Z} \otimes 1_{\mathcal{Z}}.$$

By Corollary 16.1.18 with $A = \mathcal{D} = \mathcal{Z}$, we see that ψ is approximately inner. Thus

$$\mathrm{id}_A \otimes \psi : A \otimes \mathcal{Z} \otimes \mathcal{Z} \to A \otimes \mathcal{Z} \otimes 1_{\mathcal{Z}}$$

is also approximately inner. Let $(u_n)_{n \in \mathbb{N}} \subset A \cong A \otimes \mathcal{Z}$ be a sequence of unitaries satisfying

$$\|u_n a u_n^* - (\mathrm{id}_A \otimes \psi)(a)\| \to 0, \text{ as } n \to \infty.$$

In particular, for every $\epsilon > 0$ there exists n such that $\|u_n a u_n^* - (\mathrm{id}_A \otimes \psi)(a)\| < \epsilon$. Then by Proposition 14.1.7, we have $((\mathrm{id}_A \otimes \psi)(a) - \epsilon)_+ \precsim u_n a u_n^* \sim a$ so by Theorem 14.1.9 (ii), $(\mathrm{id}_A \otimes \psi)(a) \precsim a$. Similarly $a \precsim (\mathrm{id}_A \otimes \psi)(a)$. Since $\mathrm{id}_A \otimes \psi(a) \in A \otimes 1_{\mathcal{Z}}$, this proves the claim.

Now, by Lemma 15.4.2, for every $n \in \mathbb{N}$ there is a positive element $e_n \in \mathcal{Z}$ such that $n[e_n] \le [1_{\mathcal{Z}}] \le (n+1)[e_n] \in W(\mathcal{Z})$. Let $a \in A_+$. Then, by Lemma 15.4.3 we have $n[a \otimes e_n] \le [a \otimes 1_{\mathcal{Z}}] \le (n+1)[a \otimes e_n]$. $\qquad\square$

16.3 Exercises

16.3.1. Let A and B be C*-algebras. Show that any approximately central sequence of *-homomorphisms $(\varphi_n : A \to B)_{n \in \mathbb{N}}$ induces a *-homomorphism from the central sequence algebra of A to B, that is $\varphi : A^\infty \cap A' \to B$

16.3.2. Let A be a unital C*-algebra.

(i) Suppose that A has approximately inner flip. Let $n \in \mathbb{N} \setminus \{0\}$ and i, j with $1 \le i < j \le n$. Define a *-homomorphism

$$\sigma_{i,j} : A^{\otimes n} \to A^{\otimes n}$$

on simple tensors by

$$a_1 \otimes \cdots \otimes a_i \otimes \cdots \otimes a_j \otimes \cdots \otimes a_n \mapsto a_1 \otimes \cdots \otimes a_j \otimes \cdots \otimes a_i \otimes \cdots \otimes a_n.$$

Show that $\sigma_{i,j}$ is approximately inner.

(ii) Define

$$\iota_i : A \hookrightarrow A^{\otimes n} : a \mapsto 1_A \otimes \cdots \otimes 1_A \otimes a \otimes 1_A \otimes \cdots \otimes 1_A,$$

where a is in the ith tensor factor. Show that if A has approximately inner half-flip, then ι_i is approximately unitarily equivalent to ι_j.

16.3.3. Fill in the missing final details of the proof of Proposition 16.2.2.

16.3.4. Let $\mathcal{U}$ be a UHF algebra. Show that every automorphism of $\mathcal{U}$ is approximately inner, that is, if $\varphi : \mathcal{U} \to \mathcal{U}$ is *-automorphism, then there is a sequence of unitaries $(u_n)_{n\in\mathbb{N}}$ such that $\|u_n a u_n - \varphi(a)\| \to 0$ as $n \to \infty$, for every $a \in \mathcal{U}$.

16.3.5. Let $\mathcal{U}$ be a UHF algebra. Show that $\mathcal{A}$ is strongly self-absorbing if and only if $\mathcal{U}$ is of infinite type.

16.3.6. Let $\mathcal{Q}$ denote the UHF algebra whose associated supernatural number is $\mathfrak{p} = \prod_{p \text{ prime}} p^\infty$ (see 8.3.5). The C*-algebra $\mathcal{Q}$ is called the *universal UHF algebra*.

(i) Show that $M_n \otimes \mathcal{Q} \cong \mathcal{Q}$ for every natural number $n \geq 1$.

(ii) Show that $\mathcal{U} \otimes \mathcal{Q} \cong \mathcal{Q}$ for every UHF algebra $\mathcal{U}$.

16.3.7. Let A and $\mathcal{D}$ be unital separable C*-algebras with $\mathcal{D}$ strongly self-absorbing. Use Theorem 16.2.3 to show that there is an isomorphism $\varphi : A \to A \otimes \mathcal{D}$ if and only if there is a unital *-homomorphism

$$\psi : A \otimes \mathcal{D} \to A^\infty,$$

satisfying $\psi(a \otimes 1_\mathcal{D}) = a$ for every $a \in A$. Show that in this case, the maps φ and $\mathrm{id}_A \otimes 1_\mathcal{D}$ are approximately unitarily equivalent.

17 Nuclear dimension

Theorem 7.2.2 says that C*-algebras with the completely positive approximation property (c.p.a.p.) are exactly those C*-algebras that are nuclear. In 13.3.11 we saw that we need to restrict our target for classification to nuclear C*-algebras, since otherwise when passing to the less complicated setting of von Neumann algebras there is already little hope for any reasonable classification theorem. When a C*-algebra is nuclear, its enveloping von Neumann algebra is injective, or equivalently *hyperfinite* [26, 56]. A von Neumann algebra is hyperfinite if it is generated by an increasing net of matrix algebras. Thus one might view the AF algebras of Chapter 8 as the C*-analogue of a hyperfinite von Neumann algebra, however, as we have seen, this excludes many interesting nuclear C*-algebras. The c.p.a.p. guarantees good approximation properties via finite-dimensional algebras which one might hope are tractable in the manner of hyperfiniteness in von Neumann algebras. However, thanks to the existence of simple separable nuclear C*-algebras which are not $\mathcal{Z}$-stable, we know that the c.p.a.p. is not enough.

Another interpretation of the c.p.a.p. is as a noncommutative partition of unity, as can be seen in Exercise 7.3.11. The covering dimension (Definition 8.6.2), if finite, asks that a partition of unity can be refined so that there is a bound on how many functions have a nonzero value at the same point. The subject of this chapter, the nuclear dimension, is a refinement of the c.p.a.p. akin to passing from partitions of unity to covering dimension. Just as infinite-dimensional topological spaces are generally quite wild animals, we would like to leave those C*-algebras with infinite nuclear dimension in the operator algebra jungle rather than invite them home. It turns out that what is required to ensure that a simple separable unital C*-algebra is nuclear and $\mathcal{Z}$-stable is precisely that it have finite nuclear dimension. Thus instead of restricting the scope of classification by insisting on the rather mysterious property of $\mathcal{Z}$-stability, we can equivalently restrict to those C*-algebras with finite "noncommutative covering dimension", a property that seems quite natural. This is one of the most remarkable results in the classification program and will be discussed more in the next and final chapter.

In this chapter, we begin our discussion with c.p.c. maps that are *order zero*; these will be key to refining the c.p.a.p. to define the nuclear dimension. In the second section we define the nuclear dimension and the closely related *decomposition rank* and compare the nuclear dimension and decomposition rank of a commutative C*-algebra $C(X)$ to the covering dimension of X. Next, we look at permanence properties of the nuclear dimension. In the final section, we show that the C*-algebras with nuclear dimension zero are precisely those which are AF and we show that the minor difference in their definition is enough to imply that any C*-algebra with finite decomposition rank is automatically quasidiagonal, whereas this need not be true for C*-algebras that only have finite nuclear dimension.

© The Author(s), under exclusive license to Springer Nature Switzerland AG 2021
K. R. Strung, *An Introduction to C*-Algebras and the Classification Program*, Advanced
Courses in Mathematics - CRM Barcelona, https://doi.org/10.1007/978-3-030-47465-2_17

## 17.1	Order zero maps

Nuclear dimension is a refinement of the c.p.a.p. To define it, we need to define order zero maps, which were first introduced in [131]. These are maps between C*-algebras which are stronger than c.p.c. maps – in the sense that they preserve more of the C*-algebraic structure – but not quite as strong as *-homomorphisms.

17.1.1. Definition. Let φ be a completely positive contractive map $\varphi : A \to B$. We say that φ is *order zero* if for any $a, b \in A_+$ with $ab = ba = 0$, we have $\varphi(a)\varphi(b) = 0$.

In the literature what we have called an order zero map is sometimes called a c.p.c. order zero map.

We will mostly be concerned with order zero maps whose domains are finite-dimensional C*-algebras. Any *-homomorphism is of course an order zero map since *-homomorphisms are completely positive and contractive (see Exercise 3.4.5) and are multiplicative. In fact, an order zero map is quite close to being a *-homo-morphism, as we will see.

17.1.2. Recall that the definition of the support projection of an operator $a \in \mathcal{B}(H)$ was given in Definition 5.2.10.

Lemma. *Let A be a unital C*-algebra, H a Hilbert space, and $\varphi : A \to \mathcal{B}(H)$ a completely positive map. Put $h := \varphi(1_A)$, and let p be the support projection of h. Let ξ be a unit vector in H. Then, for every $n \in \mathbb{N} \setminus \{0\}$ and $a \in A$, define*

$$\sigma_n : A \to \mathcal{B}(H),$$
$$a \mapsto (h + n^{-1}1_{\mathcal{B}(H)})^{-1/2}\varphi(a)(h + n^{-1}1_{\mathcal{B}(H)})^{-1/2} + \langle a\xi, \xi\rangle(1_{\mathcal{B}(H)} - p).$$

Then σ_n converges in the strong operator topology to a unital completely positive map

$$\sigma : A \to \mathcal{B}(H), \quad a \mapsto \lim_{n \to \infty} \sigma_n(a).$$

Proof. Since φ is linear, it is enough to assume that the element a is positive and that $\|a\| \le 1$. First, observe that h positive implies that $h + n^{-1}1_{\mathcal{B}(H)}$ is also positive and that $0 \notin \mathrm{sp}(h + n^{-1}1_{\mathcal{B}(H)})$ so it is moreover invertible. By Exercise 5.4.7 we have

$$h^{1/2}(h + n^{-1}1_{\mathcal{B}(H)})^{-1/2} \xrightarrow{\text{SOT}} p, \qquad n \to \infty.$$

Similarly,

$$(h + n^{-1}1_{\mathcal{B}(H)})^{-1/2}h^{1/2} \xrightarrow{\text{SOT}} p, \qquad n \to \infty,$$

and finally, since $(h + n^{-1}1_{\mathcal{B}(H)})^{-1/2}h(h + n^{-1}1_{\mathcal{B}(H)})^{-1/2}$ is also bounded and increasing, it follows that it converges (also to p) in the strong operator topology.

Now since $0 \le a \le 1_A$, positivity of the map φ implies $0 \le \varphi(a) \le \varphi(1_A) = h$ and so $(h + n^{-1}1_{\mathcal{B}(H)})^{-1/2}\varphi(a)(h + n^{-1}1_{\mathcal{B}(H)})^{-1/2}$ is also increasing and bounded as $n \to \infty$ and thus is SOT-convergent. It follows that $\sigma_n(a)$ is SOT-convergent to some $\sigma(a) \in \mathcal{B}(H)$.

From the above, we have

$$\sigma(1_A) = \lim \sigma_n(1_A) = p + \langle \xi, \xi \rangle (1_{\mathcal{B}(H)} - p) = 1_{\mathcal{B}(H)},$$

so σ is unital. Each σ_n is evidently linear, thus σ is linear. It is also easy to see (using, for example Exercise 3.4.5 (c)) that each σ_n is completely positive. Since $\sigma_n(a)$ is increasing in n if a is positive, σ is also completely positive.   □

17.1.3. For a C*-algebra A, the *cone* over A is the C*-algebra

$$C_0((0,1]) \otimes A.$$

Note that we do not need to specify the norm on the tensor product, since commutative C*-algebras are nuclear (Exercise 7.3.11). The next proposition says that order zero maps from $F \to A$ are in one-to-one correspondence with *-homomorphisms from the cone over F to A. For an order zero map φ we call the map π_φ in the proposition below the *supporting *-homomorphism* of φ.

17.1.4. Proposition (cf. [131, Proposition 3.2], [133, Proposition 1.2.1]). *Let F and A be C*-algebras with F finite-dimensional. Then given any *-homomorphism $\rho : C_0((0,1]) \otimes F \to A$, the map $\varphi : F \to A$ given by $\varphi(x) = \rho(\mathrm{id}_{(0,1]} \otimes x)$ is order zero. Moreover, for any order zero map $\varphi : F \to A$, there is a unique *-homomorphism*

$$\pi_\varphi : C_0((0,1]) \otimes F \to \mathrm{C}^*(\varphi(F)) \subset A$$

such that

$$\pi_\varphi(\mathrm{id}_{(0,1]} \otimes x) = \varphi(x),$$

for every $x \in F$.

Proof. It is straightforward to check that, given ρ, the map φ is order zero. Now suppose $\mu : F \to A$ is order zero. Since F is finite-dimensional, by Theorem 8.1.2 it can be decomposed as the direct sum of matrix algebras $F = \bigoplus_{i=0}^m M_{n_i}$ for some $m, n_1, \ldots, n_m \in \mathbb{N} \setminus \{0\}$. The map φ is order zero, so

$$\mathrm{C}^*(\varphi(F)) = \bigoplus_{i=0}^m \mathrm{C}^*(\varphi(M_{n_i})).$$

Thus it is enough to prove the result in the case that F is a single matrix algebra, say $F = M_r$. Furthermore, we may assume that $A = \mathrm{C}^*(\varphi(M_r))$.

Let $A \subset \mathcal{B}(H)$ be faithfully and nondegenerately represented on a Hilbert space H. Denote by $h := \varphi(1_F)$. Suppose $p \in \mathcal{B}(H)$ is the support projection of h (Definition 5.2.10). If $a \in F$ is any self-adjoint element with $\|a\| \leq 1$, positivity of φ implies that $\varphi(a) \leq \varphi(1_F) \leq p$ so $p\varphi(a) = \varphi(a)p = \varphi(a)$. Since a was an arbitrary self-adjoint element, it follows that p is a unit for A, and so by nondegeneracy, $p = 1_{\mathcal{B}(H)}$. Let

$$\sigma_n(a) := (h + n^{-1}1_{\mathcal{B}(H)})^{-1/2}\varphi(a)(h + n^{-1}1_{\mathcal{B}(H)})^{-1/2}, \quad a \in F.$$

By Lemma 17.1.2 and the fact that $p = 1_{\mathcal{B}(H)}$, this converges in the strong operator topology to a unital c.p. map $\sigma : F \to \mathcal{B}(H)$ which satisfies

$$h^{1/2}\sigma(a)h^{1/2} = p\varphi(a)p = \varphi(a),$$

for every $a \in F$.

Next, we claim that h commutes with $\sigma(F)$. Since it is obvious that h commutes with $(h + n^{-1}1_{\mathcal{B}(H)})^{-1/2}$, it is enough to show h commutes with $\varphi(a)$ for every positive $a \in F$. Since a is positive, it is diagonalisable. Let $e_1, \ldots, e_n$ be an orthonormal basis of $\mathbb{C}^n$ such that $a = \sum_{i=1}^n \lambda_i e_i$ and denote by $e_{ii} \in M_r = F$ the rank one projection onto the span of e_i. Since φ is order zero, we have $\varphi(e_{ii})\varphi(e_{jj}) = 0$ when $i \neq j$, so

$$\varphi(1_F)\varphi(a) = \sum_{i=1}^n \varphi(e_{ii})\left(\sum_{j=1}^n \lambda_j \varphi(e_{jj})\right) = \sum_{j=1}^n \lambda_j \varphi(e_{ii})^2$$

and similarly,

$$\varphi(a)\varphi(1_F) = \sum_{j=1}^n \lambda_j \varphi(e_{ii})^2,$$

so $h\varphi(F) = \varphi(F)h$.

Let us now show that σ is in fact a *-homomorphism. Let $e \in F$ be a rank one projection. Then

$$\sigma(e)\sigma(1_F - e) = 1_{\mathcal{B}(H)}\sigma(e)\sigma(1_F - e)1_{\mathcal{B}(H)} = \mathrm{SOT}\lim_n h^{1/2n}\sigma(e)\sigma(1_F - e)h^{1/2n}.$$

We see that

$$\sigma(e)\sigma(1_f 0e) = 0 \text{ if and only if } h\sigma(e)\sigma(1_F - e)h = 0.$$

Since h commutes with $\sigma(F)$, so too does $h^{1/2}$, thus

$$h\sigma(e)\sigma(1_F - e)h = h^{1/2}\sigma(e)h^{1/2}h^{1/2}\sigma(1_F - e)h^{1/2} = \varphi(e)\varphi(1_F - e) = 0,$$

since φ is order zero. It follows that $\sigma(e)^2 = \sigma(e)$. Since σ is positive, $\sigma(e)$ is a positive element, so in particular satisfies $\sigma(e)^* = \sigma(e)$. Thus $\sigma(e)$ is a projection. Now, using Exercise 7.3.5, we have $\sigma(ae) = \sigma(a)\sigma(e)$ for every $a \in M_r$. It is easy to show that σ is multiplicative on all positive elements, and hence on all of M_r. Thus σ is a *-homomorphism.

Since h is a positive element with $\|h\| < 1$, the Gelfand Theorem provides us with a *-homomorphism

$$\pi : C_0((0, 1]) \twoheadrightarrow C_0(\mathrm{sp}(h)) \cong C^*(h) \subset \varphi(F),$$

which maps $f \in C_0((0,1])$ to its restriction $f|_{\mathrm{sp}(h)}$. Let $\iota : \sigma(F) \hookrightarrow \varphi(F)$. Then π and $\iota \circ \sigma$ have commuting images in $\varphi(F)$ so, by Theorem 6.3.2, we obtain a *-homomorphism

$$\pi_\varphi : C_0((0,1]) \otimes M_r \to \varphi(F),$$

mapping $\mathrm{id}_{(0,1]} \otimes x \mapsto \varphi(x)$. $\qquad\qquad\qquad\qquad\qquad\qquad\qquad\qquad\square$

17.1.5. Proposition. *An order zero map φ is a *-homomorphism if and only if $\varphi(1_A)$ is a projection.*

Proof. Exercise. $\qquad\qquad\qquad\qquad\qquad\qquad\qquad\qquad\qquad\qquad\qquad\qquad\square$

17.1.6. The functional calculus for order zero maps whose domain is a finite-dimensional C*-algebra was introduced in [134].

Theorem (Functional calculus for order zero maps)**.** *Let $\varphi : F \to A$ be an order zero map between C*-algebras A and F, where F is finite-dimensional, and let π_φ denote the supporting *-homomorphism of φ. Suppose $f \in C_0((0,1])$ is a positive function. Then the map*

$$f(\varphi)(x) = \pi_\varphi(f(\mathrm{id}_{(0,1]}) \otimes x), \quad x \in X,$$

is a well-defined order zero map.

Proof. Exercise. $\qquad\qquad\qquad\qquad\qquad\qquad\qquad\qquad\qquad\qquad\qquad\qquad\square$

In fact the functional calculus holds for more general domains, but we will not need this. The interested reader can find the result in [139, Corollary 3.2].

The next lemma says that projections that add up to an element of norm strictly greater than one are close to orthogonal projections. It is similar in flavour to results in Section 8.4, so the proof is left as an exercise.

17.1.7. Lemma. *Let $K \in \mathbb{N}$ and $\beta > 0$. There exists $\alpha > 1$ such that the following holds. If A is a C*-algebra and $q_0, \ldots, q_k$, $k \leq K$ are projections in A satisfying $\|q_0 + \cdots + q_k\| \leq \alpha$, then there are pairwise orthogonal projections $p_1, \ldots, p_k \in A$ such that $\|p_i - q_i\| \leq \beta$.*

Proof. Exercise. $\qquad\qquad\qquad\qquad\qquad\qquad\qquad\qquad\qquad\qquad\qquad\qquad\square$

17.1.8. We saw that an order zero map φ whose domain is a finite-dimensional C*-algebra F corresponds to a *-homomorphism whose domain is the cone over F, and that if $\varphi(1_F)$ is a projection, then in fact φ is itself a *-homomorphism. This next proposition tells us that if $\varphi(1_F)$ is not a projection, we can measure how far φ is from *-homomorphism.

Proposition. *Let A and F be C*-algebras with F finite-dimensional, and suppose that $\varphi : F \to A$ is an order zero map. For any $\epsilon > 0$ there is a $\delta > 0$ such that the following holds. If*

$$\|\varphi(1_F) - \varphi(1_F)^2\| < \delta,$$

then there is a *-homomorphism $\varphi' : F \to A$ with

$$\|\varphi - \varphi'\| < \epsilon.$$

Proof. Write $F = M_{r_0} \oplus \cdots \oplus M_{r_k}$. To construct φ', we will find a suitable $c \in C^*(\varphi(F)) = \bigoplus_{i=0}^{k} C^*(\varphi(M_{r_i}))$ such that our *-homomorphism φ' satisfies $\varphi'(x) = c\varphi(x)c$. In that case, $\|\varphi' - \varphi\| = \max_{0 \le i \le k} \|\varphi'|_{M_{r_i}} - \varphi|_{M_{r_i}}\|$. Moreover, we may write $c = c_0 + \cdots + c_k$, $c_i \in \varphi(M_{r_i})$, such that $\varphi'|_{M_{r_i}} = c_i \varphi|_{M_{r_i}} c_i$ for every $1 \le i \le k$. Thus we may assume that $F = M_r$ for some $r \in \mathbb{N}$.

Let $\delta < \min\{(\epsilon/14)^2, 1/4\}$. Then, since $\|\varphi(1_F) - \varphi(1_F)^2\| < \delta$, Exercise 8.7.23 tells us there is a projection $p \in C^*(\varphi(1_F)) \subset A$ such that

$$\|p - \varphi(1_F)\| < 2\delta,$$

and moreover that $p\varphi(1_F)p$ is invertible in $pC^*(\varphi(1_F))p$. Set

$$c := (p\varphi(1_F)p)^{-1/2}.$$

Then $c\varphi(1_F)c = (p\varphi(1_F)p)^{-1/2}p\varphi(1_F)p(p\varphi(1_F)p)^{-1/2} = p$, and $\|p - c\| < 4\delta$. Define

$$\varphi' : F \to pAp, \quad x \mapsto c\varphi(x)c.$$

First, let us show that φ' is close to φ. Let $x \in F_+$ and $\|x\| \le 1$. Recall from the proof of Proposition 17.1.4 that $\varphi(1_F)$ commutes with $\varphi(F)$. Thus

$$\begin{aligned}
\|\varphi(x) - \varphi(1_F)\varphi(x)\|^2 &= \|(1 - \varphi(1_F))\varphi(x)^2(1 - \varphi(1_F))\| \\
&\le \|(1 - \varphi(1_F))\varphi(1_F)^2(1 - \varphi(1_F))\| \\
&< \delta.
\end{aligned}$$

Since $\|p - c\| < 4\delta$, we have $\|c - \varphi(1_F)\| < 6\delta$, whence

$$\begin{aligned}
\|\varphi'(x) - \varphi(x)\| = \|c\varphi(x)c - \varphi(x)\| \\
\le \|c\varphi(x)c - \varphi(1_F)\varphi(x)\varphi(1_F)\| + \|\varphi(1_F)\varphi(x)\varphi(1_F) - \varphi(x)\| \\
< 12\delta + 2\delta^{1/2} < \epsilon.
\end{aligned}$$

Thus $\|\varphi' - \varphi\| < \epsilon$.

Now, let us show that φ' is indeed a *-homomorphism. Since $\varphi'(1_F) = p$ is a projection, it is enough to show that φ' is order zero (Proposition 17.1.5). Let $e_1, \ldots, e_r$ denote rank one pairwise orthogonal projections so that $\sum_{i=1}^{r} e_i = 1_F$. Let $f \in C_0((0,1])$. By the functional calculus for order zero maps, $f(\varphi)$ is order zero, and so $f(\varphi(e_i))f(\varphi(e_j)) = 0$ for every $1 \le i \ne j \le r$ and also that $f(\varphi(1_F)) = \sum_{i=1}^r f(\varphi(e_i))$. By the above, we have

$$\|\varphi(e_i) - \varphi(e_i)^2\| \le \|\varphi(e_i) - \varphi(e_i)\varphi(1_F)\| + \|\varphi(1_F)\varphi(e_i) - \varphi(e_i)^2\|$$
$$< \delta^{1/2} + 0.$$

As above, we can find projections p_i such that $\|\varphi(e_i) - p_i\| < 2\delta^{1/2}$, and $c_i := (p_i\varphi(e_i)p_i)^{-1/2}$. Since each $p_i \in C^*(\varphi(e_i))$, these projections are pairwise orthogonal.

Let

$$f(t) := \begin{cases} 0 & t \in [0, 1/2), \\ 1 & t \in [1/2, 1]. \end{cases}$$

Then, by the choice of δ we have $f(\varphi(e_i)) = p_i$ and

$$\sum_{i=1}^r p_i = \sum_{i=1}^r f(\varphi(e_i))$$
$$= f\left(\sum_{i=1}^r \varphi(e_i)\right)$$
$$= f(\varphi(1_F))$$
$$= p.$$

Setting $d := \sum_{i=1}^r c_i$, we have $d\varphi(1_F)d = p$. Observe that each c_i commutes with $\varphi(1_F)$, which implies that d and $\varphi(1_F)$ commute. Then $d^2 = (p\varphi(1_F)p)^{-1}$ where the inverse is taken in $pC^*(\varphi(1_F))p$, which means $d^2 = c^2$. Since positive square roots are unique, $d = c$. Thus

$$\varphi'(e_i) = c\varphi(e_i)c = d\varphi(e_i)d = c_i\varphi(e_i)c_i = p_i.$$

It follows that $\varphi'(e_i)\varphi'(e_j) = 0$ for $i \ne j$. Since we chose the e_i arbitrarily, we have $\varphi'(a)\varphi'(b) = 0$ whenever $a, b \in F_+$ and $ab = 0$. Thus φ' is order zero and hence a *-homomorphism. □

17.2 Nuclear dimension and decomposition rank of commutative C*-algebras

The nuclear dimension and the closely related decomposition rank refine the definition of the completely positive approximation property by asking that the map from the finite-dimensional approximating algebra is the sum of order zero maps.

17.2.1. Definition. Let A be a separable C*-algebra. We say that A has *nuclear dimension* d, written $\dim_{\mathrm{nuc}} A = d$, if d is the least integer satisfying the following: For every finite subset $\mathcal{F} \subset A$ and every $\epsilon > 0$ there are a finite-dimensional,

C*-algebra with $d+1$ ideals, $F = F_0 \oplus \cdots \oplus F_d$, and completely positive maps $\psi : A \to F$ and $\varphi : F \to A$ such that ψ is contractive, $\varphi|_{F_n}$ are completely positive contractive order zero maps and

$$\|\varphi \circ \psi(a) - a\| < \epsilon \text{ for every } a \in \mathcal{F}.$$

If no such d exists, then we say $\dim_{\mathrm{nuc}} A = \infty$.

   If the φ can always be chosen to be contractive, then we say that A has *decomposition rank* d, written $\mathrm{dr}\, A = d$.

   It is clear that if a C*-algebra has finite decomposition rank then it is nuclear. This is also true for finite nuclear dimension (exercise). The converse does not hold, even in the simple case. The examples in [120, 50], among others, are simple separable unital and nuclear but do not have finite nuclear dimension.

17.2.2. Both the decomposition rank and nuclear dimension can be thought of as a noncommutative version of covering dimension (Definition 8.6.2).

Lemma. *Let X be a metrisable space and $\mathcal{U}$ a finite open cover of X such that $\sum_{U \in \mathcal{U}'} \chi_U(x) \le d+1$ for every $x \in X$. Then $\mathcal{U}$ has a finite subcover $\mathcal{U}'$ such that for every $U \in \mathcal{U}'$ there are at most d distinct $V \in \mathcal{U}'$, $V \ne U$, such that $U \cap V \ne \emptyset$.*

Proof. Exercise.   □

17.2.3. The next two theorems establish that the decomposition rank and nuclear dimension of $C(X)$ agree and are both bounded by $\dim X$. In fact, they are both equal to $\dim X$, but we will not prove this. Establishing lower bounds for decomposition rank and nuclear dimension is often quite difficult and technical. In applications to simple nuclear C*-algebras, it is usually enough to show that the nuclear dimension is bounded from above. To see that $\dim X \le \mathrm{dr}\, C(X)$, (in fact, more generally this holds for $C_0(X)$ when X is locally compact Hausdorff and second countable) the reader is referred to [70, Proposition 3.3] and [131, Section 3].

17.2.4. Theorem. *Let X be a compact metrisable space. Then*

$$\mathrm{dr}\, C(X) \le \dim X.$$

Proof. Let $\mathcal{F} \subset C(X)$ be a finite subset, $k \in \mathbb{N}$ and $\epsilon > 0$ be given. For every $x \in X$ there is an open set U_x such that

$$|f(y) - f(z)| < \epsilon/3,$$

whenever $y, z \in U_x$, and $f \in \mathcal{F}$. Let $\tilde{\mathcal{U}} = \{U_x \mid x \in X\}$. Since X is compact we can find a finite subcover $\mathcal{U} \subset \tilde{\mathcal{U}}$. Then, since $\dim X = d$ we may moreover assume that any $x \in X$ is contained in at most $d+1$ of the sets in $\mathcal{U}$.

Refining $\mathcal{U}$ if necessary, using Lemma 17.2.2, we can assume that $\mathcal{U} = \{U_i\}_{i \in I}$, $|I| = s < \infty$, such that, for every U_i, $i \in I$, there are at most d distinct $U \subset \mathcal{U}$, $U \neq U_i$, such that $U_i \cap U \neq \emptyset$. Thus we can partition I into $d + 1$ subsets $I_0, \ldots, I_d$ that satisfy $U_i \cap U_j = \emptyset$ for every $i \neq j \in I_k$, $0 \leq k \leq d$.

Let $g_i \in C(X)$, $1 \leq i \leq s$ be a partition of unity subordinate to $\mathcal{U}$, that is, g_i, $1 \leq i \leq s$ such that $0 \leq g_i \leq 1_{C(X)}$, $\mathrm{supp}(g_i) \subset U_i$ and $\sum_{0 \leq i \leq s} g_i = 1_{C(X)}$. For each i, let $x_i \in U_i$ be a point satisfying $g_i(x_i) = 1$.

For $0 \leq k \leq d$, let

$$F_k := \mathbb{C}^{|I_k|},$$

and

$$F = F_0 \oplus \cdots \oplus F_d.$$

Define $\psi : C(X) \to F$ by

$$\psi(a) := (\oplus_{i \in I_0} a(x_i), \ldots, \oplus_{i \in I_d} a(x_i)).$$

It is easy to see that ψ is a *-homomorphism, hence in particular, completely positive and contractive.

Define $\varphi : F \to C(X)$ by

$$\varphi((\lambda_1, \ldots, \lambda_s)) = \sum_{i=1}^{s} \lambda_i g_i.$$

Then φ is completely positive (Exercise 7.3.10), and since $U_i \cap U_j = \emptyset$ for every $i \neq j \in I_k$, it is easy to see that $\varphi|_{F_k}$ is order zero. Moreover, $\varphi(1_F) = 1_{C(X)}$, so φ is contractive.

Finally, for any $x \in X$ and any $f \in \mathcal{F}$

$$|\varphi \circ \psi(f)(x) - f(x)| = \left| \sum_{i=1}^{s} f(x_i) g_i(x) - \sum_{i=1}^{s} f(x) g_i(x) \right|$$

$$\leq \sum_{i=1}^{s} |f(x_i) - f(x)| g_i(x)$$

$$< \epsilon \cdot \sum_{i=1}^{s} g_i(x)$$

$$= \epsilon.$$

Thus $\|\varphi \circ \psi(f) - f\| < \epsilon$. $\square$

17.2.5. Theorem. *Let X be a compact metrisable space. Then*

$$\mathrm{dr}\, C(X) = \dim_{\mathrm{nuc}} C(X).$$

Proof. It is clear from their definitions that the nuclear dimension is bounded by the decomposition rank, as the decomposition rank asks that a stronger condition hold. So let us prove the inequality $\operatorname{dr} C(X) \leq \dim_{\mathrm{nuc}} C(X)$.

If $\dim_{\mathrm{nuc}} C(X) = \infty$, there is nothing to show, so let us assume that

$$\dim_{\mathrm{nuc}} C(X) = d < \infty.$$

Let $\mathcal{F} \subset C(X)$ be a finite subset and let $\epsilon > 0$. Since $\dim_{\mathrm{nuc}} C(X) = d$, we can find a finite-dimensional C*-algebra $F = F_0 \oplus \cdots \oplus F_d$, a c.p.c. map $\psi : C(X) \to F$ and a c.p. map $\varphi : F \to C(X)$ such that $\varphi|_{F_i}$ is order zero, for every $0 \leq i \leq d$ and such that

$$\|\varphi \circ \psi(f) - f\| < \epsilon/2, \text{ for every } f \in \mathcal{F} \cup \{1_{C(X)}\}.$$

Using Exercise 3.4.9, by cutting down to a hereditary subalgebra of F if necessary, we may assume that $h := \psi(1_{C(X)})$ is invertible in F.

Define

$$\hat{\psi} : C(X) \to F$$

by

$$\hat{\psi}(f) := h^{-1/2} \psi(f) h^{-1/2}, \quad f \in C(X).$$

Notice that $\hat{\psi}$ is c.p.c. In fact, we have $\hat{\psi}(1_{C(X)}) = 1_F$. Next, define

$$\hat{\varphi} : F \to C(X)$$

by

$$\hat{\varphi}(x) := (1 - \epsilon/2) \cdot \varphi(h^{1/2} x h^{1/2}), \quad x \in F.$$

Since $0 \leq h^{1/2} \leq 1_F$ and φ is positive, we have $0 \leq \varphi(h^{1/2} x h^{1/2}) \leq \varphi(x)$ for every $x \geq 0$, and so if $x, y \in F_+$ we have $\hat{\varphi}(x)\hat{\varphi}(y) = 0$ whenever $\varphi(x)\varphi(y) = 0$. In particular, $\hat{\varphi}|_{F_i}$ is order zero for every $0 \leq i \leq d$. Furthermore,

$$\hat{\varphi}(1_F) = \hat{\varphi}(\hat{\psi}(1_{C(X)})) = (1 - \epsilon/2) \cdot \varphi \circ \psi(1_{C(X)}).$$

By definition of φ and ψ, we have that

$$\varphi \circ \psi(1_{C(X)}) \leq (1 + \epsilon/2) 1_{C(X)}.$$

Thus

$$\hat{\varphi}(1_F) \leq 1_{C(X)},$$

so we see that $\hat{\varphi}$ is contractive.

Finally,

$$\begin{aligned} \|\hat{\varphi} \circ \hat{\psi}(f) - f\| &= \|\hat{\varphi} \circ \hat{\psi}(f) - \varphi \circ \psi(f) + \varphi \circ \psi(f) - f\| \\ &< \|\hat{\varphi} \circ \hat{\psi}(f) - \varphi \circ \psi(f)\| + \epsilon/2 \\ &< \|(1 - \epsilon/2)\varphi \circ \psi(f) - \varphi \circ \psi(f)\| + \epsilon/2 \\ &\leq \epsilon. \end{aligned}$$

Thus we have shown that $\operatorname{dr} C(X) \leq d = \dim_{\mathrm{nuc}} C(X)$, as required. $\qquad\square$

17.2.6. Corollary. *Let X be a compact metric space with covering dimension* $\dim X$. *Then* $\dim_{\mathrm{nuc}} C(X) \leq \dim X$.

17.3 Permanence properties of the nuclear dimension

Nuclear dimension has good permanence properties, which we show in the next series of propositions.

17.3.1. Proposition. *Let A and B be separable unital* C*-*algebras with B nuclear, and suppose that $\pi : A \twoheadrightarrow B$ is a surjective *-homomorphism. Then*

$$\dim_{\mathrm{nuc}} B \leq \dim_{\mathrm{nuc}} A.$$

Proof. If $\dim_{\mathrm{nuc}} A = \infty$ there is nothing to show. So assume $\dim_{\mathrm{nuc}} A = d < \infty$. Since B is nuclear we can apply Corollary 7.2.8 to find a c.p.c. map $\rho : B \to A$ satisfying $\pi \circ \rho = \mathrm{id}_B$. Then, for any $\mathcal{F} \subset B$ and $\epsilon > 0$, we can find a finite-dimensional subalgebra $F := F_0 \oplus \cdots \oplus F_d$, a c.p.c. map $\psi : A \to F$ and a c.p. map $\varphi : F \to A$ such that $\varphi|_{F_i}$ is order zero for every $0 \leq i \leq d$ and such that

$$\|\varphi \circ \psi(\rho(b)) - \rho(b)\| < \epsilon \text{ for every } b \in \mathcal{F}.$$

It is easy to see that the composition of an order zero map with a *-homomorphism is order zero. Since the composition of c.p.c. maps is c.p.c. (Exercise 7.3.4) putting $\tilde{\psi} := \psi \circ \rho$ and $\tilde{\varphi} := \pi \circ \varphi$, allows us to approximate any element in $\mathcal{F}$ up to ϵ through F. Thus $\dim_{\mathrm{nuc}} B \leq d$. $\qquad\square$

17.3.2. Proposition. *Let A and B be separable unital* C*-*algebras. Then*

$$\dim_{\mathrm{nuc}}(A \oplus B) = \max\{\dim_{\mathrm{nuc}} A, \dim_{\mathrm{nuc}} B\}.$$

Proof. We will show that the nuclear dimension of both A and B is finite and leave it as an easy exercise to show that the proposition holds if either A or B has infinite nuclear dimension.

To show that $\max\{\dim_{\mathrm{nuc}} A, \dim_{\mathrm{nuc}} B\} \leq \dim_{\mathrm{nuc}}(A \oplus B)$, we simply observe that there are surjective *-homomorphisms $\pi_A : A \oplus B \to A$ and $\pi_B : A \oplus B \to B$ and apply Proposition 17.3.1.

Now let $\dim_{\mathrm{nuc}} A = d_A$ and $\dim_{\mathrm{nuc}} B = d_B$ and without loss of generality, assume that $d_A \geq d_B$. Let $\mathcal{F} = \{(a_1, b_1), \ldots, (a_n, b_n)\} \subset A \oplus B$ and $\epsilon > 0$ be given. There are c.p. approximations $(F^A = \oplus_{k=0}^{d_A} F_k^A, \psi^A, \varphi^A)$ and $(F^B = \oplus_{k=0}^{d_B} F_k^B, \psi^B, \varphi^B)$ approximating $\{a_1, \ldots, a_n\}$ and $\{b_1, \ldots, b_n\}$, respectively, up to ϵ, where ψ^A and ψ^B are contractive, and $\varphi^A|_{F_k^A}$, $0 \leq k \leq d_A$ and $\varphi^B|_{F_k^B}$, $0 \leq k \leq d_B$ are order zero.

Let $F = \bigoplus_{k=0}^{d_A} F_k$ where $F_k = (F_k^A \oplus F_k^B)$ for $0 \leq k \leq d_B$ and $F_k = F_k^A$ for $d_B + 1 \leq k \leq d_A$. Define $\psi : A \oplus B \to F$ by

$$\psi((a, b)) := (\psi^A(a)_0, \psi^B(b)_0, \ldots, \psi^A(a)_{d_B}, \psi^B(b)_{d_B}, \psi^A(a)_{d_B+1}, \ldots, \psi^A(a)_{d_A}),$$

where $\psi^A(a)_k$ denotes the kth coordinate of $\psi^A(a)$, and similarly for $\psi^B(b)_k$. Then ψ is a c.p.c. map. Now define $\varphi : F \to A \oplus B$ by

$$\varphi((\lambda_0, \mu_0, \ldots, \lambda_{d_B}, \mu_{d_B}, \lambda_{d_B+1}, \ldots, \lambda_{d_A})) = (\varphi^A(\lambda_0, \ldots, \lambda_{d_A}), \varphi^B(\mu_0, \ldots, \mu_{d_B})).$$

It is easy to check that $\varphi|_{F_k}$ is order zero for each $0 \le k \le d_A$. Furthermore

$$
\begin{aligned}
\|\varphi \circ \psi((a,b)) - (a,b)\| &= \|(\varphi^A \circ \psi^A(a), \varphi^B \circ \psi^B(b)) - (a,b)\| \\
&= \max\{\|\varphi^A \circ \psi^A(a) - a\|, \|\varphi^B \circ \psi^B(b) - b\|\} \\
&< \epsilon,
\end{aligned}
$$

for every $(a,b) \in \mathcal{F}$.

Thus $\dim_{\mathrm{nuc}}(A \oplus B) \le \max\{\dim_{\mathrm{nuc}} A, \dim_{\mathrm{nuc}} B\}$. $\qquad \square$

17.3.3. Proposition. *Let (A_n, ρ_n) be an inductive system with limit $A := \varinjlim A_n$. Then*

$$\dim_{\mathrm{nuc}} A \le \liminf_{n \to \infty} \dim_{\mathrm{nuc}} A_n.$$

Proof. Let $\mathcal{F} = \{a_1, \ldots, a_k\}$ be a finite subset of A and let $\epsilon > 0$. By Exercise 8.2.7 $A = \overline{\bigcup_{n \in \mathbb{N}} \rho^{(n)} A_n}$. There is $N \in \mathbb{N}$ such that, for every $n \ge N$, there exist $b_1, \ldots, b_k \in \rho^{(n)}(A_n)$ satisfying

$$\|b_i - a_i\| < \epsilon/3, \quad 1 \le i \le k.$$

Let $m \ge N$ satisfy

$$d = \dim_{\mathrm{nuc}} A_m \le \liminf_{n \to \infty} \dim_{\mathrm{nuc}} A_n.$$

Then, there are a finite-dimensional C*-algebra $F = F_0 \oplus \cdots \oplus F_d$, a c.p.c. map $\psi : \rho^{(m)}(A_m) \to F$, and a c.p. map $\varphi : F \to \rho^{(m)}(A_m) \subset A$, with each $\varphi|_{F_i}$, $1 \le i \le d$ order zero, such that

$$\|\varphi \circ \psi(b_i) - b_i\| < \epsilon/3, \quad 1 \le i \le k.$$

Using the Arveson Extension Theorem (Theorem 7.1.12), there is a c.p.c. map $\tilde{\psi} : A \to F$ extending ψ. Then, for $1 \le i \le k$ we have,

$$
\begin{aligned}
\|\varphi \circ \tilde{\psi}(a_i) - a_i\| &< \|\varphi \circ \psi(b_i) - b_i\| + 2\epsilon/3 \\
&< \epsilon.
\end{aligned}
$$

So $\dim_{\mathrm{nuc}} A \le d \le \liminf_{n \to \infty} \dim_{\mathrm{nuc}} A_n$. $\qquad \square$

17.3.4. Lemma. *Let A and B be unital C*-algebras and suppose $\varphi_1 : M_{n_1} \to A$ and $\varphi_2 : M_{n_2} \to B$ are order zero maps. Then the induced map*

$$\varphi_1 \otimes \varphi_2 : M_{n_1} \otimes M_{n_2} \to A \otimes_{\mathrm{min}} B$$

is order zero.

Proof. Let h denote the identity function on $(0, 1]$ and let $\pi_1 : C_0((0, 1]) \otimes M_{n_1} \to A$ and $\pi_2 : C_0((0, 1]) \otimes M_{n_2} \to B$ be the supporting *-homomorphisms of φ_1 and φ_2 (17.1.3), respectively. By Theorem 7.1.6, the induced map $\varphi_1 \otimes \varphi_2 : M_{n_1} \otimes M_{n_2} \to A \otimes_{\min} B$ is the unique c.p.c. map satisfying $\varphi_1 \otimes \varphi_2(x \otimes y) = \varphi_1(x) \otimes \varphi(y)$. Define

$$\Phi : M_{n_1} \otimes M_{n_2} \to A \otimes B$$

on simple tensors by

$$\Phi(x \otimes y) = \pi_1(h \otimes x) \otimes \pi_2(h \otimes y).$$

It is straightforward to check that this is a c.p.c. order zero map. Moreover,

$$\Phi(x \otimes y) = \pi_1(h \otimes x) \otimes \pi_2(h \otimes y) = \varphi_1(x) \otimes \varphi_1(y),$$

so by uniqueness, $\varphi_1 \otimes \varphi_2 = \Phi$ is order zero. □

17.3.5. Theorem. *Let A and B be separable unital C^*-algebras. Then,*

$$\dim_{\mathrm{nuc}}(A \otimes_{\min} B) \leq (\dim_{\mathrm{nuc}} A + 1)(\dim_{\mathrm{nuc}} B + 1) - 1.$$

Proof. Suppose $\dim_{\mathrm{nuc}} A = d$ and $\dim_{\mathrm{nuc}} B = e$. We may assume $d, e < \infty$, since otherwise the result is obvious. Let $\mathcal{F} \subset A \otimes B$ be a finite subset and let $\epsilon > 0$. Without loss of generality (shrinking ϵ if necessary), we may assume that $\mathcal{F}$ is of the form $\mathcal{F}_A \otimes \mathcal{F}_B$ for finite subsets $\mathcal{F}_A \subset A$ and $\mathcal{F}_B \subset B$.

Let $(\psi^{(A)}, F^{(A)}, \varphi^{(A)})$ and $(\psi^{(B)}, F^{(B)}, \varphi^{(B)})$ be $(\mathcal{F}_A, \epsilon)$ and $(\mathcal{F}_B, \epsilon)$ c.p. approximations for the identity map on A and B respectively, with $\psi^{(A)}$, respectively $\psi^{(B)}$, contractive and $\varphi^{(A)}|_{F_i^{(A)}}$, $1 \leq i \leq d$ and $\varphi^{(B)}|_{F_j^{(B)}}$, $1 \leq j \leq e$, order zero maps. Both

$$\psi^{(A)} \otimes \psi^{(B)} : A \otimes B \to F^{(A)} \otimes F^{(B)}, \quad \varphi^{(A)} \otimes \varphi^{(B)} : F^{(A)} \otimes F^{(B)} \to A \otimes B$$

are well-defined c.p. maps (Theorem 7.1.6). It is easy to check that their composition approximates the identity on $A \otimes B$ up to ϵ on the finite subset $\mathcal{F}$. Moreover, $\psi^{(A)} \otimes \psi^{(B)}$ is contractive and the restrictions

$$\varphi^{(A)} \otimes \varphi^{(B)}|_{F_i^{(A)} \otimes F_j^{(B)}}, \quad 0 \leq i \leq d, \ 0 \leq j \leq e,$$

are order zero by Lemma 17.3.4. Thus, we can write $F^{(A)} \otimes F^{(B)}$ as the direct sum of $(d+1)(e+1)$ many ideals where the map $\phi^{(A)} \otimes \phi^{(B)}$ restricts to an order zero map. Thus,

$$\dim_{\mathrm{nuc}} A \otimes B \leq (d+1)(e+1) - 1 = (\dim_{\mathrm{nuc}} A + 1)(\dim_{\mathrm{nuc}} B + 1) - 1,$$

as required. □

17.4   Nuclear dimension of AF algebras and quasidiagonality

It was mentioned in 8.6.7 that AF algebras should be thought of as "zero-dimensional" objects and we saw that they have real rank zero and stable rank one. Here we see that they also have nuclear dimension zero. In fact, the *only* C*-algebras with nuclear dimension zero are AF algebras. Contrast this with the case of real rank: we saw that simple unital TAF C*-algebras have real rank zero and include things like AT-algebras with nontrivial K_1. The idea of seeing simple unital nuclear C*-algebras which are *not* AF as having positive dimension has been one of the most useful concepts in classification, particularly as a conceptual approach to generalising von Neumann algebra techniques. Simple AF algebras, on the other hand, being zero-dimensional, can be treated more similarly to the hyperfinite von Neumann factors. A good discussion of these ideas can be found in [12].

17.4.1. Theorem. *A separable unital* C*-*algebra has nuclear dimension zero if and only if it is AF.*

Proof. We leave it as an exercise to show that if A is AF then it has nuclear dimension zero. Let $\epsilon > 0$ and a finite subset $\mathcal{F} \subset A$ be given. Let $\delta > 0$ be the δ given by Proposition 17.1.8 with respect to $\epsilon/2$. Suppose then that A is a separable unital C*-algebra and $\dim_{\mathrm{nuc}} A = 0$. Let F be a finite-dimensional C*-algebra, $\psi : A \to F$ a c.p.c. map and $\varphi : F \to A$ an order zero map such that

$$\|\varphi \circ \psi(a) - a\| < \min\{\delta/2, \epsilon/2\}, \quad \text{for every } a \in \mathcal{F} \cup \{1_A\}.$$

Replacing ψ with $\psi + 1_F - \psi(1_A)$, we may assume that ψ is unital. Then, in particular,

$$\|\varphi(1_F) - 1_A\| < \delta/2.$$

By Proposition 17.1.8, we can approximate φ by a *-homomorphism $\varphi' : F \to A$ such that $\|\varphi' - \varphi\| < \epsilon/2$. Let $B = \varphi'(F)$, which is a finite-dimensional C*-subalgebra of A. Then, for every $a \in \mathcal{F}$ we have

$$\begin{aligned}
\mathrm{dist}(B, a) &\leq \|\varphi' \circ \psi(a) - a\| \\
&\leq \|\varphi' \circ \psi(a) - \varphi \circ \psi(a)\| + \|\varphi \circ \psi(a) - a\| \\
&< \epsilon.
\end{aligned}$$

Since $\mathcal{F}$ and ϵ were arbitrary, A is locally finite-dimensional and thus, by Theorem 8.5.2, A is AF.   □

17.4.2. Lemma. *Let A be a* C*-*algebras, let $s, d_1, \ldots, d_2 \in \mathbb{N} \setminus \{0\}$ and let F be the finite-dimensional* C*-*algebra $F := M_{r_1} \oplus \cdots \oplus M_{r_s}$. Suppose that for $\varphi : F \to A$ there is, for some $d \leq s$, a decomposition of F into $F = F_0 \oplus \cdots \oplus F_d$, such that $\varphi|_{F_i}$ is order zero. Then, for any $I \subset \{1, \ldots, s\}$ we have*

$$\left\| \sum_{i \in I} \varphi(1_{M_{r_i}}) \right\| \leq (d+1) \cdot \max_{i \in I} \|\varphi(1_{M_{r_i}})\|.$$

Proof. We may decompose I into $I = \bigsqcup_{j=0}^{d} I_j$ so that $F_j = \oplus_{i \in I_j} M_{r_i}$. Then

$$\left\| \sum_{i \in I} \varphi(1_{M_{r_i}}) \right\| \leq (d+1) \cdot \max_{j \in \{0,\ldots,d\}} \left\| \sum_{i \in I_j} \varphi(1_{M_{r_i}}) \right\|$$
$$= (d+1) \cdot \max_{j \in \{0,\ldots,d\}} \max_{i \in I_j} \| \varphi(1_{M_{r_i}}) \|.$$

Thus $\varphi(1_{M_{r_i}})\varphi(1_{M_{r_{i'}}}) = \varphi|_{F_j}(1_{M_{r_i}})\varphi|_{F_j}(1_{M_{r_{i'}}}) = 0$ for $i \neq i' \in I_j$. □

17.4.3. Theorem. *Let A be a separable C^*-algebra with finite decomposition rank. Then A is quasidiagonal.*

Proof. We will show that $\mathrm{dr}\, A \leq d < \infty$ if and only if the following holds: For any finite subset $\mathcal{F} \subset A$ and any $\epsilon > 0$ there exist a finite-dimensional C^*-algebra $F = F_0 \oplus \cdots \oplus F_d$ and c.p.c. maps $\psi : A \to F$, $\varphi : F \to A$ satisfying

(i) $\|\varphi \circ \psi(a) - a\| < \epsilon$ for every $a \in \mathcal{F}$,

(ii) $\|\psi(a)\psi(b) - \psi(ab)\| < \epsilon$ for every $a, b \in \mathcal{F}$,

(iii) $\varphi|_{F_i}$ is order zero for every $0 \leq i \leq d$.

From (i) and (ii), it follows that A is quasidiagonal by Proposition 11.1.5.

The "if" statement is clear. Let $\mathrm{dr}\, A \leq d < \infty$ and suppose we are given a finite subset $\mathcal{F} \subset A$ and $\epsilon > 0$. Without loss of generality we may assume that $\mathcal{F}$ is composed of positive elements of norm one. Let

$$\mathcal{F}^2 := \{a^2 \mid a \in \mathcal{F}\}.$$

It follows from Exercise 7.3.5 that for (ii) to hold, it is enough to show that, for every $a \in A$, the inequality $\|\psi(a)\psi(a) - \psi(a^2)\| < \epsilon$ holds. Choose ϵ_1 satisfying

$$0 < \epsilon_1 < \epsilon^4/(6(n+1)).$$

Using the fact that $\mathrm{dr}\, A \leq d$, there are a finite-dimensional $F' = F_0 \oplus \cdots \oplus F_d$, a c.p.c. map $\psi' : A \to F'$ and a c.p.c. map $\varphi' : A \to F'$ such that

$$\|\varphi' \circ \psi'(a) - a\| < \epsilon_1, \text{ for every } a \in \mathcal{F} \cup \mathcal{F}^2,$$

and such that $\varphi'|_{F_i}$ is order zero. Since F' is finite-dimensional, we can write F' as the direct sum of matrix algebras, say $F' = M_{r_0} \oplus \cdots \oplus M_{r_k}$, $k \geq d$. Since each $M_i \in F_j$ for some $0 \leq j \leq d$, the restriction $\varphi'_i := \varphi'|_{M_{r_i}}$ must be order zero. Let π_i denote the supporting *-homomorphism of φ'_i. Note that π_i is injective and hence isometric. Then

$$\|\varphi'_i(x)\| = \|\pi_i(\mathrm{id}_{(0,1]} \otimes x)\| = \|x\| \|\varphi'_i(1_F)\|,$$

for every $x \in M_{r_i}$, for every $0 \le i \le k$. Denote by ψ_i' the restriction of ψ' to $(\psi')^{-1}(M_{r_i})$ and define

$$I := \{i \in \{1, \ldots, s\} \mid \text{ there exists } a \in \mathcal{F} \text{ such that } \|\psi_i'(a)^2 - \psi_i'(a)\| \ge \epsilon^2\}.$$

Then, for every $i \in I$ we have, for suitable $a \in \mathcal{F}$, that

$$
\begin{aligned}
\epsilon^2 \cdot \|\varphi_i'(1_{M_{r_i}})\| &\le \|\varphi_i'(\psi_i'(a)^2 - \psi_i'(a^2))\| \\
&\le \left\| \sum_{i=1}^{k} \varphi_i'(\psi_i'(a)^2 - \psi_i'(a^2)) \right\| \\
&= \|\varphi'(\psi'(a)^2 - \psi(a^2))\| \\
&\le \|\varphi' \circ \psi'(a^2) - \varphi' \circ \psi'(a)^2\| \\
&\le \|\varphi' \circ \psi'(a^2) - a^2\| + \|a^2 - \varphi' \circ \psi'(a)^2\| \\
&< 3\epsilon_1.
\end{aligned}
$$

So $\|\varphi_i'(1_{M_{r_i}})\| < 3\epsilon_1/\epsilon^2$. By Lemma 17.4.2, we have

$$\left\| \sum_{i \in I} \varphi'(1_{M_{r_i}}) \right\| \le (d+1)3\epsilon_1/\epsilon^2.$$

Define a finite-dimensional algebra

$$F := \bigoplus_{i \in \{1, \ldots, s\} \setminus I} M_{r_i} \subset F',$$

a c.p.c. map

$$\psi : A \to F, \quad a \mapsto 1_F \psi'(a) 1_F,$$

and a c.p.c. map

$$\varphi : F \to A, \quad x \mapsto \varphi'|_F(x).$$

Then

$$
\begin{aligned}
\|\varphi \circ \psi(a)^2 - a^2\| &< (1 + 3(d+1)/\epsilon^2)\epsilon_1 \\
&= \epsilon^4/(6(d+1)) + \epsilon^2/2 \\
&< \epsilon^2,
\end{aligned}
$$

and also $\|\varphi \circ \psi(a) - a\| < \epsilon_1$, showing both (i) and (ii). We have already observed that $\varphi_i' := \varphi'|_{M_{r_i}}$ is order zero, so (iii) holds. □

This last theorem highlights the main difference between the decomposition rank and the nuclear dimension: while finiteness of decomposition rank implies quasidiagonality, there are many purely infinite C*-algebras with finite nuclear dimension. As we saw in Theorem 11.1.7, purely infinite C*-algebras are never quasidiagonal. The Cuntz algebras of Chapter 10, for example, all have finite nuclear dimension [140, Section 7]. Conversely, if A has finite nuclear dimension and is quasidiagonal, then A will also have finite decomposition rank; this is a result of the main theorem of [118].

17.5 Exercises

17.5.1. Let A and F be C*-algebras, with F finite-dimensional. If $\varphi : F \to B$ is an order zero map, show that φ is a *-homomorphism if and only if $\varphi(1_F)$ is a projection.

17.5.2. Let $\varphi : F \to A$ be an order zero map between C*-algebras A and F, where F is finite-dimensional, and let π_φ denote the supporting *-homomorphism of φ. Suppose $f \in C_0((0,1])$ is a positive function. Show that

$$f(\varphi)(x) = \pi_\varphi(f(\mathrm{id}_{(0,1]}) \otimes x), \quad x \in X,$$

is a well-defined order zero map.

17.5.3. Let F, A be separable unital C*-algebras. Show that an order-zero map $\varphi : F \to A$ induces a **Cu**-morphism (see Definition 14.4.2) $\varphi_* : \mathrm{Cu}(F) \to \mathrm{Cu}(A)$.

17.5.4. Prove Lemma 17.1.7 (Hint: use Lemma 10.3.4.)

17.5.5. Let X be a metric space and $\mathcal{U}$ a finite open cover of X such that

$$\sum_{U \in \mathcal{U}'} \chi_U(x) \leq d + 1 \text{ for every } x \in X.$$

Show that $\mathcal{U}$ has a finite subcover $\mathcal{U}'$ such that for every $U \in \mathcal{U}'$ there are at most d distinct $V \in \mathcal{U}'$, $V \neq U$, such that $U \cap V \neq \emptyset$.

17.5.6. Let A be AF. Show that A has nuclear dimension zero.

17.5.7. Determine the nuclear dimension and decomposition rank of the Jiang–Su algebra $\mathcal{Z}$.

17.5.8. Show that the C*-algebras of Theorem 11.3.12 have finite decomposition rank.

17.5.9. Let $n \in \mathbb{N}$. Show that $C_r^*(\mathbb{Z}^n)$ has finite decomposition rank.

18 The Classification Theorem and the Toms–Winter Theorem

In this final chapter, we will connect the material in previous chapters using the framework of the classification program for simple separable nuclear C*-algebras.

An approximately circle (A$\mathbb{T}$) algebra is an inductive limits of direct sums of algebras of the form $C(\mathbb{T}, M_n) \oplus M_m$. These were classified by Elliott [37]. Shortly after establishing this, Elliott conjectured that it might be possible to classify all simple, separable, unital, nuclear C*-algebras up to isomorphism using an invariant consisting of K-theory and traces. The invariant is of course what we now call the Elliott invariant (Definition 13.3.10).

18.0.1. Notation. We denote by $\mathcal{E}$ the class of all simple, separable, unital, nuclear, infinite-dimensional C*-algebras.

While the conjecture seemed quite bold, there were many subclasses of $\mathcal{E}$ for which it held: the AF algebras [36], and then the AH algebras with real rank zero [41], followed by AH algebras with very slow dimension growth [42]. These were all examples of inductive limits and their classification used some sort of variation of the intertwining arguments in 13.2.4. Later, Lin came up with tracial approximation which no longer required a specific inductive limit structure to obtain classification results [72, 75]. Also, in the case of purely infinite, simple, unital, nuclear C*-algebras, Kirchberg and Phillips had established the remarkable classification theorem which showed that, under the assumption of the UCT (see 12.5.14), all such C*-algebras are classified by their K-theory [65, 93].

Nevertheless, some counterexamples did pop up. Perhaps the most notable of these we already mentioned at the beginning of Chapter 14: Toms' construction in [120] of a simple separable unital nuclear C*-algebra A for which Ell$(A) \cong$ Ell$(A \otimes \mathcal{U})$ for a UHF algebra $\mathcal{U}$, and such that A and $A \otimes \mathcal{U}$ furthermore have the same real rank and stable rank. Previously, when it appeared one invariant would not suffice for a more complicated class (for example, K_0 is enough to classify AF algebras, but not A$\mathbb{T}$ algebras), one simply added more to the invariant: tracial states, K_1, and so forth. Toms' example effectively showed that the only possibility for a classification of all simple separable nuclear unital C*-algebras was to extend the Elliott invariant to (at least) include the Cuntz semigroup.

However, even if the Cuntz semigroup together with the Elliott invariant can be shown to be enough, a drawback is that in general Cuntz semigroups can be quite wild. Unlike in K-theory, there simply are not enough tools to make the invariant computable in general. Indeed, it remains unclear as to whether or not it is in fact easier to show that there is an isomorphism of Cuntz semigroups together will Ell$(\cdot)$ than it is to give a direct proof of a *-isomorphism at the level of C*-algebras. The level of complexity of the classification problem can be measured using the language of descriptive set theory by determining its *Borel complexity* (see for example [106, 40]). In [45], it was shown that separable nuclear C*-algebras

K. R. Strung, *An Introduction to C*-Algebras and the Classification Program*, Advanced Courses in Mathematics - CRM Barcelona, https://doi.org/10.1007/978-3-030-47465-2_18

cannot be *classified by countable structures*, which rules out the possibility of a computable, countable invariant as soon as one moves beyond AF algebras (the tracial state simplex already makes things too complicated). Nevertheless, the possibility that the Cuntz semigroup is the missing piece to a complete invariant is still an active area of research.

Setting aside the potential utility of the Cuntz semigroup for classification, the existence of counterexamples to the original Elliott conjecture raises the following interesting question: what is the largest subclass $\mathcal{E}_0 \subset \mathcal{E}$ for which one can expect classification by the Elliott invariant?

The construction of the Jiang–Su algebra $\mathcal{Z}$ suggested that, at the very least, one would have to restrict to those C*-algebras which are $\mathcal{Z}$-stable. How can such C*-algebras be characterised? In [123, Remarks 3.5] Toms and Winter made the following conjecture:

Conjecture (Toms–Winter I). *Let $A \in \mathcal{E}$ be a C*-algebra which is moreover stably finite. Then the following are equivalent.*

(i) $A \cong A \otimes \mathcal{Z}$,

(ii) $\operatorname{dr} A < \infty$,

(iii) *A has strict comparison of positive elements.*

Furthermore, it was conjectured that the class of stably finite C*-algebras satisfying these conditions were exactly those which could be classified by $\operatorname{Ell}(\cdot)$. Let $\mathcal{E}_{dr}$ denote this class of C*-algebras and let $\mathcal{E}_{pi}$ denote the simple separable unital nuclear purely infinite C*-algebras. From this it would follow that the correct class $\mathcal{E}_0 \subset \mathcal{E}$ would be $\mathcal{E}_0 = \mathcal{E}_{dr} \sqcup \mathcal{E}_{pi}$, at least under the assumption of the UCT.

The fact that those C*-algebras $\mathcal{E}_{pi}$ were not covered by the Toms–Winter conjecture mirrored what occurred throughout the literature: the cases of stably finite C*-algebras and purely infinite C*-algebras were often treated completely separately using different techniques. To attempt to gather all classification results under one single umbrella was a main motivation behind the generalisation of the decomposition rank to the nuclear dimension in [140]. As we saw, finite decomposition rank implies quasidiagonality and a purely infinite C*-algebra can never be quasidiagonal. However, as mentioned at the end of the previous chapter, they can have finite nuclear dimension, as is the case for the Cuntz algebras. The revised Toms–Winter conjecture [140, Conjecture 9.3] then became the following:

Conjecture (Toms–Winter II). *Let $A \in \mathcal{E}$ be a C*-algebra. Then the following are equivalent.*

(i) $A \cong A \otimes \mathcal{Z}$,

(ii) $\dim_{\mathrm{nuc}} A < \infty$,

(iii) *A has strict comparison of positive elements.*

Moreover, the C-algebras satisfying these conditions form the largest class of C*-algebras for which $\operatorname{Ell}(\cdot)$ is a complete invariant.*

We saw in Chapter 15 that if A is $\mathcal{Z}$-stable, then A has strict comparison of positive elements, or, to put it another way, a well-behaved Cuntz semigroup. Well behaved enough, in fact, that for these C*-algebras we can compute $\mathrm{Cu}(A)$. In this case, however, $\mathrm{Cu}(A)$ gives us no extra information. In fact, it was shown in [3] (based on earlier work of [19, 117]), that for every $\mathcal{Z}$-stable $A \in \mathcal{E}$ we can recover $\mathrm{Ell}(A)$ from $\mathrm{Cu}(A \otimes C(\mathbb{T}))$, and symmetrically, recover $\mathrm{Cu}(A \otimes C(\mathbb{T}))$ from $\mathrm{Ell}(A)$.

18.0.2. Theorem. *For the class of simple, separable, unital, nuclear, $\mathcal{Z}$-stable C*-algebra, the Elliott invariant and the Cuntz semigroup of any such algebra tensored with $C(\mathbb{T})$ are equivalent functors.*

The fact that finite decomposition rank implies $\mathcal{Z}$-stability was shown by Winter in [134] and then this was generalised to show that finite nuclear dimension implies $\mathcal{Z}$-stability in [136]. This establishes that (ii) implies (i) implies (iii) in Toms–Winter I and Toms–Winter II above.

Completing the cycle and showing the implications in the other direction proved significantly trickier. It was a breakthrough by Matui and Sato in [82] that allowed progress towards the conjecture finally being settled, up to the minor restriction that the extreme boundary of the tracial state space, denoted $\partial_e T(A)$, has a finite topological dimension.

18.0.3. Theorem. *Let $A \in \mathcal{E}$ be a C*-algebra and suppose $\partial_e T(A)$ is finite-dimensional. Then the following are equivalent.*

(i) $A \cong A \otimes \mathcal{Z}$,

(ii) $\dim_{\mathrm{nuc}} A < \infty$,

(iii) *A has strict comparison of positive elements.*

The restriction to the case where the dimension of $\partial_e T(A)$ is finite is only required for (iii) implies (i) [69, 107, 121]. We saw in Theorem 15.4.6 that (i) implies (iii). For (i) implies (ii), see [21], which was based on earlier work [12, 108, 82]. That (ii) implies (i) was shown in [136].

18.0.4. Definition. A tracial state τ on a unital C*-algebra A is *quasidiagonal* if for every finite subset $\mathcal{F} \subset A$ and every $\epsilon > 0$ there exist a matrix algebra M_n and a u.c.p. map $\psi : A \to M_n$ such that

(i) ψ is $(\mathcal{F}, \epsilon)$-approximately multiplicative, that is, $\|\psi(ab) - \psi(a)\psi(b)\| < \epsilon$ for every $a, b \in \mathcal{F}$;

(ii) ψ is $(\mathcal{F}, \epsilon)$-approximately trace-preserving, that is, $|\mathrm{tr}_{M_k} \circ \psi(a) - \tau(a)| < \epsilon$ for every $a \in \mathcal{F}$.

18.0.5. Theorem ([118, Proposition 1.4]). *Let A be a separable unital nuclear C*-algebra. If A has a faithful tracial state, then A is quasidiagonal.*

Of course, if A is simple, then every tracial state is automatically faithful. And, as it turns out, if A is also separable and nuclear then every tracial state is quasidiagonal, so long as we assume that A satisfies the UCT.

18.0.6. Theorem (Tikuisis–White–Winter, [118]). *Let A be a separable nuclear C^*-algebra which satisfies the UCT. Then every faithful tracial state is quasidiagonal.*

18.0.7. The Tikusis–White–Winter Theorem gives us a partial converse to Theorem 17.4.3, which also serves as a complete characterisation of the difference, among those C^*-algebras with finite nuclear dimension and which satisfy the UCT, between having finite or infinite decomposition rank. A simplified proof was later given by Schafhauser [109], and a generalisation was given by Gabe [48] which requires only that A is exact, rather than nuclear (recall that a C^*-algebra is *exact* if it is the C^*-subalgebra of a nuclear C^*-algebra).

Corollary. *Let $A \in \mathcal{E}$ and suppose that A has finite nuclear dimension and satisfies the UCT. Then A is quasidiagonal if and only if A has finite decomposition rank.*

While the first successful classifications were variations on Elliott's original AF classification, establishing the full classification theorem, Theorem 18.0.12 below, required new machinery. Lifting maps from the Elliott invariant to apply an intertwining argument already becomes incredibly complicated for AH algebras, the classification of which either requires the assumption of real rank zero [41] or so-called "very slow dimension growth" which asks that the size of the matrices grow much faster than the dimension of the spaces [42]. While AH algebras did not exhaust all possible Elliott invariants (AH algebras always have Riesz decomposition in K_0), Elliott showed in [38] that this can be done by approximately subhomogeneous (ASH) C^*-algebras (a *subhomogeneous* C^*-algebra is a C^*-subalgebra of a homogeneous C^*-algebra). Apart from some less complicated examples, such as inductive limits of dimension drop algebras, an Elliott intertwining-type classification for ASH algebras has been elusive. Furthermore, even if such a classification theorem were available, its use would be limited to those C^*-algebras for which we already know such an inductive limit structure exists. It is far from obvious, for example, whether or not a simple crossed product can be realised as an inductive limit.

Huaxin Lin's introduction of tracial approximation provided a big step forward for classification because it no longer required finding a specific inductive limit representation of a given C^*-algebra. Lin's program began with the classification of simple, separable, unital, nuclear, tracially approximately finite C^*-algebras satisfying the UCT [75]. Further progress was made when Winter showed that in some cases one could take classification up to UHF-stability to imply classification up to $\mathcal{Z}$-stability [137]. For example, if A and B are two simple separable unital nuclear C^*-algebras such that $A \otimes \mathcal{Q}$ and $B \otimes \mathcal{Q}$ are both tracially approximately finite, then one would be able to conclude that $A \otimes \mathcal{Z} \cong B \otimes \mathcal{Z}$ if and only if $\mathrm{Ell}(A \otimes \mathcal{Z}) \cong \mathrm{Ell}(B \otimes \mathcal{Z})$. This meant that showing a given C^*-algebra, or class of C^*-algebras, fits into a classification theorem could be broken up into two simpler steps: show that the class of C^*-algebras is covered by a classification theorem after tensoring with a UHF algebra, and show that the C^*-algebras in that class are $\mathcal{Z}$-stable. In general, it is much easier to work with a C^*-algebra which has been tensored with a UHF algebra because many structural properties – such as strict comparison and property (SP) (Definition 11.2.12) – hold automatically [99, 100].

18.0.8. Definition. Let F_1 and F_2 be two finite-dimensional C*-algebras and suppose there are two unital *-homomorphisms $\varphi_0, \varphi_1 : F_1 \to F_2$. Let

$$A = A(F_1, F_2, \varphi_0, \varphi_1)$$
$$= \{(f, g) \in C([0, 1], F_2) \oplus F_1 \mid f(0) = \varphi_0(g) \text{ and } f(1) = \varphi_1(g)\}.$$

A C*-algebra of this form is called an *Elliott–Thomsen building block*.

18.0.9. Definition ([54, Definition 9.2]). Let A be a simple unital C*-algebra. Then A has *generalised tracial rank at most one* if the following holds: For any $\epsilon > 0$, any nonzero $c \in A_+$ and any finite subset $\mathcal{F} \subset A$ there exists a nonzero projection p and a C*-subalgebra B which is an Elliott–Thomsen building block with $1_B = p$ satisfying

(i) $\|pa - ap\| < \epsilon$ for every $a \in \mathcal{F}$,

(ii) $\text{dist}(pap, B) < \epsilon$ for every $a \in \mathcal{F}$,

(iii) $1_A - p$ is Murray–von Neumann equivalent to a projection in $\overline{cAc}$.

In the language of Chapter 11, such a C*-algebra would be called "tracially approximately Elliott–Thomsen".

Building on earlier classification results for TAF algebras (Definition 11.3.1), tracially approximately interval algebra (TAI algebras for short, also called algebras with *tracial rank one*; here the building blocks are direct sums of matrices and C*-algebras of the form $C([0, 1], M_n))$ and others (for example, [77, 88]), Gong, Lin and Niu proved the following.

18.0.10. Theorem ([54]). *Let A and B be simple, separable, unital, nuclear C*-algebras. Suppose that $A \otimes Q$ and $B \otimes Q$ have generalised tracial rank at most one. Then $A \otimes \mathcal{Z} \cong B \otimes \mathcal{Z}$ if and only if $\text{Ell}(A \otimes \mathcal{Z}) \cong \text{Ell}(B \otimes \mathcal{Z})$. Moreover, any isomorphism between $\text{Ell}(A \otimes \mathcal{Z})$ and $\text{Ell}(B \otimes \mathcal{Z})$ can be lifted to an isomorphism of $A \otimes \mathcal{Z}$ and $B \otimes \mathcal{Z}$.*

In the same paper, they show that those C*-algebras with generalised tracial rank at most one after tensoring with the universal UHF algebra can exhaust the range of the Elliott invariant for simple separable unital nuclear stably finite C*-algebras with weakly unperforated K_0. Thus one expects that any simple separable unital stably finite $\mathcal{Z}$-stable C*-algebra should belong to this class. From [43], we have the following.

18.0.11. Theorem. *Let A be a simple separable unital C*-algebra with finite decomposition rank and which satisfies the UCT. Suppose that all tracial states of A are quasidiagonal. Then $A \otimes Q$ has generalised tracial rank at most one.*

Of course, we have already seen above that the if A is such a C*-algebra, the restriction on the tracial states is in fact redundant. Thus, putting all of these pieces together we arrive at the classification of *all* separable, unital simple, infinite-dimensional C*-algebras with finite nuclear dimension and which satisfy the UCT.

18.0.12. Theorem. *Let A and B be separable, unital simple, infinite-dimensional C*-algebras with finite nuclear dimension and which satisfy the UCT. Suppose there is an isomorphism*

$$\varphi : \mathrm{Ell}(A) \to \mathrm{Ell}(B).$$

*Then there is a *-isomorphism*

$$\Phi : A \to B,$$

which is unique up to approximate unitary equivalence and satisfies $\mathrm{Ell}(\Phi) = \phi$.

Bibliography

[1] Charles A. Akemann, Joel Anderson, and Gert K. Pedersen, *Excising states of* C*-*algebras*, Canad. J. Math. **38** (1986), no. 5, 1239–1260. MR 869724

[2] Claire Anantharaman-Delaroche, *Systèmes dynamiques non commutatifs et moyennabilité*, Math. Ann. **279** (1987), no. 2, 297–315. MR 919508

[3] Ramon Antoine, Marius Dadarlat, Francesc Perera, and Luis Santiago, *Recovering the Elliott invariant from the Cuntz semigroup*, Trans. Amer. Math. Soc. **366** (2014), no. 6, 2907–2922. MR 3180735

[4] Pere Ara, Francesc Perera, and Andrew S. Toms, *K-theory for operator algebras. Classification of* C*-*algebras*, Aspects of operator algebras and applications, Contemp. Math., vol. 534, Amer. Math. Soc., Providence, RI, 2011, pp. 1–71. MR 2767222

[5] William B. Arveson, *Subalgebras of* C*-*algebras*, Acta Math. **123** (1969), 141–224. MR 0253059

[6] Selçuk Barlak and Xin Li, *Cartan subalgebras and the UCT problem*, Adv. Math. **316** (2017), 748–769. MR 3672919

[7] Bruce Blackadar, *Comparison theory for simple* C*-*algebras*, Operator algebras and applications, Vol. 1, London Math. Soc. Lecture Note Ser., vol. 135, Cambridge Univ. Press, Cambridge, 1988, pp. 21–54. MR 996438

[8] _____, *K-theory for operator algebras*, second ed., Mathematical Sciences Research Institute Publications, vol. 5, Cambridge University Press, Cambridge, 1998. MR 1656031

[9] Bruce Blackadar and David Handelman, *Dimension functions and traces on* C*-*algebras*, J. Funct. Anal. **45** (1982), no. 3, 297–340. MR 650185

[10] Bruce Blackadar and Eberhard Kirchberg, *Generalized inductive limits of finite-dimensional* C*-*algebras*, Math. Ann. **307** (1997), no. 3, 343–380. MR 1437044

[11] Bruce Blackadar and Mikael Rørdam, *Extending states on preordered semigroups and the existence of quasitraces on* C*-*algebras*, J. Algebra **152** (1992), no. 1, 240–247. MR 1190414

[12] Joan Bosa, Nathanial P. Brown, Yasuhiko Sato, Aaron Tikuisis, Stuart White, and Wilhelm Winter, *Covering dimension of* C*-*algebras and 2-*

coloured classification, Mem. Amer. Math. Soc. **257** (2019), no. 1233, vii+97. MR 3908669

[13] Ola Bratteli, *Inductive limits of finite-dimensional C*-algebras*, Trans. Amer. Math. Soc. **171** (1972), 195–234. MR 0312282

[14] Ola Bratteli, George A. Elliott, and Richard H. Herman, *On the possible temperatures of a dynamical system*, Comm. Math. Phys. **74** (1980), no. 3, 281–295. MR 578045

[15] Emmanuel Breuillard, Mehrdad Kalantar, Matthew Kennedy, and Narutaka Ozawa, *C*-simplicity and the unique trace property for discrete groups*, Publ. Math. Inst. Hautes Études Sci. **126** (2017), 35–71. MR 3735864

[16] Lawrence G. Brown and Gert K. Pedersen, *C*-algebras of real rank zero*, J. Funct. Anal. **99** (1991), no. 1, 131–149. MR 1120918

[17] Nathanial P. Brown, *Excision and a theorem of Popa*, J. Operator Theory **54** (2005), no. 1, 3–8. MR 2168856

[18] Nathanial P. Brown and Narutaka Ozawa, *C*-algebras and finite-dimensional approximations*, Graduate Studies in Mathematics, vol. 88, American Mathematical Society, Providence, RI, 2008. MR 2391387

[19] Nathanial P. Brown, Francesc Perera, and Andrew S. Toms, *The Cuntz semigroup, the Elliott conjecture, and dimension functions on C*-algebras*, J. Reine Angew. Math. **621** (2008), 191–211. MR 2431254

[20] Nathanial P. Brown and Andrew S. Toms, *Three applications of the Cuntz semigroup*, Int. Math. Res. Not. IMRN (2007), no. 19, Art. ID rnm068, 14. MR 2359541

[21] Jorge Castillejos, Samuel Evington, Aaron Tikuisis, Stuart White, and Wilhelm Winter, *Nuclear dimension of simple C*-algebras*, Preprint arXiv: 1901.05853v2, 2019.

[22] Man Duen Choi and Edward G. Effros, *The completely positive lifting problem for C*-algebras*, Ann. of Math. (2) **104** (1976), no. 3, 585–609. MR 0417795

[23] _____, *Separable nuclear C*-algebras and injectivity*, Duke Math. J. **43** (1976), no. 2, 309–322. MR 0405117

[24] _____, *Nuclear C*-algebras and injectivity: the general case*, Indiana Univ. Math. J. **26** (1977), no. 3, 443–446. MR 0430794

[25] _____, *Nuclear C*-algebras and the approximation property*, Amer. J. Math. **100** (1978), no. 1, 61–79. MR 0482238

[26] Alain Connes, *Classification of injective factors. Cases II_1, II_∞, III_λ, $\lambda \neq$ 1*, Ann. of Math. (2) **104** (1976), no. 1, 73–115. MR 0454659

[27] Kristofer T. Coward, George A. Elliott, and Cristian Ivanescu, *The Cuntz semigroup as an invariant for C*-algebras*, J. Reine Angew. Math. **623** (2008), 161–193. MR 2458043

[28] Joachim Cuntz, *Simple C*-algebras generated by isometries*, Comm. Math. Phys. **57** (1977), no. 2, 173–185. MR 0467330

[29] _____, *The structure of multiplication and addition in simple C*-algebras*, Math. Scand. **40** (1977), no. 2, 215–233. MR 0500176

[30] _____, *Dimension functions on simple C*-algebras*, Math. Ann. **233** (1978), no. 2, 145–153. MR 0467332

[31] _____, *K-theory for certain C*-algebras*, Ann. of Math. (2) **113** (1981), no. 1, 181–197. MR 604046

[32] Kenneth R. Davidson, *C*-algebras by example*, Fields Institute Monographs, vol. 6, American Mathematical Society, Providence, RI, 1996. MR 1402012

[33] Marius Dădărlat, *Nonnuclear subalgebras of AF algebras*, Amer. J. Math. **122** (2000), no. 3, 581–597. MR 1759889

[34] Marius Dădărlat and Søren Eilers, *Approximate homogeneity is not a local property*, J. Reine Angew. Math. **507** (1999), 1–13. MR 1670254

[35] Edward G. Effros and Jonathan Rosenberg, *C*-algebras with approximately inner flip*, Pacific J. Math. **77** (1978), no. 2, 417–443. MR 510932

[36] George A. Elliott, *On the classification of inductive limits of sequences of semisimple finite-dimensional algebras*, J. Algebra **38** (1976), no. 1, 29–44. MR 0397420

[37] _____, *On the classification of C*-algebras of real rank zero*, J. Reine Angew. Math. **443** (1993), 179–219. MR 1241132

[38] _____, *An invariant for simple C*-algebras*, Canadian Mathematical Society. 1945–1995, Vol. 3, Canadian Math. Soc., Ottawa, ON, 1996, pp. 61–90. MR 1661611

[39] George A. Elliott and David E. Evans, *The structure of the irrational rotation C*-algebra*, Ann. of Math. (2) **138** (1993), no. 3, 477–501. MR 1247990

[40] George A. Elliott, Ilijas Farah, Vern I. Paulsen, Christian Rosendal, Andrew S. Toms, and Asger Törnquist, *The isomorphism relation for separable C*-algebras*, Math. Res. Lett. **20** (2013), no. 6, 1071–1080. MR 3228621

[41] George A. Elliott and Guihua Gong, *On the classification of C*-algebras of real rank zero. II*, Ann. of Math. (2) **144** (1996), no. 3, 497–610. MR 1426886

[42] George A. Elliott, Guihua Gong, and Liangqing Li, *On the classification of simple inductive limit C*-algebras. II. The isomorphism theorem*, Invent. Math. **168** (2007), no. 2, 249–320. MR 2289866

[43] George A. Elliott, Guihua Gong, Huaxin Lin, and Zhuang Niu, *On the classification of simple amenable C*-algebras with finite decomposition rank II*, arXiv preprint math.OA/1507.03437v2, 2015.

[44] Ilijas Farah and Takeshi Katsura, *Nonseparable UHF algebras I: Dixmier's problem*, Adv. Math. **225** (2010), no. 3, 1399–1430. MR 2673735

[45] Ilijas Farah, Andrew S. Toms, and Asger Törnquist, *Turbulence, orbit equivalence, and the classification of nuclear* C*-*algebras*, J. Reine Angew. Math. **688** (2014), 101–146. MR 3176617

[46] Brian E. Forrest, Nico Spronk, and Matthew Wiersma, *Existence of tracial states on reduced group* C*-*algebras*, arXiv preprint math.OA/1706.05354v2.

[47] Bent Fuglede, *A commutativity theorem for normal operators*, Proc. Nat. Acad. Sci. U. S. A. **36** (1950), 35–40. MR 0032944

[48] James Gabe, *Quasidiagonal traces on exact* C*-*algebras*, J. Funct. Anal. **272** (2017), no. 3, 1104–1120. MR 3579134

[49] Israel Gelfand and Mark Neumark, *On the imbedding of normed rings into the ring of operators in Hilbert space*, Rec. Math. [Mat. Sbornik] N.S. **12(54)** (1943), 197–213. MR 0009426

[50] Julien Giol and David Kerr, *Subshifts and perforation*, J. Reine Angew. Math. **639** (2010), 107–119. MR 2608192

[51] James G. Glimm, *On a certain class of operator algebras*, Trans. Amer. Math. Soc. **95** (1960), 318–340. MR 0112057

[52] ———, *A Stone–Weierstrass theorem for* C*-*algebras*, Ann. of Math. (2) **72** (1960), 216–244. MR 0116210

[53] Guihua Gong, Xinhui Jiang, and Hongbing Su, *Obstructions to* $\mathcal{Z}$-*stability for unital simple* C*-*algebras*, Canad. Math. Bull. **43** (2000), no. 4, 418–426. MR 1793944

[54] Guihua Gong, Huaxin Lin, and Zhuang Niu, *Classification of finite simple amenable* $\mathcal{Z}$-*stable* C*-*algebras*, preprint math.OA/arXiv:1501.00135v6, 2015.

[55] Kenneth R. Goodearl and David Handelman, *Rank functions and* K_O *of regular rings*, J. Pure Appl. Algebra **7** (1976), no. 2, 195–216. MR 0389965

[56] Uffe Haagerup, *A new proof of the equivalence of injectivity and hyperfiniteness for factors on a separable Hilbert space*, J. Funct. Anal. **62** (1985), no. 2, 160–201. MR 791846

[57] ———, *Connes' bicentralizer problem and uniqueness of the injective factor of type* III$_1$, Acta Math. **158** (1987), no. 1-2, 95–148. MR 880070

[58] ———, *Every quasi-trace on an exact* C*-*algebra is a trace*, Preprint, 1991.

[59] ———, *Quasitraces on exact* C*-*algebras are traces*, C. R. Math. Acad. Sci. Soc. R. Can. **36** (2014), no. 2-3, 67–92. MR 3241179

[60] Alfred Haar, *Der Massbegriff in der Theorie der kontinuierlichen Gruppen*, Ann. of Math. (2) **34** (1933), no. 1, 147–169. MR 1503103

[61] Nigel Higson and John Roe, *Analytic K-homology*, Oxford Mathematical Monographs, Oxford University Press, Oxford, 2000, Oxford Science Publications. MR 1817560

[62] Xinhui Jiang and Hongbing Su, *On a simple unital projectionless* C*-*algebra*, Amer. J. Math. **121** (1999), no. 2, 359–413. MR 1680321

[63] Richard V. Kadison, *Irreducible operator algebras*, Proc. Nat. Acad. Sci. U.S.A. **43** (1957), 273–276. MR 0085484

[64] Matthew Kennedy and Sven Raum, *Traces on reduced group* C*-*algebras*, Bull. Lond. Math. Soc. **49** (2017), no. 6, 988–990. MR 3743482

[65] Eberhard Kirchberg, *The classification of purely infinite* C*-*algebras using Kasparov's theory*, in preparation; to appear in Fields Institute Communications.

[66] ———, C*-*nuclearity implies CPAP*, Math. Nachr. **76** (1977), 203–212. MR 0512362

[67] Eberhard Kirchberg and N. Christopher Phillips, *Embedding of exact* C*-*algebras in the Cuntz algebra* $\mathcal{O}_2$, J. Reine Angew. Math. **525** (2000), 17–53. MR 1780426

[68] Eberhard Kirchberg and Mikael Rørdam, *Non-simple purely infinite* C*-*algebras*, Amer. J. Math. **122** (2000), no. 3, 637–666. MR 1759891

[69] ———, *Central sequence* C*-*algebras and tensorial absorption of the Jiang–Su algebra*, J. Reine Angew. Math. **695** (2014), 175–214. MR 3276157

[70] Eberhard Kirchberg and Wilhelm Winter, *Covering dimension and quasidiagonality*, Internat. J. Math. **15** (2004), no. 1, 63–85. MR 2039212

[71] Mark Krein and David Milman, *On extreme points of regular convex sets*, Studia Math. **9** (1940), 133–138. MR 0004990

[72] Huaxin Lin, *Classification of simple tracially AF* C*-*algebras*, Canad. J. Math. **53** (2001), no. 1, 161–194. MR 1814969

[73] ———, *The tracial topological rank of* C*-*algebras*, Proc. London Math. Soc. (3) **83** (2001), no. 1, 199–234. MR 1829565

[74] ———, *Tracially AF* C*-*algebras*, Trans. Amer. Math. Soc. **353** (2001), no. 2, 693–722. MR 1804513

[75] ———, *Classification of simple* C*-*algebras of tracial topological rank zero*, Duke Math. J. **125** (2004), no. 1, 91–119. MR 2097358

[76] ———, *Simple nuclear* C*-*algebras of tracial topological rank one*, J. Funct. Anal. **251** (2007), no. 2, 601–679. MR 2356425

[77] ———, *Asymptotic unitary equivalence and classification of simple amenable* C*-*algebras*, Invent. Math. **183** (2011), no. 2, 385–450. MR 2772085

[78] ———, *Locally AH algebras*, Mem. Amer. Math. Soc. **235** (2015), no. 1107, vi+109. MR 3338301

[79] ———, *Crossed products and minimal dynamical systems*, J. Topol. Anal. **10** (2018), no. 2, 447–469. MR 3809595

[80] Huaxin Lin and N. Christopher Phillips, *Crossed products by minimal homeomorphisms*, J. Reine Angew. Math. **641** (2010), 95–122. MR 2643926

[81] Terry A. Loring, *Lifting solutions to perturbing problems in* C*-*algebras*, Fields Institute Monographs, vol. 8, American Mathematical Society, Providence, RI, 1997. MR 1420863

[82] Hiroki Matui and Yasuhiko Sato, *Decomposition rank of UHF-absorbing* C*-*algebras*, Duke Math. J. **163** (2014), no. 14, 2687–2708. MR 3273581

[83] Gerard J. Murphy, C*-*algebras and operator theory*, Academic Press, Inc., Boston, MA, 1990. MR 1074574

[84] Francis J. Murray and John von Neumann, *On rings of operators*, Ann. of Math. (2) **37** (1936), no. 1, 116–229. MR 1503275

[85] _____, *On rings of operators. II*, Trans. Amer. Math. Soc. **41** (1937), no. 2, 208–248. MR 1501899

[86] _____, *On rings of operators. IV*, Ann. of Math. (2) **44** (1943), 716–808. MR 0009096

[87] Jun-iti Nagata, *Modern dimension theory*, North-Holland, Amsterdam, 1965.

[88] Zhuang Niu, *A classification of tracially approximate splitting interval algebras. III. Uniqueness theorem and isomorphism theorem*, C. R. Math. Acad. Sci. Soc. R. Can. **37** (2015), no. 2, 41–75. MR 3363480

[89] Gert K. Pedersen, C*-*algebras and their automorphism groups*, London Mathematical Society Monographs, vol. 14, Academic Press, Inc. [Harcourt Brace Jovanovich, Publishers], London-New York, 1979. MR 548006

[90] _____, *Unitary extensions and polar decompositions in a* C*-*algebra*, J. Operator Theory **17** (1987), no. 2, 357–364. MR 887230

[91] _____, *Analysis now*, Graduate Texts in Mathematics, vol. 118, Springer-Verlag, New York, 1989. MR 971256

[92] Francesc Perera and Andrew S. Toms, *Recasting the Elliott conjecture*, Math. Ann. **338** (2007), no. 3, 669–702. MR 2317934

[93] N. Christopher Phillips, *A classification theorem for nuclear purely infinite simple* C*-*algebras*, Doc. Math. **5** (2000), 49–114. MR 1745197

[94] Mihai Pimsner and Dan Virgil Voiculescu, *Exact sequences for* K-*groups and Ext-groups of certain cross-product* C*-*algebras*, J. Operator Theory **4** (1980), no. 1, 93–118. MR 587369

[95] Sorin Popa, *On local finite-dimensional approximation of* C*-*algebras*, Pacific J. Math. **181** (1997), no. 1, 141–158. MR 1491038

[96] Ian F. Putnam, *On the topological stable rank of certain transformation group* C*-*algebras*, Ergodic Theory Dynam. Systems **10** (1990), no. 1, 197–207. MR 1053808

[97] Iain Raeburn, *On crossed products and Takai duality*, Proc. Edinburgh Math. Soc. (2) **31** (1988), no. 2, 321–330. MR 989764

[98] Marc A. Rieffel, *Dimension and stable rank in the* K-*theory of* C*-*algebras*, Proc. London Math. Soc. (3) **46** (1983), no. 2, 301–333. MR 693043

[99] Mikael Rørdam, *On the structure of simple C*-algebras tensored with a UHF-algebra*, J. Funct. Anal. **100** (1991), no. 1, 1–17. MR 1124289

[100] ———, *On the structure of simple C*-algebras tensored with a UHF-algebra. II*, J. Funct. Anal. **107** (1992), no. 2, 255–269. MR 1172023

[101] ———, *Classification of certain infinite simple C*-algebras*, J. Funct. Anal. **131** (1995), no. 2, 415–458. MR 1345038

[102] ———, *The stable and the real rank of $\mathcal{Z}$-absorbing C*-algebras*, Internat. J. Math. **15** (2004), no. 10, 1065–1084. MR 2106263

[103] Mikael Rørdam, Flemming Larsen, and Niels Jakob Laustsen, *An introduction to K-theory for C*-algebras*, London Mathematical Society Student Texts, vol. 49, Cambridge University Press, Cambridge, 2000. MR 1783408

[104] Jonathan Rosenberg and Claude Schochet, *The Künneth theorem and the universal coefficient theorem for Kasparov's generalized K-functor*, Duke Math. J. **55** (1987), no. 2, 431–474. MR 894590

[105] Walter Rudin, *Functional analysis*, second ed., International Series in Pure and Applied Mathematics, McGraw-Hill, Inc., New York, 1991. MR 1157815

[106] Marcin Sabok, *Completeness of the isomorphism problem for separable C*-algebras*, Invent. Math. **204** (2016), no. 3, 833–868. MR 3502066

[107] Yasuhiko Sato, *Trace spaces of simple nuclear C*-algebras with finite-dimensional extreme boundary*, arXiv preprint math.OA/1209.3000, 2012.

[108] Yasuhiko Sato, Stuart White, and Wilhelm Winter, *Nuclear dimension and $\mathcal{Z}$-stability*, Invent. Math. **202** (2015), no. 2, 893–921. MR 3418247

[109] Christopher Schafhauser, *A new proof of the Tikuisis–White–Winter Theorem*, to appear in J. Reine Angew. Math., doi.org/10.1515/crelle-2017-0056.

[110] Claude Schochet, *Topological methods for C*-algebras. II. Geometric resolutions and the Künneth formula*, Pacific J. Math. **98** (1982), no. 2, 443–458. MR 650021

[111] Irving E. Segal, *Irreducible representations of operator algebras*, Bull. Amer. Math. Soc. **53** (1947), 73–88. MR 20217

[112] Karen R. Strung, *On classification, UHF-stability, and tracial approximation of simple nuclear C*-algebras*, Ph.D. thesis, Universität Münster, 2013.

[113] ———, *C*-algebras of minimal dynamical systems of the product of a Cantor set and an odd-dimensional sphere*, J. Funct. Anal. **268** (2015), no. 3, 671–689. MR 3292350

[114] Karen R. Strung and Wilhelm Winter, *Minimal dynamics and $\mathcal{Z}$-stable classification*, Internat. J. Math. **22** (2011), no. 1, 1–23. MR 2765440

[115] Hiroshi Takai, *On a duality for crossed products of C*-algebras*, J. Functional Analysis **19** (1975), 25–39. MR 0365160

[116] Masamichi Takesaki, *On the cross-norm of the direct product of C*-algebras*, Tôhoku Math. J. (2) **16** (1964), 111–122. MR 0165384

[117] Aaron Tikuisis, *The Cuntz semigroup of continuous functions into certain simple C*-algebras*, Internat. J. Math. **22** (2011), no. 8, 1051–1087. MR 2826555

[118] Aaron Tikuisis, Stuart White, and Wilhelm Winter, *Quasidiagonality of nuclear C*-algebras*, Ann. of Math. (2) **185** (2017), no. 1, 229–284. MR 3583354

[119] Thomas Timmermann, *An invitation to quantum groups and duality*, EMS Textbooks in Mathematics, European Mathematical Society (EMS), Zürich, 2008, From Hopf algebras to multiplicative unitaries and beyond. MR 2397671

[120] Andrew S. Toms, *On the classification problem for nuclear C*-algebras*, Ann. of Math. (2) **167** (2008), no. 3, 1029–1044. MR 2415391

[121] Andrew S. Toms, Stuart White, and Wilhelm Winter, *$\mathcal{Z}$-stability and finite-dimensional tracial boundaries*, Int. Math. Res. Not. IMRN (2015), no. 10, 2702–2727. MR 3352253

[122] Andrew S. Toms and Wilhelm Winter, *Strongly self-absorbing C*-algebras*, Trans. Amer. Math. Soc. **359** (2007), no. 8, 3999–4029. MR 2302521

[123] ———, *The Elliott conjecture for Villadsen algebras of the first type*, J. Funct. Anal. **256** (2009), no. 5, 1311–1340. MR 2490221

[124] ———, *Minimal dynamics and K-theoretic rigidity: Elliott's conjecture*, Geom. Funct. Anal. **23** (2013), no. 1, 467–481. MR 3037905

[125] Léon Van Hove, *Von Neumann's contributions to quantum theory*, Bull. Amer. Math. Soc. **64** (1958), 95–99. MR 92587

[126] Jesper Villadsen, *The range of the Elliott invariant*, J. Reine Angew. Math. **462** (1995), 31–55. MR 1329901

[127] Dan-Virgil Voiculescu, *A non-commutative Weyl–von Neumann theorem*, Rev. Roumaine Math. Pures Appl. **21** (1976), no. 1, 97–113. MR 0415338

[128] Niels Erik Wegge-Olsen, *K-theory and C*-algebras*, Oxford Science Publications, The Clarendon Press, Oxford University Press, New York, 1993, A friendly approach. MR 1222415

[129] J.H.C. Whitehead, *On incidence matrices, nuclei and homotopy types*, Ann. of Math. (2) **42** (1941), 1197–1239. MR 0005352

[130] Dana P. Williams, *Crossed products of C*-algebras*, Mathematical Surveys and Monographs, vol. 134, American Mathematical Society, Providence, RI, 2007. MR 2288954

[131] Wilhelm Winter, *Covering dimension for nuclear C*-algebras*, J. Funct. Anal. **199** (2003), no. 2, 535–556. MR 1971906

[132] ———, *On the classification of simple $\mathcal{Z}$-stable C*-algebras with real rank zero and finite decomposition rank*, J. London Math. Soc. (2) **74** (2006), no. 1, 167–183. MR 2254559

[133] ———, *Covering dimension for nuclear C*-algebras. II*, Trans. Amer. Math. Soc. **361** (2009), no. 8, 4143–4167. MR 2500882

[134] _____, *Decomposition rank and Z-stability*, Invent. Math. **179** (2010), no. 2, 229–301. MR 2570118

[135] _____, *Strongly self-absorbing C*-algebras are Z-stable*, J. Noncommut. Geom. **5** (2011), no. 2, 253–264. MR 2784504

[136] _____, *Nuclear dimension and Z-stability of pure C*-algebras*, Invent. Math. **187** (2012), no. 2, 259–342. MR 2885621

[137] _____, *Localizing the Elliott conjecture at strongly self-absorbing C*-algebras*, J. Reine Angew. Math. **692** (2014), 193–231. MR 3274552

[138] _____, *QDQ vs. UCT*, Operator algebras and applications – the Abel Symposium 2015, Abel Symp., vol. 12, Springer, [Cham], 2017, pp. 327–348. MR 3837604

[139] Wilhelm Winter and Joachim Zacharias, *Completely positive maps of order zero*, Münster J. Math. **2** (2009), 311–324. MR 2545617

[140] _____, *The nuclear dimension of C*-algebras*, Adv. Math. **224** (2010), no. 2, 461–498. MR 2609012

[141] Stanisław L. Woronowicz, *Compact matrix pseudogroups*, Comm. Math. Phys. **111** (1987), no. 4, 613–665. MR 901157

[142] _____, *Differential calculus on compact matrix pseudogroups (quantum groups)*, Comm. Math. Phys. **122** (1989), no. 1, 125–170. MR 994499

[143] _____, *Compact quantum groups*, Symétries quantiques (Les Houches, 1995), North-Holland, Amsterdam, 1998, pp. 845–884. MR 1616348

[144] Georges Zeller-Meier, *Produits croisés d'une C*-algèbre par un groupe d'automorphismes*, J. Math. Pures Appl. (9) **47** (1968), 101–239. MR 0241994

Index

© The Author(s), under exclusive license to Springer Nature Switzerland AG 2021
K. R. Strung, *An Introduction to C*-Algebras and the Classification Program*, Advanced
Courses in Mathematics - CRM Barcelona, https://doi.org/10.1007/978-3-030-47465-2

Printed in the United States
By Bookmasters